CHAOS

Chaos

The Science of Predictable Random Motion

Richard Kautz

Formerly of the National Institute of Standards and Technology, Boulder, CO, USA

OXFORD
UNIVERSITY PRESS

OXFORD
UNIVERSITY PRESS

Great Clarendon Street, Oxford OX2 6DP

Oxford University Press is a department of the University of Oxford.
It furthers the University's objective of excellence in research, scholarship,
and education by publishing worldwide in

Oxford New York

Auckland Cape Town Dar es Salaam Hong Kong Karachi
Kuala Lumpur Madrid Melbourne Mexico City Nairobi
New Delhi Shanghai Taipei Toronto

With offices in

Argentina Austria Brazil Chile Czech Republic France Greece
Guatemala Hungary Italy Japan Poland Portugal Singapore
South Korea Switzerland Thailand Turkey Ukraine Vietnam

Oxford is a registered trade mark of Oxford University Press
in the UK and in certain other countries

Published in the United States
by Oxford University Press Inc., New York

British Library Cataloguing in Publication Data

Data available

Library of Congress Cataloging in Publication Data

Kautz, Richard.

Chaos : the science of predictable random motion / Richard Kautz.

p. cm.

ISBN 978–0–19–959458–0

1. Prediction theory. 2. Chaotic behavior in systems. I. Title.

QA279.2.K38 2011

003'.857—dc22 2010030953

Typeset by SPI Publisher Services, Pondicherry, India
Printed in Great Britain
on acid-free paper by
CPI Antony Rowe, Chippenham, Wiltshire

ISBN 978–0–19–959457–3 (Hbk)
978–0–19–959458–0 (Pbk)

1 3 5 7 9 10 8 6 4 2

In memory of my father, Ralph Kautz,
Who taught me the common sense of mechanics

Preface

I unknowingly encountered chaos for the first time in 1977 when I observed noisy behavior in a superconducting tunnel junction (STJ). This unremarkable observation occurred during the course of research at the National Institute of Standards and Technology (NIST) in Boulder, Colorado. Three years later I learned that the noise was truly astonishing. It didn't result from the usual thermal agitation of electrons but from the apparently noise-free equation that governs the dynamics of an STJ. Lest you think that a superconducting device might be complicated, I hasten to add that an STJ is modeled by the same equation that describes the motion of a simple pendulum. But how can a mathematically perfect pendulum driven by an exactly repeating oscillatory force produce motion that is best described as noisy? My initial response to this question, like many scientists', was that complex, noisy behavior couldn't possibly arise in such a simple system. And yet it does.

As I explored chaos further, it proved to be a fascinating mathematical and physical phenomenon. Soon I began to publish papers and present lectures on chaos in the pendulum. In addition, as part of a science outreach program at NIST, I developed a presentation on chaos suitable for high school students, including a brief history of the science of dynamics and computer animations of chaotic motion. Many students "got it" and saw that understanding ideas at the frontier of science was within their reach. With *Chaos: The Science of Predictable Random Motion*, I hope to bring this message to other motivated students. Science in general and chaos in particular aren't impossibly difficult, and you can truly understand much of current research with elementary mathematics.

In the 30 years since chaotic motion began to make a splash in the popular press, more than 20 books on the subject have been written for general audiences. These books, listed in the bibliography, are excellent introductions to chaos but usually avoid mathematics. My strategy in writing *Chaos* has been to explore the topic using elementary algebra, vectors, and trigonometry, while avoiding calculus. Although a little math makes the book more challenging, it also affords deeper insights into the nature and origins of chaotic behavior. In translating my original lecture into a book, I've tried to retain some of the lighter elements, like historical sketches and computer animations. (The latter are available on the companion CD.) However, the final chapters take the reader into the mathematical realm of state space, where neither my lecture nor most introductory books dare to tread. These chapters rely on graphics rather than algebra to introduce the homoclinic tangle, a topological monster that lies at the heart of chaotic motion.

In writing *Chaos*, I was inspired by the Science Study Series (SSS), a collection of 73 pocket-sized paperbacks published by Anchor Doubleday between 1959 and 1974. The series grew out of efforts by the Physical Science Study Committee to reform the high school physics curriculum by emphasizing physical principles rather than collected facts. As a student, I was never good at memorization and found the SSS exactly to my taste. Written by experts but pitched to high school students, the best of the series are like intellectual rocket sleds, taking you from a standing start (no prior knowledge) to supersonic speeds (the research frontier) in almost no time at all. I still remember an "Aha!" moment from my teens that occurred while reading *Electrons and Waves* by John R. Pierce (number 38 of the series). Suddenly I understood that light is an electromagnetic wave in which the magnetic field is generated by the changing electric field and the electric field is generated by the changing magnetic field—a pair of dynamic, mutually self-sustaining fields. My debt to an earlier generation of scientists will be repaid in part if a few younger readers experience an "Aha!" moment while reading *Chaos*.

Acknowledgements

It is my pleasure to thank the many people who have helped in the preparation of this book. For reading all or part of the manuscript, I thank Kathleen Danna, Jimmy Fuller, Clark Hamilton, Tim Kautz, Alice Levine, Keagan McAllister, Don McDonald, Jim Meiss, Bob Phelan, and Matt Young. For impromptu translations of source materials in French and Italian, I thank Nicolas Hadacek and Luca Lorini. For the resolution of minor historical points, I thank Ralph Abraham, Francis Everitt, David Ruelle, Tatyana Shaposhnikova, and James Yorke. For technical support, I thank Gene Hilton (computer), Peter Kuemple (music), and Dave Wortendyke (video). For locating source materials, no matter how obscure, I thank the staff of the library at NIST Boulder, especially Katherine Day, Carol Gocke, Heather McCullough, Michael Robinson, and Joan Segal. Special thanks go to Jim Meiss for reviewing the technical accuracy of some chapters, to the Quantum Devices Group, especially Dave Rudman, for hosting my continuing stay at NIST as a guest researcher, and to Don McDonald for his advice and encouragement throughout the writing process.

Contents

Part I
Introduction

Chaos everywhere

What is chaos? To a three-year-old, chaos is altogether ordinary. Chaos happens. A glass tumbles to the floor and shatters into a dozen pieces, splattering orange juice far and wide. In everyday life, such events are no more remarkable than the rising and setting of the Sun. Scientists, on the other hand, have traditionally distinguished between random, unpredictable events and those, like the movements of the Sun and planets, that obey simple, mathematical rules. Prediction is the essence of science, and physicists take pride in having honed the laws of motion, the science of dynamics, to the point where the movements of planets, moons, and space probes can be foretold with precision.

For more than three centuries, physicists thought that their mathematical laws described only regular motion, leaving random events in the realm of the unpredictable. It made sense. Simple laws should predict simple motion, and random events are neither simple nor predictable. Thus, it came as a shock, a lightning bolt out of the clear blue sky, when scientists of the 1960s began to realize that the same equations that predict regular behavior can also predict motion that is aptly described as chaotic. How could this be? Predictable random motion is a paradox, an apparent self-contradiction. Chaos doesn't make much sense, unless perhaps you are three years old.

As scientists further explored the chaos predicted by their equations, they discovered additional paradoxes. Not only is chaotic motion predictable and random at the same time, but it generates complexity from simplicity, combines stability with instability, and, in the long term, is predictable only in principle, not in practice. Given these apparent contradictions, perhaps it isn't too surprising that chaotic motion also yields geometric objects called fractals with dimensionalities that aren't whole numbers—something between a 2-D sheet of paper and a 3-D brick. For the physicist, chaos is altogether extraordinary, a riddle that challenges previous notions of order and defies understanding.

Even so, physicists have learned that chaotic motion is quite ordinary. Knowing what to look for, they discovered that simple rules can explain a wide variety of seemingly random patterns in everyday life. In a mathematical sense, chaos can be found in the flutter of a falling leaf, in the dripping of a leaky faucet, or in the irregular beat of a distressed heart. In each of these cases, scientists discovered that apparently random motion can be explained by a mathematical model that does not include random forces. The leaf falls through still air, the faucet is fed by a uniform flow of water, and the heart is stimulated by exactly periodic nerve impulses, yet the resulting behavior is irregular. While not all instances of random motion are explained by simple sets of equations (many result from intrinsic complexity), a surprising number are well described by what is now called chaos theory.

1.1   Tilt-A-Whirl

To get a better idea of what the science of chaos is all about, let's look at another example. Just for fun, we consider an amusement-park ride called the Tilt-A-Whirl, a devilish contraption that you may well have chanced to ride. As shown in Fig. 1.1, the Tilt-A-Whirl consists of several cars, each mounted on a circular platform that moves evenly along a circular track with three identical hills. The detailed geometry of the Tilt-A-Whirl will be explained in a later chapter, but for now it suffices to understand that the motion of each platform is entirely regular. Every time a platform goes over a hill, it moves through the same sequence of tilt angles. If the cars were rigidly attached to the platforms, the Tilt-A-Whirl would be a totally boring ride. Instead, the cars are free to rotate about a central pivot point, and this freedom allows chaos to creep into the machine. To see what I mean, pop the companion CD into your computer and check out the film clip of a Tilt-A-Whirl in action.

Passengers on the Tilt-A-Whirl experience chaos first hand. The cars whirl clockwise and then counterclockwise, pausing and reversing direction in an apparently random fashion that keeps you guessing what will happen next. Far from boring, the Tilt-A-Whirl may make you wish that you hadn't eaten a hot dog before boarding. Not surprisingly, the inventor of the Tilt-A-Whirl, Herbert Sellner, was well aware of the ride's chaotic nature. In his 1926 patent application, Sellner wrote, "A further object is to provide amusement apparatus wherein the riders will be

Fig. 1.1 The Tilt-A-Whirl, a chaotic amusement park ride. (Courtesy of Sellner Manufacturing.)

moved in general through an orbit and will unexpectedly swing, snap from side to side or rotate without in any way being able to figure what movement may next take place in the car." The random motion of the cars is what makes the Tilt-A-Whirl fun.

Is the Tilt-A-Whirl chaotic in the scientific sense? To answer this question, I teamed up with Bret Huggard, then a sophomore in the Physics Department at Northern Arizona University, and we set about constructing a mathematical model of the Tilt-A-Whirl. Adding in the pull of gravity, the centrifugal force of rotation, the regular tilting of the platform, and a little friction, we arrived at an equation that describes the motion of a single car in the absence of random forces. Since our equation assumes that the passengers sit perfectly still in their seats, the motion it predicts is due entirely to the periodic tilting of the platform as it traverses the successive hills. Given this periodicity, we might have presumed that the car would also settle into periodic motion, rotating in exactly the same way every time it goes over a hill. Indeed, before the 1960s, this would have been the expectation of almost any physicist. Bret and I were pleased to discover, however, that our equation instead predicted irregular motion, aptly described as chaotic and quite like that of the real machine.

Exactly what kind of chaos do you experience when riding the Tilt-A-Whirl? Figure 1.2 shows the motion of a car, predicted by our mathematical model, as it traverses 500 hills in succession. In this figure, each bar records the net rotation of the car with respect to the platform while the platform moves from the bottom of one valley to the bottom of the next. The model predicts that the car will rotate about one fifth of a revolution in the reverse (clockwise) direction while going over the first hill. On the second hill it rotates one third of a revolution forward and on the third hill almost a full revolution in reverse. The car never rotates more than about one and a half revolutions in either direction, but the sequence of rotations is apparently without pattern, alternating in an irregular way between large and small, forward and reverse. This mixed-up motion never ends: it continues for at least a million hills without any sign of repetition.

Although the reader might easily imagine motion that is much more irregular than that shown in Fig. 1.2, the Tilt-A-Whirl provides a good example of chaos in the scientific sense. The scientist's chaos is constrained but persistently random. Watching the cars circle around the Tilt-A-Whirl, we don't see anything particularly surprising. The cars never jerk or lurch: they move in a smooth, continuous fashion, accelerating and decelerating according to the laws of physics. But there is a surprise. In the long term, this very predictable motion is apparently random. Given the series of rotations shown in Fig. 1.2 for the first 100 hills, who would venture to predict the pattern for the next 100 hills? And yet we know that the second 100 hills are predicted by continuing to apply the same equation that predicted the first. This is the essential paradox of chaotic motion: it's predictable but random.

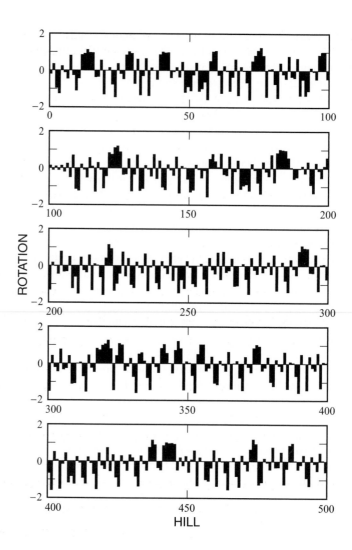

Fig. 1.2 The computed rotation of a Tilt-A-Whirl car as it traverses 500 identical hills at a platform speed of 5.9 rpm. Counterclockwise rotations are considered positive.

1.2 Digits of π

But wait, doesn't this paradox sound familiar? Aren't the digits of π, the famous mathematical constant, supposed to be random, even though it's specified by a simple formula? Perhaps physicists shouldn't have been quite so surprised when they discovered chaos.

Defined as the ratio of a circle's circumference to its diameter, π has fascinated mathematicians since ancient times. While various approximate values of π were tabulated by the Babylonians and Egyptians, Archimedes of Syracuse (287–212 B.C.) was the first to find a method of computing π to arbitrary accuracy. Fitting a circle with inscribed and circumscribed polygons having 96 sides, Archimedes showed that $3\frac{10}{71} < \pi < 3\frac{1}{7}$ and established its value within 0.1%. However, it wasn't until the advent of the calculus in the seventeenth century that π was expressed in terms of infinite series, allowing routine computation. For example, mathematicians discovered that

$$\frac{\pi}{4} = 1 - \frac{1}{3} + \frac{1}{5} - \frac{1}{7} + \frac{1}{9} - \frac{1}{11} + \cdots$$

and

$$\frac{\pi}{6} = \frac{1}{2} + \frac{1}{2}\left(\frac{1}{3 \cdot 2^3}\right) + \frac{1 \cdot 3}{2 \cdot 4}\left(\frac{1}{5 \cdot 2^5}\right) + \frac{1 \cdot 3 \cdot 5}{2 \cdot 4 \cdot 6}\left(\frac{1}{7 \cdot 2^7}\right) + \cdots .$$

A calculator suffices to confirm that these series converge toward $\pi/4$ and $\pi/6$ as more terms are included. Try it! More important, such expressions make it clear that π is a definite number that can be calculated to as many digits as desired.

What is the result of such a calculation? Using a series like those given above, we can compute the first 500 digits of π, simply by summing the series until the remaining terms are smaller than 10^{-500}. The result is

$\pi = 3.14159$ 26535 89793 23846 26433 83279 50288 41971 69399 37510

58209 74944 59230 78164 06286 20899 86280 34825 34211 70679

82148 08651 32823 06647 09384 46095 50582 23172 53594 08128

48111 74502 84102 70193 85211 05559 64462 29489 54930 38196

44288 10975 66593 34461 28475 64823 37867 83165 27120 19091

45648 56692 34603 48610 45432 66482 13393 60726 02491 41273

72458 70066 06315 58817 48815 20920 96282 92540 91715 36436

78925 90360 01133 05305 48820 46652 13841 46951 94151 16094

33057 27036 57595 91953 09218 61173 81932 61179 31051 18548

07446 23799 62749 56735 18857 52724 89122 79381 83011 94912

A quick inspection of the tabulated digits suggests that π is a complete jumble, even if it is the sum of a regular series. This impression is confirmed by Fig. 1.3, where we plot the first 500 digits of π without revealing any special order in its decimal expansion.

Mathematicians began to suspect that the digits of π might be random during the nineteenth century, when its decimal expansion was first extended to several hundred digits. This suspicion was recently confirmed in spectacular fashion by two mathematicians from the former Soviet Union, the brothers David and Gregory Chudnovsky, who calculated π to over 8 billion (8×10^9) decimal places. Their statistical analysis reveals that the digits of π are as random as possible. Not only do the digits 0 to 9 each occur with the same frequency (within statistical expectations), but no correlations of any kind are found between successive digits. Thus, the sequence "27" is as likely to appear in π as any other two-digit sequence, the sequence "835" is as likely

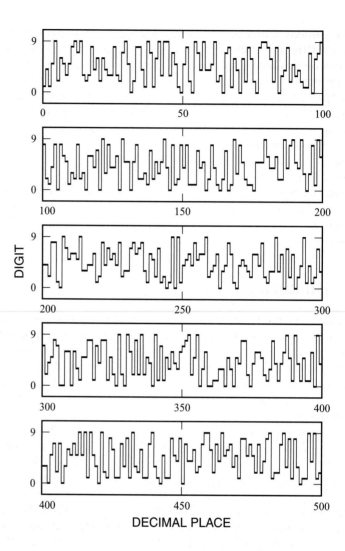

Fig. 1.3 The first 500 decimal digits of π.

as any other three-digit sequence, and so forth. In short, the decimal expansion of π is thoroughly mixed up.

Mathematicians often describe the digits of π as "pseudorandom" (literally "fake random") to indicate that they are random according to statistical tests, even though their pattern is completely specified by a simple formula. Chaos in the Tilt-A-Whirl is analogous. Although the sequence of rotations shown in Fig. 1.2 exhibit correlations over the short term, rotations separated by just a few hills are nearly statistically independent. All the same, the motion of the Tilt-A-Whirl is completely specified by the laws of dynamics, so we know there is an underlying pattern, even if we can't guess what it is by observing the motion. Like the digits of π, chaos can be described as pseudorandom: it's statistically random but predictable.

Is the paradox found in the random but predictable nature of chaos simply a matter of semantics? Perhaps we are surprised to discover that things can be jumbled and predictable at the same time just because the word "random" encompasses both "jumbled" and "unpredictable." Of course, "pseudorandom" was introduced precisely to allow for the possibility that "jumbled" doesn't imply "unpredictable." However, making the distinction between random and pseudorandom doesn't lessen our surprise that order can result in a jumbled mess. Gregory Chudnovsky once said, "Pi is a damned good fake of a random number," expressing his frustration that the digits of π give no hint of the underling order. Similarly, the pseudorandom behavior of a chaotic machine is paradoxical because the rule describing its motion shows no sign of disorder and the observed motion gives no sign of an ordered rule. Before the end of the book, we'll try to resolve this paradox.

1.3 Butterfly effect

The random character of chaos is its most obvious feature, apparent even to a casual observer of the Tilt-A-Whirl. A more subtle but equally important feature of chaos is known to scientists as "extreme sensitivity to initial conditions" or, more poetically, the "butterfly effect." As we'll see, extreme sensitivity explains why chaos is predictable in principle but not in practice and how chaos can be both stable and unstable at the same time.

What could scientists possibly mean by extreme sensitivity to initial conditions? While the phrase is a mouthful, the effect is easily demonstrated with our mathematical model of the Tilt-A-Whirl. To show the effect, we'll perform what physicists call a numerical experiment. In particular, we'll use a computer to calculate the motion of the Tilt-A-Whirl using several slightly different assumptions about the initial position of the car. For the calculation shown in Fig. 1.2, the machine was started with the platform at the bottom of a hill and the car oriented with the passengers facing the center of the machine. If we always begin the calculation with the car in this position, the pattern of rotations is invariably that shown in Fig. 1.2.

Now suppose that a passenger bumps the car while boarding, shifting its initial position by a millimeter. Experience with other machines suggests that a millimeter offset won't make much difference: the sequence of rotations should be about the same, although a small offset in the car's position may persist. Performing the numerical experiment, however, proves that a silly millimeter can make a dramatic difference. The first two frames of Fig. 1.4 compare the rotations calculated for the first 100 hills with and without an initial 1 mm offset. At first, the offset doesn't appear to make any difference: the rotations recorded for the first nine hills are nearly identical to those with no offset. Beginning with the tenth hill, however, differences suddenly appear. Given an initial offset, the car rotates -0.3 rather than -1.1 revolutions on the tenth hill,

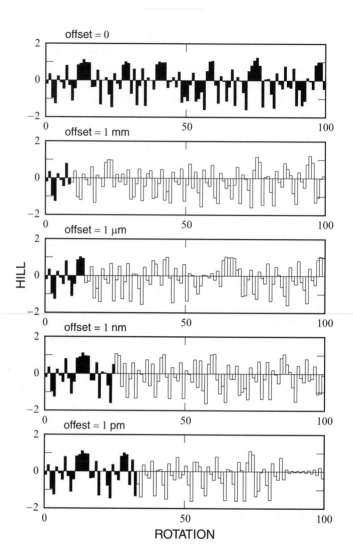

Fig. 1.4 The computed rotation of a Tilt-A-Whirl car as it traverses 100 hills at a platform speed of 5.9 rpm, assuming various initial positions of the car. In all cases the platform is initially at the bottom of a hill and the car is at rest with respect to the platform. Rotations plotted in the first frame assume that the car is positioned exactly toward the outside of the machine, such that the passengers face the center. Other frames show the rotations that result when the car is offset from this position by 1 mm, 1 μm, 1 nm, and 1 pm, measured at the outer edge of the platform in the counterclockwise direction. In each plot, rotations are shown by open bars after they differ from the zero-offset case by more than 0.2 revolutions.

+0.5 rather than −0.4 revolutions on the eleventh hill, and −1.3 rather than +0.9 revolutions on the twelfth hill. Once the effect of the offset becomes obvious, such differences persist indefinitely, and the offset leads to a completely new sequence of random rotations. In Fig. 1.4, the new portion of the sequence is shown with open bars.

Since one random sequence is much like another, the passengers will never guess the effect of the bump that shifted the car's position. After ten hills or so, they may be rotating forward instead of backward, but they'll have just as much fun. For a physicist trying to predict the future, on the other hand, the millimeter offset spells disaster. Since the Tilt-A-Whirl passes over a hill every 3 seconds, his predictions will be grossly wrong after just 30 seconds, if he doesn't account for the offset. This kind of error is particularly embarrassing when you consider that the

Tilt-A-Whirl is governed by the same tried and true laws of dynamics that landed a man on the Moon. What is a physicist to do?

As our numerical experiment demonstrates, accurate prediction of the Tilt-A-Whirl's motion requires, at minimum, a precise knowledge of the car's initial position. Pulling out all stops, suppose we use a laser interferometer to locate the car within one micrometer ($1 \mu m = 10^{-6}$ m), roughly 1/50 of the diameter of a human hair. Another numerical experiment, shown in the third frame of Fig. 1.4, reveals that an offset of $1 \mu m$ leads to rotations that match the pattern with no offset for the first 14 hills. As expected, improving our knowledge of the initial position of the car allows us to predict farther into the future. But after 14 hills, the pattern of rotations with the $1 \mu m$ offset is again grossly different from that with no offset! What's happened to our ability to predict motion? Measuring the location of the car within $1 \mu m$ is at the limit of what is practically possible, and even with this accuracy we can't predict the motion of the Tilt-A-Whirl for more that 40 seconds.

And things get worse. As the last two frames of Fig. 1.4 show, even if the offset is reduced to one nanometer (1 nm $= 10^{-9}$ m), about the diameter of a sugar molecule, or one picometer (1 pm $= 10^{-12}$ m), about 70 times the diameter of a uranium nucleus, accurate prediction is extended only to 80 or 100 seconds.

Long-term prediction is simply impossible. Because the Tilt-A-Whirl exhibits extreme sensitivity to initial conditions, there is no practical way of predicting its motion even a minute into the future. Our model ignores all noise, all irregularities in the machine, all outside disturbances, and yet, even under these ideal conditions, long-term prediction depends on knowing the initial position of the car with inconceivable precision. Clearly, sensitivity to initial conditions leads to motion that is predictable in principle but not in practice.

Although stated in terms of initial conditions, the extreme sensitivity of the Tilt-A-Whirl is a persistent effect. No matter when it occurs, the slightest bump will, after a brief time, cause a car to rotate clockwise when it would otherwise have rotated counterclockwise and completely alter the succeeding random sequence of rotations. In this sense, the Tilt-A-Whirl is highly unstable. Even so, bumping a car doesn't cause it to suddenly stop or rotate extraordinarily fast. In fact, the perturbed motion has the same statistical properties as the unperturbed motion, as suggested by the open bars in Fig. 1.4. In this sense, small disturbances make almost no difference, and the Tilt-A-Whirl is extremely stable. Surprisingly, chaotic motion is unstable and stable at the same time: unstable in its exact trajectory but stable in its statistical properties.

Sensitivity to initial conditions also provides a reply to those who object that chaotic motion is merely pseudorandom because it's exactly predicted by a mathematical model. In the real world, chaotic systems are always subject to minute perturbations that prevent us from predicting their future. Since the motion of an actual Tilt-A-Whirl isn't predictable in the long term, the machine can be considered truly random, not just pseudorandom. To be precise, we might say that the mathematical model produces exactly predictable, pseudorandom motion, while the real-world machine produces pseudopredictable, truly random motion. Thus, to be precise, the final phrase of this book's subtitle should be read as either *"Predictable Pseudorandom Motion"* or *"Pseudopredictable Random Motion"*—take your pick.

For simplicity, in all that follows we'll use the word "random" when we mean "pseudorandom" and make the distinction only when it's warranted.

1.4   Weather prediction

Extreme sensitivity to initial conditions is a hallmark of chaotic motion. It severely limits our ability to predict the future, and it implies that even the smallest disturbance will eventually change the course of history. Of course, predicting the future of the Tilt-A-Whirl is of no consequence: even the passengers don't care whether they're rotating clockwise or counterclockwise. On the other hand, meteorologists believe that the Earth's weather is also chaotic, and accurate weather prediction is of interest to almost everyone.

Compared to our mathematical model of the Tilt-A-Whirl, the equations that describe the Earth's weather are highly complex. The principle is the same, however. Given the present state of the atmosphere around the world, meteorologists use powerful computers to solve the equations of fluid dynamics and forecast the weather in days to come. If these equations describe a chaotic system, then we can understand some basic features of the weather.

As you've probably noticed, the weather simply isn't the same from day to day. Instead, we experience sunshine, clouds, heat, cold, wind, and rain in patterns that have a certain familiarity but aren't the same from week to week or year to year. Indeed, weather patterns are much like those plotted in Fig. 1.2 for the rotation of the Tilt-A-Whirl: persistently random but limited in their variety. It would make perfect sense if weather were the product of chaos.

Now consider weather prediction. If the weather is sensitive to initial conditions, then the accuracy of a forecast will depend critically on how well we know the present state of the atmosphere. Using the Tilt-A-Whirl as a guide, we can understand why meteorologists are seldom able to predict the weather more than a few days in advance. Any deficiency in our knowledge of the present state of the atmosphere will inevitably lead to predictions that are grossly in error a short time later. Indeed, it's likely that the state of the atmosphere has never been recorded with the precision needed to make an accurate forecast more than a few days in advance. Moreover, if the Earth's weather really is chaotic, the weatherman is likely to remain the butt of jokes for the foreseeable future. As with the Tilt-A-Whirl, longer-range weather forecasts will require heroic efforts just to pin down the initial conditions with sufficient accuracy.

Perhaps you won't be surprised to learn that a meteorologist gave the name "butterfly effect" to the property of extreme sensitivity to initial conditions. In 1972, while alerting scientists to the possibility that the weather is chaotic, Edward Lorenz posed the question, "Does the flap of a butterfly's wings in Brazil set off a tornado in Texas?" Of course, we can't expect the butterfly to have an immediate effect, but, just like a small disturbance in the Tilt-A-Whirl, we can expect the butterfly's action to set the atmosphere on a new path that will lead, months or years later, to a completely altered pattern of weather. A butterfly can't change the statistics of the weather, but it can determine whether a tornado occurs at a particular time and place. With butterflies involved, it's little wonder that long-range forecasts are so difficult to get right.

1.5   Inward spiral

In this chapter, we have begun to unfold the paradox of chaotic motion. The essential surprise is that random motion is predicted by the same laws of dynamics that were developed to explain regular motion. Equally surprising, chaotic systems are so sensitive to perturbations that predicting a real-world trajectory over an extended time is practically impossible, even if the equation of motion is known exactly. Thus, although we describe chaos as predictable random motion, it's entirely predictable only within mathematical models, and it's truly random, as opposed to pseudorandom, only in the real world. Chaos is strange indeed.

But chaos is also exciting. For three centuries the laws of dynamics were successfully applied to system after system, until it was believed that the future was entirely predictable, if we only knew all the forces involved. This picture of a clockwork universe began to lose credibility first in the atomic domain with the advent of quantum mechanics during the 1920s. At that time, physicists discovered that microscopic systems are irreducibly random and unpredictable. With the discovery of chaos in the 1960s, however, unpredictability was extended to the macroscopic world. According to chaos theory, the motion of the Tilt-A-Whirl, a gross machine, cannot be foretold a minute in advance, even in the absence of quantum uncertainty. Chaos is a strange new effect that forces us to revise downward our estimate of what science can do. Scientists now realize that when chaos is present their knowledge of the future may be extremely limited even in the macroscopic world. During the last 50 years, this new paradigm has overturned much of what was once held sacred in the science of dynamics, and its consequences continue to ripple throughout scientific thought.

In this book our primary goal is to discover what makes chaos tick. In pursuing this goal we follow an inward spiral that proceeds from easy to difficult. Thus, we repeatedly circle the topic of chaotic motion, presenting the simplest ideas first and only slowly make our way toward the deeply mathematical core of chaos theory. To stay in touch with reality, we focus primarily on one example: the motion of a pendulum. Long a symbol of regularity, the pendulum is also one of the simplest mechanical systems to display chaotic behavior. Of course, we won't find chaos if we confine our pendulum inside the mahogany case of a grandfather clock. But, when driven with enough force to propel it full circle, a pendulum frequently goes chaotic.

Indeed, the Tilt-A-Whirl has already provided our first example of a chaotic pendulum. Although the cars rotate about an axis that is nearly vertical rather than horizontal, the equation that describes the motion of a car on a fixed, tilted platform is exactly that of a pendulum. The passengers just add weight to the pendulum bob and tag along for a wild ride.

Before we begin to delve into the paradox of chaos, however, we need to take a giant step backward and learn more about the physics of motion. Fortunately, what we need to know won't take us beyond the mathematical realms of elementary algebra, vectors, and trigonometry. Using these tools, Part II of the book introduces the science of dynamics as it was developed by Galileo Galilei and Isaac Newton in the sixteenth and seventeenth centuries. Here and throughout the book, examples guide the way, and the motion, presented initially in terms of equations, is also shown graphically.

In Part III, chaotic motion itself takes center stage, and we begin by exploring its randomness. As we learn here, chaos can mimic many classic random processes, from a drunkard's walk, to the hiss of radio static, to the decay of radioactive atoms. Similarly, Part IV explores the consequences of the butterfly effect in several chaotic systems, including the weather, the Tilt-A-Whirl, and the atoms of a gas.

Fig. 1.5 *"All in all, I'd say it's a pretty convincing explication of chaos theory."* (© Gahan Wilson—The New Yorker Collection—www.cartoonbank .com, by permission.)

While Parts III and IV describe the basic phenomena of chaotic motion, behavior that is directly observable, Part V penetrates to the mathematical heart of chaos. As we approach the center of our inward spiral, the concepts become more abstract, and we'll rely on analogies and pictures to reveal the nature of the underlying mathematics. In the process, you may feel that you're gradually slipping from left to right in Gahan Wilson's cartoon about chaos theory (Fig. 1.5). In the end, however, you'll glimpse something close to the frontier of mathematical dynamics, where the topology of an abstract space is of central importance.

As you spiral inward through the coming chapters, please take time to try the experiments presented in the Dynamics Lab on the companion CD. The experiments animate many of the dynamic systems used as examples and demonstrate how simple equations can capture realistic motion, whether it's regular or chaotic.

Further reading

Readings that use advanced mathematics are marked with a solid bullet (●).

Butterfly effect

○ Lorenz, E. N., "The butterfly effect", in *The Essence of Chaos* (University of Washington Press, Seattle, 1993) Appendix 1.

Digits of π

○ Beckmann, P., *A History of π* (St. Martin's Press, New York, 1971).

○ Blatner, D., *The Joy of π* (Walker, New York, 1997).

○ Petersen, I., "Pi á la mode", *Science News* **160**, 136–137 (2001).

○ Posamentier, A. S. and Lehmann, I., *π: A Biography of the World's Most Mysterious Number* (Prometheus, Amherst, New York, 2004).

○ Preston, R., "Mountains of pi", in *Life Stories: Profiles from the New Yorker*, edited by D. Remnick (Random House, New York, 2000).

○ Schumer, P. D., "Episodes in the calculation of pi", in *Mathematical Journeys* (Wiley, Hoboken, 2004) chapter 11.

○ Wrench, J. W., "The evolution of extended decimal approximations to π", *The Mathematics Teacher* **53**, 644–650 (1960).

Dripping faucet

● Ambravaneswaran, B., Phillips, S. D., and Basaran, O. A., "Theoretical analysis of a dripping faucet", *Physical Review Letters* **85**, 5332–5335 (2000).

○ Carlson, S., "Falling into chaos", *Scientific American* **281**(5), 120–121 (November 1999).

● Martien, P., Pope, S. C., Scott, P. L., and Shaw, R. S., "The chaotic behavior of the leaky faucet", *Physics Letters* A **110**, 399–404 (1985).

● Shaw, R., *The Dripping Faucet as a Model Chaotic System* (Aerial Press, Santa Cruz, 1984).

Falling leaves

● Tanabe, Y. and Kaneko, K., "Behavior of a falling paper", *Physical Review Letters* **73**, 1372–1375 (1994).

○ Weiss, P., "The puzzle of Flutter and Tumble", *Science News* **154**, 285–287 (1998).

Heart attack

○ Glass, L., Shrier, A., and Bélair, J., "Chaotic cardiac rhythms", in *Chaos*, Holden, A. V., editor (Princeton University Press, Princeton, 1986) pp. 237–256.

○ Glass, L., "Dynamics of cardiac arrhythmias", *Physics Today* **49**(8), 40–45 (August 1996).

○ Peterson, I., "Identifying chaos in heart quakes", *Science News* **151**, 52 (1997).

Tilt-A-Whirl

● Kautz, R. L. and Huggard, B. M., "Chaos at the amusement park: Dynamics of the Tilt-A-Whirl", *American Journal of Physics* **62**, 59–66 (1994).

○ Page, D., "Formula for fun: The physics of chaos", *Funworld* **12**(3), 42–46 (March 1996).

○ Sellner, H. W., "Amusement device", U.S. Patent number 1,745,719, February 4, 1930 (Filed April 24, 1926).

Weather prediction

○ Lorenz, E. N., "Our chaotic weather", in *The Essence of Chaos* (University of Washington Press, Seattle, 1993) chapter 3.

○ Marchese, J., "Forecast: Hazy", *Discover* **22**(6), 44–51 (June 2001).

○ Matthews, R., "Don't blame the butterfly", *New Scientist* **171**(2302), 24–27 (August 4, 2001).

● Read, P. L., "Applications of chaos to meteorology and climate", in *The Nature of Chaos*, Mullin, T., editor (Oxford University Press, Oxford, 1993) pp. 222–260.

○ Rosenfeld, J., "The butterfly that roared", *Scientific American Presents* **11**(1), 22–27 (Spring 2000).

Part II
Dynamics

Galileo Galilei—Birth of a new science

In 1583, while still a teenager, Galileo Galilei (1564–1642) became interested in the science of motion. At the time, he was a student of philosophy and medicine at the University of Pisa, and a simple observation piqued his interest. While contemplating a chandelier swinging back and forth in the Cathedral at Pisa, it occurred to Galileo that, contrary to intuition, the time required for the chandelier to complete a full swing might not depend on the size of the swing. He guessed that the extra speed gained when the chandelier was released from a greater height would compensate for the longer distance it had to travel.

Perhaps daydreaming in church isn't entirely a bad thing. Galileo's speculation led him first to check the period of the chandelier's motion against his own pulse. Sure enough, a large swing was completed in about the same time as a small one. But Galileo wasn't satisfied with this crude measurement. Later, he constructed two identical pendulums, consisting of lead balls suspended from strings. When he released the balls simultaneously, it was easy to see that the pendulums moved in synchrony, even if one was initially displaced from vertical much farther than the other. The period of oscillation was nearly independent of the size of the swing. Further experiments revealed another simple regularity: the period of a pendulum is proportional to the square root of its length. Thus, even when the pendulum first entered the realm of science, it was a model of regularity, and, given the constancy of its swing, Galileo soon advocated building a pendulum clock.

Galileo wasn't the first to contemplate motion, just the first to understand it correctly. In Galileo's day, the writings of Aristotle (384–322 B.C.) were the unquestioned authority on motion and were part of the curriculum at the University of Pisa. However, Galileo knew intuitively that the only real authority in science is experiment, and he learned to distrust the pronouncements of ancient Greeks. While Aristotle said that the natural state of any body is at rest, Galileo came to understand that a body in motion will continue in motion unless acted on by a force. At first, Galileo's idea seems crazy: no matter how hard you throw a rock it eventually stops moving. But Galileo was well aware that even the air provides a resistance that acts to slow a rock, not to mention the friction imparted by the ground when it falls back to Earth. To Galileo,

Fig. 2.1 Galileo Galilei. Portrait by Domenico Tintoretto, circa 1605–1607. (© National Maritime Museum, Greenwich, London, by permission.)

these frictional forces explained why bodies in motion tend to stop, not a universal tendency of bodies to seek a state of rest.

Moreover, Galileo argued, if all bodies seek rest, then how can you explain the behavior of a ball tossed upward in the cabin of a speeding ship? If a ball always seeks rest, it would immediately begin to lose its speed relative to the land and end up hitting the stern wall of the cabin. Instead, the ball behaves as if the ship were stationary, falling right back into your hand. The natural tendency of the ball is to continue with the same speed as the ship, and it slows only when there are forces opposing its motion. Can you imagine the pandemonium aboard a jetliner if Aristotle had been right? Everything that wasn't tied down would quickly end up at the rear of the plane.

Often called the law of inertia, Galileo's observation that a body in motion continues in motion unless acted on by a force became the cornerstone of the science of dynamics. Galileo developed dynamics as an experimental and mathematical science over a period of 20 years, from roughly 1590 to 1610, while a professor at the Universities of Pisa and Padua. However, he published his ideas only in 1638 at the age of 74. By this time, Galileo was old, infirm, and living under virtual house arrest. Why was an elderly professor under house arrest? Galileo had advocated the idea that the Earth orbits the Sun and isn't the center of the universe, contrary to Aristotle and the doctrine of the Catholic Church. Tried by the Inquisition as a heretic in 1633, Galileo escaped with his life only when he agreed to recant, and he remained sequestered until his death in 1642. Given the possibility of censorship or worse, Galileo's treatise on motion, entitled *Discourses and Mathematical Proofs, Concerning Two New Sciences Pertaining to Mechanics and Local Motions*, was smuggled out of Italy and published in Holland. Eventually the Church apologized to Galileo, but not until 1997, after more than three centuries.

In his *Discourses*, Galileo discusses three types of natural motion: the uniform inertial motion of a body with no forces acting on it, the accelerated motion of a falling body, and the trajectory of a cannonball. Even today, there is no better introduction to dynamics than a survey of these simple forms of motion.

2.1   When will we get there?

As a kid growing up in western South Dakota, I remember spending long hours in our family's '53 Chevy, driving across interminable stretches of sagebrush prairie. When I was maybe 10, Dad answered the inevitable question by handing me the road map and telling me to figure it out for myself. Okay, you add up the mileage to your destination, divide by your average speed, and, *voilà,* predict how many hours it will be before you roll into Lusk, Wyoming. The long division was a little tedious, but I thought it was neat to be able to predict our arrival time, occasionally within a minute or two. Little did I suspect then that I would make a career out of calculations that are only slightly more sophisticated.

Of course, a '53 Chevy doesn't move in the absence of an applied force, but motoring across the prairie is governed by the same mathematics as inertial motion. In both cases the velocity V is constant, and the time T required to go a distance D is simply

$$T = D/V. \qquad (2.1)$$

This is a useful formula if you want to know when you'll arrive in Lusk, and it also applies to bodies moving in the absence of a force. In either case, Eq. (2.1) tells us that traveling a distance of 120 miles at 60 miles per hour will require 2 hours, a simple fact of life on the highway. We can also rearrange Eq. (2.1) to tell us the distance traveled in a given time,

$$D = V \times T \qquad (2.2)$$

and this is the form of the equation that we'll find most useful here. In all that follows, we'll be predicting the location of a body in motion at some later time.

Inertial motion is particularly simple. In Galileo's day, it was exemplified by a ball rolling along a horizontal track or a boat drifting across the water after pushing off from shore. Because the friction is slight, the ball and boat lose their initial velocity only gradually. Today, a better example is provided by the game of air hockey. This game requires a special table with a closely spaced grid of tiny holes and a pump that forces air through the holes from below. Because the puck rides on a cushion of air, it glides along without perceptibly slowing. Air hockey provides an excellent idea of inertial motion, the best this side of strapping on a rocket pack and drifting around in outer space. Having spent considerable time experimenting with balls rolling on tracks, Galileo would be envious of such modern toys as air hockey and rocket packs.

2.2   Computer animation

We can easily create a computer animation of a hockey puck in motion. As with any motion picture, it is simply a matter of displaying a succession of still frames with the position of the moving object shifted slightly from frame to frame. While the required mathematics is given entirely by Eq. (2.2), it is useful to rewrite this equation in a form suited to animation. Because motion is all about change, we will adopt the common practice of prefacing a quantity X by a Greek capital delta (Δ) to denote a small change in that quantity. Thus, the pair of symbols ΔX, pronounced "delta-ex," represents a small change in X, not Δ multiplied by X. Now, if the change in time between animation frames is ΔT, then according to Eq. (2.2) a puck moving with velocity V will shift its position by a distance $V \times \Delta T$ from one frame to the next. That is, if the last frame was displayed at time T_{old} with the puck at position D_{old}, then the time T_{new} and the position D_{new} for the next frame are given by

$$T_{\text{new}} = T_{\text{old}} + \Delta T, \tag{2.3}$$

and

$$D_{\text{new}} = D_{\text{old}} + V \times \Delta T. \tag{2.4}$$

These equations tell us how to update the time and position incrementally, given the velocity V and time step ΔT.

How about the next frame of our movie? Once the "new" frame is displayed, we can reuse Eqs. (2.3) and (2.4) by letting the variables T_{old} and D_{old} assume the values just computed for T_{new} and D_{new}, then compute the time and position for yet another frame. Mathematicians call the repeated use of equations in this fashion "iteration," and by iterating Eqs. (2.3) and (2.4) we can predict successive times and positions of the hockey puck as far into the future as desired.

Amazingly, all of the motion discussed in this book, whether regular or chaotic, is described by sets of iterated equations similar to Eqs. (2.3) and (2.4). In each case, the future is predicted step by step, ΔT by ΔT, using simple equations that relate the new values of the motion variables to their old values. While these equations generally look more complicated than Eqs. (2.3) and (2.4), they work in the same simple way. If dynamics is really this easy, you might be thinking, it doesn't take a rocket scientist to predict the future. But of course, this is exactly how rocket scientists calculate trajectories!

An animated hockey puck is presented as the first experiment of the Dynamics Lab. To check it out, insert the companion CD into your PC, load the "Dynamics Lab" program, and select "Air Hockey" from the list of experiments. Just click the "Start" button to see the inertial motion of a puck as it bounces back and forth between the ends of the table. The program that creates this animation is simple. For each time increment, it calculates the new position of the puck using Eq. (2.4), waits until the time ΔT has elapsed, and redraws the puck at its new location. Bouncing isn't built into Eq. (2.4), but it's easily added by reversing the velocity whenever the puck reaches the end of the table. Try the experiment with different time increments ΔT to see how it affects the animation.

The strategy of describing motion in small time increments, while natural to animation, would have seemed strange to Galileo, who generally expressed his results in terms of total time and distance. Indeed, the ancient Greeks discovered that breaking time and distance into small pieces could be dangerous. According to Zeno of Elea (495–430 B.C.), doing so leads to the inevitable conclusion that motion is simply impossible. How so? Suppose our air hockey table is 1 meter long and the puck travels at 0.1 meter/second. While we immediately conclude that the puck will reach the far end in 10 seconds, Zeno argued that it would never get there. According to Zeno, the puck must first go half the distance to arrive $\frac{1}{2}$ meter from the end. Then it must travel half the remaining distance to arrive $\frac{1}{4}$ meter from the end, then half the yet remaining distance to arrive $\frac{1}{8}$ meter from the end (see Fig.2.2). Continuing in this fashion, Zeno argued that the puck can come as close to the end of the table as you'd like—$\frac{1}{16}$, $\frac{1}{32}$, or even $\frac{1}{1024}$ meter—but it can never actually reach the end. Moreover, because the end of the table is an arbitrarily chosen goal, Zeno concluded that the puck can never move anywhere at all.

Zeno's motion paradox may seem patently absurd, but it illustrates the difficulty of working with infinity. Mathematically, the interval from 0 to 1 is a continuum that includes an infinite number of points. As Zeno realized, between any two points on a line, we can always find another point, say the point midway between the two, so a line can be broken into an arbitrarily large number of segments. Thus, if our hockey puck is to traverse a line 1 meter long, it must pass through an infinite number of line segments, a seemingly impossible task. On the other hand, we know that the total length of this infinity of segments is just 1 meter. Furthermore, since the time required to traverse any segment is proportional to its length, the sum of all the time increments is also finite. Thus, while Zeno is writing out his list of all the line segments that the puck has to cross before it can travel 1 meter, the puck has long since bounced off the end of the table. Contrary to Zeno's implicit assumption, an infinite number of things can happen in a finite time. Nevertheless, from Zeno's point of view, even uniform inertial motion is truly amazing.

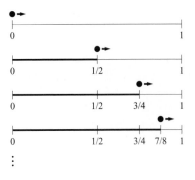

Fig. 2.2 Zeno argued that motion is impossible because a body cannot pass through an infinite number of line segments in a finite time.

2.3   Acceleration

But what happens if the velocity isn't constant? Whenever a body in motion speeds up, slows down, or changes its direction of motion, physicists say it's accelerating. Clearly, if we're going to describe motion in general, we'll have to deal with acceleration. For anyone who's ridden in a car, acceleration is a familiar concept. It happens when the driver steps on the gas and we're pushed back into our seat, when the driver slams on the brakes and we're thrown toward the dashboard, or when she steers around a corner and we're forced against the door. In each case, we're well aware that our motion is no longer inertial, and during such maneuvers we don't expect a ball tossed in the air to return neatly to our hand.

When I was a teen in the 1960s, we often talked about cars being able to go from 0 to 60 miles per hour (mph) in a few seconds, maybe 10. In that case, a physicist would say that the average acceleration is 6 miles per hour per second. Today, sports cars off the showroom floor can reach 60 mph in about 4 seconds, for an average acceleration of 15 miles per hour per second. In any case, life on the freeway has taught us not only what acceleration is but how to measure it. Acceleration is the rate of change of velocity and is measured in miles per hour per second or, if you're a physicist, in meters per second per second (abbreviated m/s/s or m/s^2). To avoid using two different units of time, in this book

Fig. 2.3 The top fuel dragster is an acceleration machine. (Photo by Major William Thurmond, courtesy of the US. Army.)

we'll measure velocities in meters per second and accelerations in meters per second per second. Since a velocity of 1 m/s is about 2.24 mph, an acceleration of 1 m/s² is equivalent to 2.24 miles per hour per second. If you want to see real acceleration, you should check out the action at a drag strip, where top fuel dragsters go from 0 to 300 mph (134 m/s) in 5 seconds. That's an average acceleration of 27 m/s². What a rush!

For motion along a line, the acceleration A tells us how fast the velocity is changing, either increasing ($A > 0$) or decreasing ($A < 0$). Thus, if we accelerate at A for a time T, our velocity will change by $A \times T$, and accelerating at 5 m/s² for 6 seconds will increase our speed by $5 \times 6 = 30$ m/s or 67 mph. Of course, the acceleration of a moving body isn't always constant. However, if we are satisfied with incremental calculations that predict the velocity V_{new} just a short time ΔT later, then we can assume that the acceleration is roughly constant at its old value A_{old} over ΔT. In this case, we arrive at the following incremental equation for the velocity.

$$V_{\text{new}} = V_{\text{old}} + A_{\text{old}} \times \Delta T. \tag{2.5}$$

When combined with Eqs. (2.3) and (2.4), this equation completes the basic set of equations we'll need to describe all the types of motion we'll encounter in this book. To predict the position of a moving body a short time ΔT in the future, we add $V_{\text{old}} \times \Delta T$ to the old position, and to predict the new velocity we add $A_{\text{old}} \times \Delta T$ to the old velocity. How do we predict the new acceleration? That's a question we won't answer in general until Chapter 3.

2.4   Free-fall

One of Galileo's greatest discoveries concerns the acceleration of falling bodies. Because free-fall occurs so rapidly, Galileo made his discovery by observing the slower motion of balls rolling down inclined planes. In the end, however, he came to the conclusion that, when wind resistance can be neglected, freely falling bodies experience constant acceleration. That is, a falling body always speeds up at exactly the same rate, whether it has just been released or has been falling for some time. Moreover, Galileo found that the acceleration is the same for all kinds of things—wooden blocks, lead balls, apples, and oranges—as long as they aren't strongly affected by wind resistance. Thus, while feathers don't obey Galileo's law of free-fall, knowing a single number, the acceleration of gravity, allows us to predict the motion of most falling bodies. A remarkably simple and powerful result!

In the case of free-fall, we know how to predict the acceleration because it's always the same. Physicists usually denote the acceleration of gravity by g, which has a value of about 9.8 m/s². Accordingly, when you drop a ball from a tall building, its speed will be 9.8 m/s 1 second after release, 19.6 m/s 2 seconds after release, and 29.4 m/s 3 seconds after release.

Of course, the faster the ball goes, the greater the wind resistance, so we can't expect the acceleration to continue indefinitely.

Called "little gee" by physicists, g is such a well-known acceleration, that aviators commonly say things like, "I pulled 6 g's on that maneuver," and know they're likely to blackout at accelerations over about 10 g. Similarly, the driver of a top-fuel dragster experiences an average acceleration of almost 3 g.

How can we use g to predict the position of a falling ball? Suppose that we take the position of the ball to be its height H above the ground and assume that it is dropped from a building of height B at time $T = 0$. Initially, then, the ball is at height $H = B$ and has velocity $V = 0$. To find its position at later times, we apply the incremental equations (2.3)–(2.5). In this case the distance D is the height H, and the acceleration A has magnitude g. However, because the the acceleration is opposite to the direction in which position is measured, we have $A = -g$. Thus, when the incremental equations are applied to a falling ball, they become,

$$T_{\text{new}} = T_{\text{old}} + \Delta T, \tag{2.6}$$

$$H_{\text{new}} = H_{\text{old}} + V_{\text{old}} \times \Delta T, \tag{2.7}$$

$$V_{\text{new}} = V_{\text{old}} - g \times \Delta T. \tag{2.8}$$

Do these equations make sense? From Eq. (2.8), we see that during the first time increment the velocity will go from zero to $-g \times \Delta T$ and thus becomes negative. A negative velocity means that, when Eq. (2.7) is used to compute the ball's height after the next time increment, the height is reduced. The ball is falling!

Perhaps the most important feature of Eqs. (2.6)–(2.8) is that they can be used repeatedly by substituting the new values of T_{new}, H_{new}, and V_{new} obtained on the right-hand side into the variables T_{old}, H_{old}, and V_{old} on the left to obtain yet newer values of the dynamic variables. Thus, thanks to Galileo's insight that the acceleration of a falling body is constant, we can compute the body's height and velocity as far into the future as desired. Indeed, the procedure is so simple that you can easily make the required computations on a hand calculator. If, for example, you consider a ball dropped from a height of 20 meters and take $\Delta T = 0.1$ second, then you obtain the results shown in the first three columns of Table 2.1. That is, beginning with $(T_{\text{old}}, V_{\text{old}}, H_{\text{old}}) = (0, 0, 20)$, the first row of Table 2.1, we obtain from Eqs. (2.6)–(2.8) the result $(T_{\text{new}}, V_{\text{new}}, H_{\text{new}}) = (0.1, -0.98, 20)$, the second row. And, beginning with $(T_{\text{old}}, V_{\text{old}}, H_{\text{old}}) = (0.1, -0.98, 20)$, we obtain $(T_{\text{new}}, V_{\text{new}}, H_{\text{new}}) = (0.2, -1.96, 19.9)$, the third row. Predicting the future is easy.

However, the predictions made by our incremental equations are only approximate. As we noted earlier, these equations assume that during each time increment ΔT, the velocity is constant at the value it assumed at the beginning of the increment. Thus, after 0.1 s, we find in Table 2.1 that the height of the ball is still 20 m, although its speed has increased to 0.98 m/s. For greater accuracy, we can use a smaller time increment,

Table 2.1 The velocity and height of a falling ball as a function of time. The ball is initially at rest and falls from a height of $B = 20$ m. The height at later times was computed using incremental equations with time increments ΔT of 0.1 s and 0.01 s and using the exact formula.

Time (s)	Velocity (m/s)	$\Delta T = 0.1$ s Height (m)	$\Delta T = 0.01$ s Height (m)	Exact Height (m)
0.0	0.0000	20.0000	20.0000	20.0000
0.1	−0.9800	20.0000	19.9559	19.9510
0.2	−1.9600	19.9020	19.8138	19.8040
0.3	−2.9400	19.7060	19.5737	19.5590
0.4	−3.9200	19.4120	19.2356	19.2160
0.5	−4.9000	19.0200	18.7995	18.7750
0.6	−5.8800	18.5300	18.2654	18.2360
0.7	−6.8600	17.9420	17.6333	17.5990
0.8	−7.8400	17.2560	16.9032	16.8640
0.9	−8.8200	16.4720	16.0751	16.0310
1.0	−9.8000	15.5900	15.1490	15.1000
1.1	−10.7800	14.6100	14.1249	14.0710
1.2	−11.7600	13.5320	13.0028	12.9440
1.3	−12.7400	12.3560	11.7827	11.7190
1.4	−13.7200	11.0820	10.4646	10.3960
1.5	−14.7000	9.7100	9.0485	8.9750
1.6	−15.6800	8.2400	7.5344	7.4560
1.7	−16.6600	6.6720	5.9223	5.8390
1.8	−17.6400	5.0060	4.2122	4.1240
1.9	−18.6200	3.2420	2.4041	2.3110
2.0	−19.6000	1.3800	0.4980	0.4000

so that the velocity is updated more frequently. If we use $\Delta T = 0.01$ s and print out every tenth computation, we obtain the result for the height shown in the fourth column of Table 2.1. This more accurate computation shows that the ball has dropped by about 4 cm after 0.1 s.

Amazingly, even in the seventeenth century, Galileo knew the mathematically exact equations for a body subject to constant acceleration. As usual, Galileo stated his results in terms of the total elapsed time, rather than incremental time. Suppose again that a ball is released at height $H = B$ with velocity $V = 0$ at time $T = 0$. Expressed in modern notation, Galileo's results for H and V at any later time T are,[1]

$$H = B - \frac{1}{2}gT^2, \tag{2.9}$$

$$V = -gT. \tag{2.10}$$

Using these formulas, we can easily fill in the exact values for the height of the ball in Table 2.1. As the table shows, the height predicted by the incremental calculation with $\Delta T = 0.1$ s is in error by almost 1 m after 2 seconds, while the calculation with $\Delta T = 0.01$ s is off by only about 10 cm. Clearly, we must be careful to choose a small enough time increment when using incremental equations.[2]

[1] Here and in subsequent equations, multiplication is indicated without a times sign, using the convention $AB = A \times B$.

[2] The calculations presented in Table 2.1 make use of what is called Euler's method, in which the dynamic variables remain fixed over each time increment. While this method is easy to understand and adequate to reveal all of the phenomena presented in this book, greater accuracy can be obtained by interpolating more carefully over the time increments. To insure accuracy, all other computations presented here and in the Dynamics Lab use an interpolation method known as fourth-order Runge–Kutta.

If Eqs. (2.9) and (2.10) give the exact result and are easy to evaluate, what is the point of introducing the incremental equations, Eqs. (2.6)–(2.8), when they are approximate and require us to evaluate the dynamic variables at many intermediate times to obtain the final position and velocity? In the case of free-fall, the incremental equations aren't very useful. However, as we consider more complex systems, we'll encounter many cases in which exact results simply don't exist. Indeed, we can never hope to write down a function that is perfectly smooth yet random over the long term, as required to describe chaotic motion. Thus, like it or not, we're stuck with solving incremental equations for the majority of dynamic systems, and we might as well get used to the idea.

Just for fun, try out the animation of a bouncing ball, presented as the second experiment of the Dynamics Lab. The animation is based on Eqs. (2.6)–(2.8), and I think you'll agree that Galileo's analysis of free-fall yields an uncanny resemblance to reality.

2.5   Reconstructing the past

When we wrote Eqs. (2.6)–(2.8), our purpose was to predict the future, but the same equations are useful for looking backwards into the past.

Table 2.2 The past velocity and height of a falling ball, computed backward in time from an initial height of $H = 0.4$ m and a (downward) velocity of $V = -19.6$ m/s. The height at earlier times is obtained from the incremental equations using time increments ΔT of -0.1 s and -0.01 s.

Time (s)	Velocity (m/s)	$\Delta T = -0.1$ s Height (m)	$\Delta T = -0.01$ s Height (m)	Exact Height (m)
0.0	−19.6000	0.4000	0.4000	0.4000
−0.1	−18.6200	2.3600	2.3159	2.3110
−0.2	−17.6400	4.2220	4.1338	4.1240
−0.3	−16.6600	5.9860	5.8537	5.8390
−0.4	−15.6800	7.6520	7.4756	7.4560
−0.5	−14.7000	9.2200	8.9995	8.9750
−0.6	−13.7200	10.6900	10.4254	10.3960
−0.7	−12.7400	12.0620	11.7533	11.7190
−0.8	−11.7600	13.3360	12.9832	12.9440
−0.9	−10.7800	14.5120	14.1151	14.0710
−1.0	−9.8000	15.5900	15.1490	15.1000
−1.1	−8.8200	16.5700	16.0849	16.0310
−1.2	−7.8400	17.4520	16.9228	16.8640
−1.3	−6.8600	18.2360	17.6627	17.5990
−1.4	−5.8800	18.9220	18.3046	18.2360
−1.5	−4.9000	19.5100	18.8485	18.7750
−1.6	−3.9200	20.0000	19.2944	19.2160
−1.7	−2.9400	20.3920	19.6423	19.5590
−1.8	−1.9600	20.6860	19.8922	19.8040
−1.9	−0.9800	20.8820	20.0441	19.9510
−2.0	0.0000	20.9800	20.0980	20.0000

Suppose, for example, that you catch sight of a falling object just before it hits the ground. Thinking quickly, you note that the object is falling straight down, then whip out your radar gun and measure its speed as 19.6 m/s when it is 0.4 m off the ground.

Was the object a meteor from outer space or a piece of trash tossed from a nearby building? This question could easily be answered by inspecting the object, but we will apply Eqs. (2.6)–(2.8). To work backward in time, we set $\Delta T = -0.1$ s so that T_{new} will be less than T_{old}. Now, plugging in $H_{\text{old}} = 0.4$ m and $V_{\text{old}} = -19.6$ m/s, we compute that 0.1 s before the radar measurement, the object's height was 2.36 m and its speed was 18.62 m/s. Continuing to work backward in this way, as shown in Table 2.2, we soon discover that 2 seconds before the measurement, the object's speed was 0 and its height was about 21 meters (closer to 20 meters if we use a smaller time increment). Clearly, the object was not a meteor but possibly something dropped from a six-story building.

While knowing the past is seldom as interesting as knowing the future, the science of dynamics allows us to determine both, and we'll occasionally find this useful. For now, let's return to forecasting the future.

2.6   Projectile motion

What's the trajectory of a rock hurled into the air? Aristotle believed that a rock follows a straight line as it rises, then a circular arc as it passes through maximum height, and a vertical line as it falls to Earth again. As usual, Galileo was suspicious of Aristotle's pronouncement, and careful experiments led him to conclude that a rock instead follows a parabola as it rises and falls. To mathematicians, a parabola is much simpler than Aristotle's hodgepodge of lines and arcs, as shown in Fig. 2.4, and it also proves to be correct. In science, simplicity has often proved to be a reliable guide to the truth, although experiment is the ultimate arbiter.

Once Galileo realized that rolling balls mimic the motion of falling objects, he was led to a simple demonstration of the parabolic trajectory. He dipped a marble in ink and tossed it onto a sheet of paper mounted on a flat surface set at an incline. Gauging his toss so the marble rolled up and across the incline then back down again, as in Fig. 2.5, Galileo immediately discovered a parabola traced in ink. What could be simpler or more convincing?

In addition to his experiments, Galileo devised a mathematical proof of the parabolic trajectory. His proof was based on a simple assumption that would later become an important principle of dynamics. In particular, Galileo assumed that the horizontal component of the motion is independent of the vertical component. Thus, if we use X and Y to denote the horizontal and vertical components of the rock's position, the incremental equations for X will duplicate those for inertial motion, while those for Y will correspond to free fall. Using U and V for the

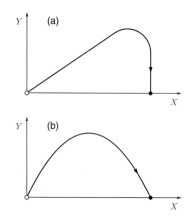

Fig. 2.4 The trajectory of a projectile according to Aristotle (a) and Galileo (b).

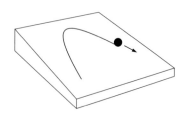

Fig. 2.5 Tossing an inked ball onto an incline reveals a parabola.

horizontal and vertical components of the velocity, we can write,

$$T_{\text{new}} = T_{\text{old}} + \Delta T, \qquad\qquad (2.11)$$

$$X_{\text{new}} = X_{\text{old}} + U\Delta T, \qquad\qquad (2.12)$$

$$Y_{\text{new}} = Y_{\text{old}} + V_{\text{old}}\Delta T, \qquad\qquad (2.13)$$

$$V_{\text{new}} = V_{\text{old}} - g\Delta T, \qquad\qquad (2.14)$$

where U is understood to be constant. Thus, X increases uniformly in time, while Y changes with an ever decreasing velocity.

The independence of the two perpendicular components of motion might seem surprising at first. Assuming wind resistance can be neglected, is the vertical motion of a rock really independent of how fast it's moving horizontally? Again, our experience in the cabin of a ship or airplane proves the point. As long as we're moving at constant velocity, we experience the same result when tossing a rock into the air, whether we're standing on the ground, cruising at 20 knots on a ship, or flying at 600 mph in an airliner. Thus, when one component of the motion is inertial, we, like Galileo, aren't surprised to find that a perpendicular component behaves independently. Later, scientists proved the general case: motion in a given direction depends only on the force applied in that direction and not on forces applied in perpendicular directions.

Although kids throwing rocks probably weren't impressed with Galileo's analysis of projectile motion, his work did find immediate use. At the time, Italy consisted of several independent city states, often at war with each other. As professor at Padua, Galileo lived within the Venetian republic, and, armed with his new science of dynamics, he became a consultant to the Venetian arsenal. Using his new parabolic trajectory, Galileo helped prepare new, more accurate gunnery tables for the republic's cannons.

The trajectory of a cannonball depends on the speed[3] S with which the ball leaves the muzzle and the angle θ of the muzzle with respect to horizontal. As shown in Fig 2.6, the initial horizontal and vertical components of the velocity are

[3]Physicists use "speed" to denote the magnitude of the velocity vector.

$$U = S\cos\theta, \qquad\qquad (2.15)$$

$$V = S\sin\theta. \qquad\qquad (2.16)$$

If we additionally assume that the cannonball begins at $X = Y = 0$, then Eqs. (2.11)–(2.14) can be used to determine its position as a function of time. The third Dynamics Lab shows off Galileo's parabolic trajectory for a cannonball, computed using incremental equations.

As usual, Galileo expressed his results in terms of the total time T rather than incremental time, and in modern notation his equations for the trajectory of a cannonball are

$$X = (S\cos\theta)T, \qquad\qquad (2.17)$$

$$Y = (S\sin\theta)T - \frac{1}{2}gT^2. \qquad\qquad (2.18)$$

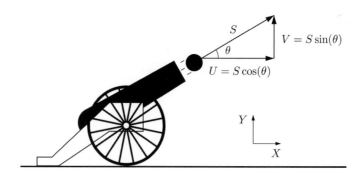

Fig. 2.6 The trajectory of a cannon-ball is determined by the muzzle speed S and the muzzle angle θ, which fix the horizontal and vertical components, U and V, of its initial velocity.

These equations are essential if you want to construct gunnery tables for a cannon or analyze the trajectory of a home-run ball. Where's the parabola, you ask? To find it, you'll need to eliminate T and express Y directly in terms of X. In addition, you can derive formulas for the maximum height, the point of impact (at $Y = 0$), and the time of flight, all in terms of S, θ, and g. Give it a try!

In 1609, the year in which Galileo devised his mathematical proof of the parabolic trajectory, he also learned of a marvelous new instrument invented in Holland: the telescope. Setting aside his treatise on dynamics, Galileo spent the next 20 years perfecting the telescope, using it to investigate the heavens, and writing about astronomy. Indeed, Galileo quickly became one of the foremost astronomers of his day, discovering the craters of the Moon, sunspots, the crescent phases of Venus, four of Jupiter's moons, and two appendages of Saturn, later recognized as the tips of its ring. Although these discoveries were certainly important, Galileo's earlier investigation of motion is undoubtedly his greatest contribution to science. His work on dynamics was entirely in the mold of modern physics, combining careful experiment and cutting-edge mathematics to resolve a problem that had stood unresolved since ancient times. Unanticipated and unwarranted in his day, Galileo's new science of dynamics, while incomplete, was a masterpiece of observation and reason that set the stage for all that would follow.

Further reading

Drag racing

- Fox, G. T., "On the physics of drag racing", *American Journal of Physics* **41**, 311–313 (1973).
- Post, R. C., *High Performance: The Culture and Technology of Drag Racing, 1950–2000* (Johns Hopkins University Press, Baltimore, 2001).

Galileo

- Drake, S., "Galileo's discovery of the law of free fall", *Scientific American* **228**(5), 84–92 (May 1973).
- Drake, S., *Galileo at Work: His Scientific Biography* (University of Chicago Press, Chicago, 1978).

○ Drake, S. and MacLachlan, J., "Galileo's discovery of the parabolic trajectory", *Scientific American* **232**(3), 102–110 (March 1975).

○ Gindikin, S. G., "Two tales of Galileo", in *Tales of Physicists and Mathematicians* (Birkhäuser, Boston, 1988) pp. 25–74.

○ Renn, J., editor, *Galileo in Context* (Cambridge University Press, Cambridge, 2001).

Projectile motion

○ Gurstelle, W., *Backyard Ballistics: Build Potato Cannons, Paper Match Rockets, Cincinnati Fire Kites, Tennis Ball Mortars, and More Dynamite Devices* (Chicago Review Press, Chicago, 2001).

○ Gurstelle, W., *The Art of the Catapult: Build Greek Ballistae, Roman Onagers, English Trebuchets, and More Ancient Artillery* (Chicago Review Press, Chicago, 2004).

Zeno

○ Mazur, J., *The Motion Pradox: The 2,500-Year-Old Puzzle Behind All the Mysteries of Time and Space* (Dutton, New York, 2007).

○ McLaughlin, W. I., "Resolving Zeno's paradoxes", *Scientific American* **271**(5), 84–89 (November 1994).

3 Isaac Newton—Dynamics perfected

When Galileo died, he left the world with a rudimentary science of dynamics. While he had demonstrated that motion could be predicted when the acceleration is constant, Galileo didn't attempt to understand more general types of motion. By chance, the man who would finish the job, Isaac Newton (1642–1727), was born in Lincolnshire, England, in the very year that Galileo died. By 1666, Newton would discover the general laws of motion that now bear his name and explain the dynamics of ordinary things, from pendulums and carnival rides to moons and planets. He was just 23 at the time.

Newton's discovery was made at the time of the Great Plague, an event that claimed the lives of one tenth of the inhabitants of London. Newton had studied mathematics and natural philosophy (science, as we would say today) at Cambridge University and received his Bachelor's degree in the spring of 1665, just before the plague turned serious. That fall, the University was closed as a precaution against the plague, and Newton returned home to his family's farm in Lincolnshire, where he would remain for two years. In his old age, Newton recalled that at home during the plague years, "... I was in the prime of my age of invention and minded mathematics and philosophy more than at any time since." Indeed, in those two years he not only discovered the basic laws of motion but single-handedly developed the calculus and performed optical experiments that led to his theory of light and color.

Fig. 3.1 Isaac Newton. Portrait by Godfrey Kneller, 1689. (Courtesy of the Trustees of the Portsmouth Estates.)

Like Galileo, Newton was slow to publish his new insight into the science of dynamics. Newton's ideas appeared in print only in 1687, more than 20 years after their initial inspiration. By then, Newton had fully developed his laws of motion with mathematical rigor, applied them to a wide variety of natural phenomena, and demonstrated their accuracy to astounding levels of precision. Newton's treatise on motion, *Philosphiae Naturalis Principia Mathematica* or the *Principia* for short, was immediately recognized as a masterpiece and is now regarded as one of the greatest scientific works of all time.

3.1 Equations of motion

For the present purpose, we need not delve deeply into the mathematical proofs contained in the *Principia*. However, we rely on Newton to solve the riddle of acceleration. For natural motion, Newton showed that when the acceleration isn't constant, it can generally be expressed in terms of a body's position and velocity. This is the key insight that completes our program for predicting the future. As you'll remember, our master equations,

$$T_{\text{new}} = T_{\text{old}} + \Delta T, \tag{3.1}$$

$$D_{\text{new}} = D_{\text{old}} + V_{\text{old}}\Delta T, \tag{3.2}$$

$$V_{\text{new}} = V_{\text{old}} + A_{\text{old}}\Delta T, \tag{3.3}$$

allow us to calculate the time T, distance D, and velocity V after a time increment ΔT but fail to specify how the acceleration A can be updated. According to Newton, however, A_{old} can be evaluated directly from T_{old}, D_{old}, and V_{old}, so we have all the information needed to iterate the equations of motion. By continuing to substitute new values of T, D, V, and A into the right-hand side of Eqs. (3.1)–(3.3), we can predict motion as far into the future as desired.

Newton chose to write the acceleration of a body in terms of its mass M and a function $F(T, D, V)$ as

$$A = \frac{F(T, D, V)}{M}. \tag{3.4}$$

What is the mysterious function F? Rewriting Eq. (3.4), we obtain

$$F(T, D, V) = MA, \tag{3.5}$$

which may help you recognize Newton's most famous equation: force equals mass times acceleration. Here, I have made Newton's equation uglier than the usual $F = MA$ to remind us that the force can generally be expressed as a function of time and the body's position and velocity. Does Newton's equation make sense? A surprisingly good answer to this question is: no, $F = MA$ is simply the definition of force. On the other hand, if we think of a force as a push or pull, it makes sense that the same push will produce less acceleration if the body is more massive. Thus, we're not surprised to find M in the denominator of Eq. (3.4). Indeed, even if $F = MA$ is no more than a definition of force, it is a clever choice because experiments show that, for a given size push, the acceleration really is inversely proportional to a body's mass.

The flow chart in Fig. 3.2 illustrates how Newton's equations are used to compute the motion of a mass M. First, the program reads the initial time and the initial position and velocity of the mass. Next, Eqs. (3.1)–(3.4) are used to calculate values of T, D, and V after the brief time interval ΔT. In Fig. 3.2, we abbreviate the subscripts "old" and "new" by "o" and "n." Incrementing the variables in this way is at the heart of Newton's method. The program next prints the new values of T, D,

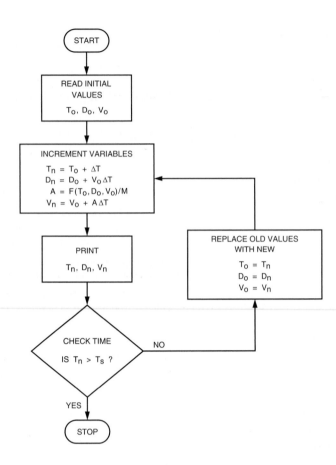

Fig. 3.2 Flow chart for solving Newton's equations of motion by computer.

and V or, if animation is the goal, draws the mass at its new position. After completing its output, the program checks to see if the new time T_n exceeds the stop time T_s. If $T_n > T_s$, the program halts, if not, it replaces the old values of T, D, and V with the new ones and loops back to increment the variables again. Depending on the chosen stop time, this loop may be executed hundreds or thousands of times, iterating the equations of motion over and over again until T_n exceeds T_s and the program halts. Predicting motion is simple, especially when iteration can be carried out by a computer.

3.2 Force laws

All in all, however, we haven't made much progress by writing Eq. (3.4), since we still don't have an explicit formula for the acceleration in terms of T, D, and V. Newton's real genius wasn't $F = MA$, but the formulas that he discovered for the forces that arise in a variety of situations. How and why does the force depend on T, D, and V? Three examples serve to illustrate.

In the simplest case, often called natural motion, the force doesn't depend explicitly on time. We'll look at this possibility first. Consider, for example, a mass attached to a wall by a spring, as shown in Fig. 3.3(a). Here the force exerted by the spring is a function only of the distance D of the mass from the wall. The force law is

$$F = -K(D - D_r), \tag{3.6}$$

where D_r is the length of the spring when it's relaxed and K is a constant that specifies the stiffness of the spring. According to Eq. (3.6), the force is zero when $D = D_r$ but acts to pull the mass back toward the wall if $D > D_r$ or push the mass away if $D < D_r$. Thus, the spring force always accelerates the mass toward $D = D_r$. With the force law specified, we now have all the information required to predict the motion of the mass and spring system. That is, combining Eqs. (3.4) and (3.6) yields

$$A = -\frac{K}{M}(D - D_r), \tag{3.7}$$

and with the acceleration expressed in terms of distance we can complete the iteration loop of Fig. 3.2 to predict the motion. Alternatively, substituting $A_{\text{old}} = -(K/M)(D_{\text{old}} - D_r)$ into Eq. (3.3) yields

$$V_{\text{new}} = V_{\text{old}} - \frac{K}{M}(D_{\text{old}} - D_r)\Delta T, \tag{3.8}$$

which eliminates acceleration from the problem and allows T, D, and V to be computed by iterating Eqs. (3.1), (3.2), and (3.8).

Suppose that we pull the mass away from the wall until the spring is tightly stretched, then let go. What's the resulting motion? Since the force goes to zero when $D = D_r$, you might guess that the mass will return to the point where the spring is relaxed and come to a stop. However, iterating Eqs. (3.1), (3.2), and (3.7) on a computer proves that the motion is oscillatory, as shown in Fig. 3.4. If you think for a minute, the oscillations aren't too surprising. When the mass is released at $T = 0$, the stretched spring accelerates the mass toward the wall, so the velocity goes from 0 to increasingly negative values (the velocity is opposite to the direction in which D is measured), and the displacement $(D - D_r)$ of the mass from its relaxed position begins to decrease. When the displacement reaches 0 at about $T = 0.7$ s, the spring is relaxed and exerts no force on the mass, but the mass has acquired a large negative velocity and, in accord with the law of inertia, it just keeps going. The mass only slows as the spring compresses, and it comes to a stop at $T = 1.4$ s when the spring is compressed as much as it was initially stretched. Now the spring is set to push the mass away from the wall, so the motion is reversed and the mass soon ends up in its original location, ready to repeat the cycle. To see the resulting oscillatory motion in animation, check out Experiment 4 of the Dynamics Lab.

All in all, a mass on a spring is an amazing example. First, it demonstrates the success of Newton's dynamics. By expressing the spring force in terms of the position, we get a closed set of equations

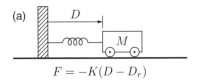

(a)

$$F = -K(D - D_r)$$

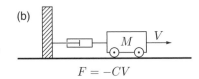

(b)

$$F = -CV$$

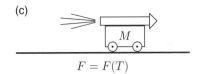

(c)

$$F = F(T)$$

Fig. 3.3 Forces can depend on (a) position, (b) velocity, and (c) time.

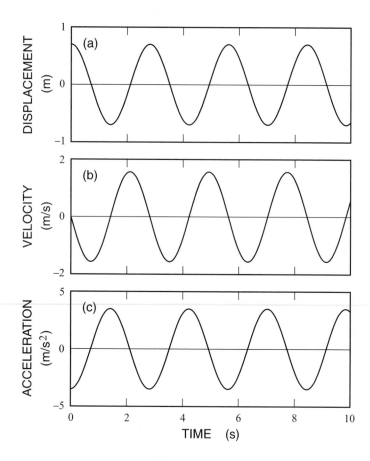

Fig. 3.4 The displacement $(D - D_r)$, velocity V, and acceleration A of a mass attached to a wall by a spring, in the absence of friction. The spring constant is $K/M = 5 \ 1/\text{s}^2$.

that can be iterated to simulate motion like that of a real mass on a spring. Newton's program works! Second, Eqs. (3.1), (3.2), and (3.8) are remarkably simple. Over a short time interval ΔT, the change in position is proportional to the velocity, and the change in velocity is proportional to the position. It's so simple that you can easily compute the motion on a programmable calculator! Finally, the resulting oscillations appear as if by magic, beautifully smooth and regular, from equations that give no hint of periodic behavior. Newton must have been pleased.

As our second force, we consider a type of a friction produced by what engineers call a dashpot. A dashpot is a fluid-filled cylinder with a loosely fitting piston. If the cylinder is connected to a mass and the piston is connected to a wall, as in Fig. 3.3(b), then the mass is free to move, but motion requires that fluid flow around the piston. In this case, the force is given by,

$$F = -CV, \qquad (3.9)$$

where C is a constant that depends on the viscosity of the fluid and V is the velocity of the mass. This force law makes sense because the force is 0 when the mass is stationary and acts in a direction opposite to the motion whenever the mass moves. Thus, friction provides an example of a force that depends on velocity.

To explore the effect of friction, suppose that a mass is connected to a wall by both a spring and a dashpot. In this case, two forces act on the mass, and, according to Newton, the motion is correctly predicted if we simply add the forces together. The total force is,

$$F = -K(D - D_r) - CV, \tag{3.10}$$

so the resulting acceleration is,

$$A = \frac{F}{M} = -\frac{K}{M}(D - D_r) - \frac{C}{M}V. \tag{3.11}$$

As before, to predict the motion, Eq. (3.11) can be used to eliminate A_{old} from Eq. (3.3) with the result,

$$V_{\text{new}} = V_{\text{old}} - \left[\frac{K}{M}(D_{\text{old}} - D_r) + \frac{C}{M}V_{\text{old}}\right]\Delta T. \tag{3.12}$$

While everything is beginning to look a little complicated, combining Eqs. (3.1), (3.2) and (3.12) provides all the information needed for iteration, and allows us to predict T, D, and V as far into the future as we wish. According to Eq. (3.12), the new velocity depends on both the old distance and the old velocity, but iteration is almost as easy as with only a spring force.

What's the motion like when a mass on a spring encounters friction? As shown in Fig. 3.5, the mass still oscillates about the point at which the spring is relaxed, but now the oscillations gradually decay. On each oscillation, friction slows the mass and prevents it from compressing the spring to the same extent that it was previously stretched. With time, the maximum displacement decreases, and the mass slowly approaches a state of rest at its relaxed position, where both the spring force and the friction force are 0. Altogether, the spring isn't as bouncy as without friction, but the predicted motion fits more closely with our experience of real springs, since there's always a little friction, even without a dashpot. Experiment 5 of the Dynamics Lab lets you play with the spring and friction forces in combination.

As a third example, consider what happens when a force is artificially applied, perhaps by pushing the mass by hand or by strapping on a rocket engine, as in Fig. 3.3(c). In such situations, it's usually impossible to relate the force to the position or velocity of the mass, but measuring the force as a function of time provides all the information we need. For example, if we record the thrust $F(T)$ of a rocket engine in a static test, then Newton's equations can be used to predict the rocket's trajectory in flight. Indeed, whenever the force can be expressed as a function of T, D, and V in any combination, we can predict the resulting motion. Almost every kind of motion encountered in everyday life is subject to Newton's laws.

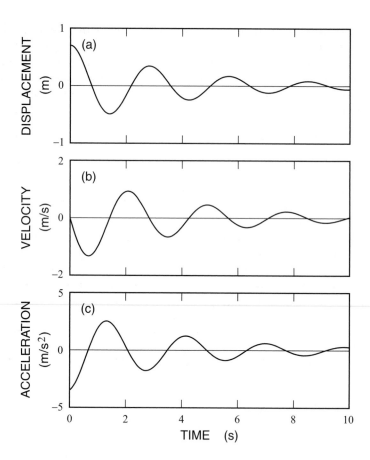

Fig. 3.5 The motion of a mass on a spring in the presence of friction. The spring constant is $K/M = 5$ $1/s^2$, and the friction constant is $C/M = 0.5$ $1/s$.

3.3   Calculus

Calculating the damped oscillation of a mass on a spring from Eqs. (3.1), (3.2), and (3.12) is almost trivial on a modern computer. In the absence of computers, Newton was forced to confront such problems with pencil and paper alone. As a result, he was led to the invention of a new mathematical technique, which he called the method of fluxions and we now call the calculus. As with most of his work, the seminal idea for the calculus came to Newton during the intense studies he undertook at home in Lincolnshire during the plague years of 1665 and 1666.

From the modern point of view, the calculus is the natural mathematical language of motion. Although this book is principally concerned with motion, we have sidestepped the calculus to keep the presentation as simple as possible. However, we now take a brief look at the calculus, just to see how close we've come to that greatly feared domain of mathematics.

Actually, we've already described the basic concepts of the calculus, even if they weren't labeled as such. The differential calculus deals with calculating the velocity given the position as a function of time, while the integral calculus deals with calculating the position given the velocity as a function of time. The procedures involved are now familiar. For example, consider what happens if you throw a rock straight up into the air and watch it climb, then fall to Earth again. Suppose that we know the height H of the rock as a function of the time T, and we wish to calculate its velocity. If at some time the rock climbs by $\Delta H = 1$ m in a time $\Delta T = 0.1$ s, then its velocity at that time is $V(T) = \Delta H/\Delta T = 10$ m/s. There's certainly nothing hard about calculating the velocity from the position, and yet this calculation is at the very heart of the differential calculus! Indeed, the name "differential calculus" comes simply from the fact that we're computing the differences, ΔH and ΔT, in position and time.

To understand more, let's look at a geometric interpretation of our difference calculation. As shown in Fig. 3.6(a), when we evaluate the velocity at a given time, we are really calculating the slope of the $H(T)$ curve. In particular, if we use a discrete grid of time points and suppose that $T = N\Delta T$, then the velocity is

$$V(N\Delta T) = \frac{H((N+1)\Delta T) - H(N\Delta T)}{\Delta T}, \tag{3.13}$$

That is, $V(N\Delta T)$ is just the slope (rise over run) of the line passing through the two points indicated in Fig. 3.6(a). With different values of N, Eq. (3.13) can be used to estimate the velocity of the rock at any point in time: it's always the slope of the $H(T)$ curve at that point. Does this interpretation makes sense? At $T = 0$, when the rock is launched, we see that the slope is large and positive, so the rock is climbing fast. By the time the rock reaches its maximum height, the $H(T)$ curve is horizontal, so the slope is zero and the rock has stopped in mid-air. Later, the slope becomes negative, indicating that the rock is falling. It's easy to read $V(T)$ from the $H(T)$ curve: just measure the slope of the curve at the time of interest.

What about the reverse—can we calculate the position from the velocity curve? As we have argued all along, if the initial height is $H(0)$ and the initial velocity is $V(0)$, then after a short time ΔT the height will be $H(\Delta T) = H(0) + V(0)\Delta T$, where we've added the short distance $V(0)\Delta T$ traveled in the time ΔT. Similarly, at the beginning of the next time interval the velocity is $V(\Delta T)$, so the height after $2\Delta T$ can be computed by adding the distance $V(\Delta T)\Delta T$ to the earlier result: $H(2\Delta T) = H(0) + V(0)\Delta T + V(\Delta T)\Delta T$. Continuing in this fashion yields

$$H(N\Delta T) = H(0) + V(0)\Delta T + V(\Delta T)\Delta T + \cdots + V((N-1)\Delta T)\Delta T$$

$$= H(0) + \sum_{n=0}^{N-1} V(n\Delta T)\Delta T \tag{3.14}$$

where the Greek capital sigma (Σ) has been used to indicate that a sum is taken over terms from $n = 0$ to $n = N - 1$. Equation (3.14) adds

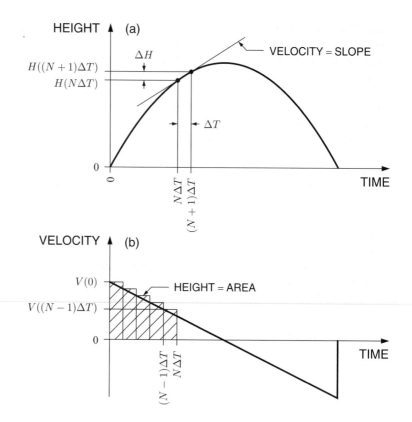

Fig. 3.6 The height H and velocity V of a rock thrown straight up provide an example of curves related by the differential and integral calculus. At any time T, the velocity $V(T)$ is the slope of the line tangent to the $H(T)$ curve, and the height $H(T)$ is the area under the $V(T)$ curve between the times 0 and T.

up the all the distances traveled in the intervals ΔT assuming that the rock's velocity remains roughly constant over each interval. This is the same approximation used previously when we obtained the position by iterating the equation $H_{\text{new}} = H_{\text{old}} + V_{\text{old}}\Delta T$. Not surprisingly, given that integration means summation, Eq. (3.14) represents the integral calculus. By addition (integration) we can undo what was done by subtraction (differentiation) and go from the velocity curve back to the position curve.

As with the differential form, the integral calculus has a simple geometric interpretation. As shown in Fig. 3.6(b), each of the terms $V\Delta T$ in Eq. (3.14) is represented by the area of a hatched rectangular box of dimension V by ΔT. The total change in the height of the rock from $T = 0$ to $T = N\Delta T$ is thus the combined area of all the boxes in this interval. By choosing N, we can determine how high the rock has climbed after any time $N\Delta T$, simply by adding the areas of the appropriate number of boxes. Does this make sense? As long as the velocity is positive, adding more boxes means that the height of the rock increases as time goes on. When the velocity goes to zero, however, the added box has no area and the height of the rock stays the same. Finally, when the velocity becomes negative, the boxes have negative area and subtract from the total, so the height of the rock begins to decrease. Thus, by summing areas of boxes, we can compute the height of the rock at any time if we know the velocity curve.

In our discussion so far, we have adopted a finite time interval ΔT as the basis of our computations. For good accuracy, ΔT must be small. For example, in Fig 3.6(a) the velocity at a given time was approximated as the average velocity over the interval ΔT following that time. Making ΔT smaller gives a more accurate evaluation of the instantaneous velocity at the time of interest. Similarly, by using finite intervals to add up the changes in height $\Delta H = V \Delta T$ in Fig. 3.6(b), we assume that the velocity is constant over each interval, although V is always decreasing. Making ΔT smaller will increase the accuracy of the computed height by accounting for the continuously changing velocity. For smaller ΔT, the stair steps in Fig. 3.6(b) become smaller, the staircase approaches the $V(T)$ curve more closely, and the sum of the hatched boxes approaches the area under the $V(T)$ curve.

Although Zeno warned against dividing time into smaller and smaller intervals, Newton discovered that, with proper care, this forbidden limit can be taken explicitly to obtain the ultimate in accuracy. Thus, Newton developed techniques for computing the instantaneous velocity by finding the slope of the $H(T)$ curve at an exact point and for computing the position by finding the exact area under the $V(T)$ curve. For example, the curves plotted in Fig. 3.6, are given by

$$H(T) = V_0 T - \tfrac{1}{2} g T^2, \tag{3.15}$$

$$V(T) = V_0 - gT, \tag{3.16}$$

where V_0 is the initial velocity and g is the acceleration of gravity. Although we won't explain how, with Newton's calculus it's a simple matter to "differentiate" Eq. (3.15) to obtain Eq. (3.16) or to "integrate" Eq. (3.16) to obtain Eq. (3.15), using only pencil and paper. In this instance, the calculus completely eliminates the need for tedious numerical computations of uncertain accuracy: the exact result is handed to us on a silver platter. Galileo would have been astonished to see motion analyzed with such precision and efficiency!

Given that the calculus is the natural mathematical language of motion, it may seem strange that the *Principia*, Newton's great work on motion, doesn't use this powerful technique. Instead, Newton presented his dynamics strictly in terms of geometry. By omitting calculus from the *Principia*, Newton probably sought to avoid any controversy that might arise from an unconventional approach and give his work the air of rigor associated with Euclid's classic treatise. Moreover, if Newton sought to claim the calculus as his own invention, he was already too late. By the time the *Principia* appeared in 1687, the calculus had been in the public domain for three years, published in 1684 by Gottfried Wilhelm Leibniz (1646–1716), a German mathematician. It's now generally accepted that Newton invented the calculus in 1666, and Leibniz came to the idea in 1675, but historians usually give them equal credit for independent discoveries. Newton can count himself lucky in this regard, since, in today's "publish-or-perish" world of science, accolades always go to the first in print. Newton belatedly published his account of the calculus in 1704, as a treatise appended to his *Opticks*, and later fought bitterly with Leibniz over priority. But whoever deserves credit for its invention, the calculus immediately became the mathematician's most important tool in the analysis of motion.

Of the various forces considered by Newton, one stands above all others in importance: the force of gravity. Unlike the spring force or friction force, which depend on material properties, gravity is fundamental in character. For Galileo, gravity was simply the force that made things fall, but for Newton gravity resulted from a universal attraction of every speck of mass in the universe for every other speck. This is the hypothesis Newton framed during the plague years while living on the family farm, and he eventually applied it to explain not only bodies falling on Earth but the shape of the Earth, ocean tides, and the motions of celestial bodies as well. Predicting celestial motion was Newton's greatest triumph, and in the next chapter we explore how Newton explained the regular motion of the Solar System, the counterpoint to all that is chaotic.

Further reading

Calculus

- Berlinski, D., *A Tour of the Calculus* (Pantheon, New York, 1996).
- Sawyer, W. W., *What Is Calculus About?* (Random House, New York, 1961).

Dynamics

- Bitter, F., *Mathematical Aspects of Physics: An Introduction* (Anchor Doubleday, Garden City, New York, 1963).
- Pólya, G., "From the history of dynamics", in *Mathematical Methods in Science* (Mathematical Association of America, Washington, 1977) chapter 3.
- Rothman, M. A., *Discovering the Natural Laws: The Experimental Basis of Physics* (Anchor Doubleday, Garden City, New York, 1972).

Newton

- Andrade, E. N. da C., *Sir Isaac Newton: His Life and Work* (Anchor Doubleday, Garden City, New York, 1964).
- Arnol'd, V. I., *Huygens and Barrow, Newton and Hooke: Pioneers in Mathematical Analysis and Catastrophe Theory from Involutes to Quasi-Crystals* (Birkhäuser, Boston, 1990).
- Aughton, P., *Newton's Apple: Isaac Newton and the English Scientific Renaissance* (Weidenfeld and Nicolson, London, 2003).
- Berlinski, D., *Newton's Gift* (Duckworth, London, 2000).
- Chandrasekhar, S., *Newton's Principia for the Common Reader* (Oxford University Press, Oxford, 1995).
- Cohen, I. B., *The Birth of a New Physics* (Anchor Doubleday, Garden City, New York, 1960).
- Drake, S., "Newton's apple and Galileo's Dialogue", *Scientific American* **243**(2), 151–156 (February 1980).
- Fauvel, J., Flood, R., Shortland, M., and Wilson, R., editors, *Let Newton Be!* (Oxford University Press, Oxford, 1988).
- Westfall, R. S., *The Life of Isaac Newton* (Cambridge University Press, Cambridge, 1993).

Celestial mechanics— The clockwork universe

Within our natural environment, the Sun and Moon provide unique examples of motion that is accurately regular. Little wonder that man has always reckoned time by days, months, and years. Celestial motion is the antithesis of the chaos that we ordinarily find in nature and has fascinated man from the beginning. To the ancients, the Sun, Moon, stars, and planets were beyond the reach of mortals and embodied otherworldly perfection. However, as our knowledge and understanding of the heavens increased, celestial phenomena began to look more and more mundane, and after Newton we knew that the heavens are governed by the same laws that apply to our everyday world. The story of this discovery is a classic of modern science and well worth retelling. We begin our account in ancient Greece.

4.1 Ptolemy

In accord with the supposed perfection of the heavens, Aristotle dictated that celestial bodies should move with constant speed along circles centered on the Earth, and for centuries thereafter astronomers tried to reconcile the observed motions with Aristotle's ideal. While stars do indeed rotate about the Earth with constant speed, as if embedded in a rigid celestial sphere, the motions of the Sun, Moon, and planets are not so simple. As man had known for centuries, these heavenly bodies wander through the constellations of the zodiac, generally progressing from west to east. However, careful observations revealed that the speed of the Sun and Moon vary noticeably, while some of the planets go so far as to reverse direction occasionally, in what is called retrograde motion.

How could the observed motions of the Sun, Moon, and planets be explained while upholding Aristotle's ideal? Ancient astronomers decided that the ideal could be stretched a little without violating its spirit. Rather than a single circle, they found that two circles were needed to explain planetary motion. The motion of Mars, for example, could be understood if it moved uniformly around a small circle, called an epicycle, while the center of the epicycle moved uniformly around a large circle, called a deferent, centered on the Earth. As shown in Fig. 4.1, this scheme occasionally creates a small loop in the trajectory of Mars

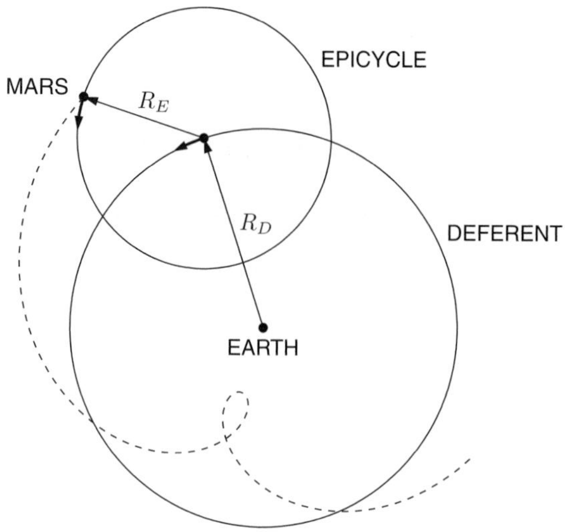

Fig. 4.1 A simplified version of Ptolemy's geocentric orbit for Mars. Mars moves uniformly around a small circle called an epicycle, the center of which moves uniformly around a large circle called a deferent, centered on the Earth. A dashed line indicates the trajectory of Mars relative to Earth, with a loop illustrating a period of retrograde motion.

as seen from Earth, explaining its retrograde motion. While Aristotle may not have been pleased, the system of deferents and epicycles was entirely workable and explained the gross motions of all the planets. Indeed, by the second century, when Claudius Ptolemy of Alexandria wrote his great work on astronomy, the positions of the known planets could be predicted centuries in advance with errors no greater than a few degrees. To obtain this accuracy, however, Ptolemy was forced to make additional compromises in the Aristotelian ideal: he displaced the center of the deferent slightly from the Earth and took the motion of the epicycle to be uniform not from the point of view of the Earth or the center of the deferent but from yet another nearby point. Nevertheless, Arab astronomers, who used and preserved Ptolemy's work during the Middle Ages, named his treatise the *Almagest*, literally "The Greatest."

4.2   Copernicus

The Ptolemaic system of the universe remained unquestioned for more than a millennium and was eventually accepted as dogma by the Catholic Church. This geocentric system, in which the Sun, Moon, and planets all rotate about the Earth, was successful in part because it appealed to man's sense of self-importance. In the sixteenth century, however, a Polish astronomer began to tinker with Ptolemy's circles and discovered a point of view that greatly simplified our picture of the heavens. The insight revealed by Nicolaus Copernicus (1473–1543) resulted when he reversed the roles of Ptolemy's deferent and epicycle. As shown in Fig. 4.2, the motion of Mars relative to the Earth is unchanged if we assume that the planet moves around a circle having the same radius as the deferent while the center of this circle moves around another having

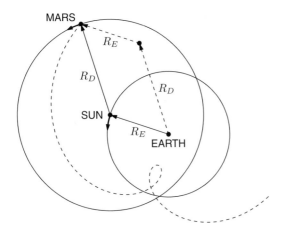

Fig. 4.2 Copernicus's geo-heliocentric orbit for Mars is obtained by reversing the roles of Ptolemy's deferent and epicycle. For comparison, the radius vectors of Ptolemy's system are shown as dashed lines. In the geo-heliocentric system, Mars moves uniformly around a large circle, the center of which moves uniformly around a small circle, which is centered on the Earth. The trajectory of Mars relative to Earth, indicated by a dashed line, is the same as in Ptolemy's system, but the Sun is located at the center of the large circle.

the same radius as the epicycle, but centered on the Earth. In both systems, the location of Mars is found by adding the radius vectors R_D and R_E, and reversing the order of addition doesn't affect the outcome.

While Copernicus might appear to have gained nothing by this interchange, he observed that, with the deferent and epicycle reversed, the radius vector R_E always points from the Earth directly toward the Sun. Is this simply a coincidence? No, Copernicus found that the same thing happens for every planet![1] He concluded that the Sun plays a special role in the movements of the planets, and this role became more obvious when he took his next step. Because only the relative sizes of the deferent and epicycle are involved in determining the location of a planet in the sky, Copernicus could scale both circles until the radius R_E matched the distance from the Earth to the Sun, placing the Sun at the center of the circle about which each planet moves. Thus, Copernicus arrived at the geo-heliocentric system for Mars shown in Fig. 4.2. When applied to all the planets, the geo-heliocentric system stipulates that every planet moves in a circular orbit about the Sun, while the Sun moves in a circular orbit about the Earth.

Copernicus never published his geo-heliocentric system because it was only a stepping stone to his ultimate theory. In a final step, Copernicus imagined how the geo-heliocentric system would look if the Sun were assumed to be stationary, rather than the Earth. From this point of view, the Earth circles the Sun and is just another planet, not the center of the universe. Thus, Copernicus transformed the geocentric Ptolemaic universe into the heliocentric Solar System now familiar to everyone. Of course, with the Sun held stationary, the Earth–Sun–Mars system becomes that shown in Fig. 4.3, with both Mars and Earth circling the Sun. In this view, the radii R_D and R_E of Mar's deferent and epicycle are just the radii of the solar orbits of Mars and Earth, while the time required for Mars to make a circuit of its epicycle in the Ptolemaic system is one Earth year, and the time required for the epicycle to circle once around its deferent is one Martian year. Thus, the geocentric system

[1] For the inner planets, Mercury and Venus, it isn't necessary to interchange the deferent and epicycle. The circles are correctly chosen in the Ptolemaic system, with the larger radius R_D pointing toward the Sun.

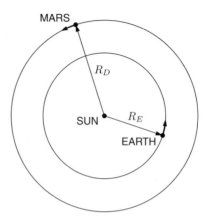

Fig. 4.3 Copernicus's heliocentric orbits for Mars and Earth with uniform circular motion. The system is identical to the geo-heliocentric system except that the Sun is assumed to be stationary, rather than the Earth.

of Ptolemy and the heliocentric system of Copernicus are essentially equivalent.

Copernicus published a short description of his heliocentric system in 1515, but a full account did not appear in print until 1543, the year of his death. His great work, entitled *De Revolutionibus*, was modeled on the *Almagest* and described in detail how to predict the motions of the planets. Using strategies analogous to those of Ptolemy to account for the fact that planetary orbits are not quite circular and the motion is not quite uniform, Corpernicus was able to predict planetary positions with similar accuracy, always within a few degrees. The real advantage of Copernicus's system wasn't its accuracy but its simplicity. By shifting the astronomer's coordinate system from one centered on the Earth to one centered on the Sun, Copernicus revealed that the apparently complex motions of the planets are basically simple. Retrograde motion, which so puzzled the ancients, is no more than an illusion created by the Earth's own motion. In the heliocentric system, Mars doesn't stop and move backwards, it only appears that way from the Earth when we speed by Mars during our closest approach.

In addition to simplifying planetary motion, Copernicus gave the Solar System its first unified scale. Because Ptolemy was concerned only with predicting the location of planets in the sky, the relative size of the deferent and epicycle were fixed for a given planet, but the relative sizes of the deferents of say Jupiter and Saturn were anyone's guess. On any given day, Ptolemy knew where to find Jupiter and Saturn in the sky, but he couldn't tell you which was further from Earth. In the Copernican system, however, the epicycles of the outer planets (Mars, Saturn, and Jupiter) and the deferents of the inner planets (Mercury and Venus) all represent the orbit of the Earth around the Sun, so their radii are all equal to the distance between the Earth and the Sun, a distance now called the astronomical unit (AU). With this equality, Copernicus could calculate all distances within the Solar System, expressed neatly in AU. This unified scale was the feature of his system that Copernicus thought most valuable, and indeed it explained why the planets vary in brightness during their travels through the heavens. More important, a century later, planetary distances would prove crucial to Newton's formulation of his theory of gravity.

The heliocentric system probably wasn't the first breakthrough in understanding gained by a careful choice of coordinate system, and it certainly wasn't the last. Time and again, scientists have gained new insights simply by adopting a coordinate system appropriate to the problem. The phenomenon of superconductivity, for example, went unexplained for decades before physicists realized that it could be explained using coordinates representing pairs of electrons rather than individual electrons. Likewise, the possibility of chaotic behavior long went unrecognized, in part because scientists adopted coordinate systems suited to periodic motion. Changing your point of view is an important tool in science.

4.3 Brahe and Kepler

The Ptolemaic and Copernican models of the Solar System adequately explain casual observations of planetary motion. However, as the Danish astronomer Tycho Brahe (1546–1601) was to discover, these models are completely inadequate when compared with careful quantitative observations. Although he lived before the invention of the telescope, Brahe perfected the art of naked-eye astronomy, and his observations became the gold standard of his day. Working with a six-foot brass quadrant at his observatory on the island of Hven, situated between Denmark and Sweden, Brahe routinely recorded the positions of stars and planets within 1 arc minute, or $\frac{1}{60}$ of a degree. Of course, the inadequacy of the Ptolemaic and Copernican models, which could be off by several degrees, was immediately apparent to Brahe, and it became his ambition to discover a new and more accurate system of the universe. To this end, Brahe accumulated meticulous observations for more than three decades, but in the end he admitted defeat. On his deathbed, Brahe pleaded, "Let me not seem to have lived in vain."

Brahe's last wish was eventually fulfilled by Johannes Kepler (1571–1630), a German mathematician and astronomer who had become Brahe's assistant a year earlier. The problem of celestial motion that Brahe willed to Kepler was a little like Galileo's ballistics problem, which was solved around the same time, except that Galileo could perform experiments, while Kepler was forced to seek a solution in the reams of data bequeathed by Brahe. Kepler could do no more than guess a possible scheme for planetary motion, laboriously calculate the consequent positions, compare with Brahe's data, then guess a new scheme. Working in Prague as the imperial mathematician of Emperor Rudolph II, Kepler began by attacking the orbit of Mars, and indeed referred to his labors as his "war on Mars." In the beginning, he employed all the tricks of Ptolemy and Copernicus to fit the orbit of Mars with various combinations of uniform circular motions, and after some 70 trials reduced the discrepancy between his model and Brahe's observations to a mere 8 arc minutes. But Kepler wasn't satisfied.

In the end, Kepler won his war on Mars only when he threw Aristotle's circles out the window and tried a whole new approach. Finally recognizing that the orbit of Mars simply wasn't circular, Kepler chanced to try fitting the orbit with an ellipse and succeeded brilliantly. He would later remark, "It was almost as if I had awakened from a sleep." By 1605, four years after Brahe's death, Kepler had transformed astronomy by establishing his first two laws of planetary motion. In particular, Kepler discovered that,

I. Planets move in elliptical orbits with the Sun located at one focus of the ellipse.

II. The line between a planet and the Sun sweeps out equal areas in equal times.

Thus, planetary motion is neither circular nor uniform! However, because the orbits of the planets in our Solar System are nearly circular, we illustrate Kepler's laws in Fig. 4.4 with a highly eccentric orbit typical of a comet. Here, the shaded areas are equal, so the comet moves through the corresponding arcs in equal times. Clearly, comets move much more rapidly near perihelion (the point of closest approach to the Sun) than at aphelion (the point farthest from the Sun). Compared to those of comets, planetary orbits are nearly circular and their motion is nearly uniform, but Brahe's careful observations forced Kepler to ferret out their true nature.

Kepler discovered his third and final law of planetary motion in 1618. In the heliocentric system, the relative distances of the planets from the Sun are known from observations, and even Copernicus noticed that the periods of the planetary orbits increase with their size. However, it was Kepler who first realized that to high accuracy,

III. The square of the periods of planetary orbits increase in proportion to the cube of their mean radii.

This law would later provide Newton with an important clue to the theory of gravity.

Kepler published his investigations of Mars as the *Astronomia Nova*, or *New Astronomy*, in 1609. His ephemeris for all of the known planets, the *Rudolphine Tables*, would not appear until 1627, just three years before his death, but it would be a factor of 30 more accurate than any previous system of the heavens. Indeed, the *Rudolphine Tables* were a remarkable accomplishment. They served as ample proof of Kepler's laws of planetary motion and justified the toil of two lifetimes: Brahe's as well as Kepler's.

But Kepler is also remembered for a planetary hypothesis that would remain unproven. While the ancients saw the planets as otherworldly, Kepler suggested that they are held in their orbits by a magnetic

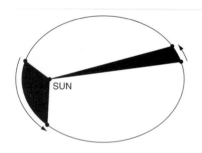

Fig. 4.4 Kepler showed that planets orbit the Sun in elliptical trajectories with the Sun at one focus. A line connecting a planet to the Sun sweeps out equal areas in equal times, as illustrated by the two sectors of equal area shown in black. Here we show a highly eccentric orbit typical of comets.

force, and this mundane point of view foreshadowed Newton's theory of celestial motion.

4.4   Universal gravitation

Once Galileo trained his telescope on the heavens and began to discover craters on the Moon, spots on the Sun, and the crescent phases of Venus, it gradually became clear that the heavens are far less perfect than the ancients imagined. With heavenly bodies beginning to look more and more like the Earth, it's not too surprising that Newton chanced to suppose, during his musings of the plague years, that the Moon might be held in its orbit by the same force that causes objects to fall to the ground on Earth. After all, if the Moon is no more than a spherical pile of rocks, then, in the absence of some force pulling it toward the Earth, it would surely move in a straight line rather than a circle. In any case, whether inspired by Kepler's notion of magnetic attraction or by the fall of a Lincolnshire apple, something clicked in Newton's mind, and he suddenly saw that gravitation might hold the Solar System together.

According to Newton's idea, every speck of mass in the universe attracts every other speck with a force,

$$F = G\frac{M_1 M_2}{R^2}, \tag{4.1}$$

where M_1 and M_2 are the masses of the two specks, R is the distance between them, and G is a universal constant that physicists call "big gee." With this simple rule, illustrated in Fig. 4.5, Newton sought to explain both Galileo's laws of free-fall and Kepler's laws of planetary motion. In the end, Eq. (4.1) would prove to be almost magic, a top-hat from which Newton could pull explanations for all and sundry to the astonishment of his audience. To get an idea of its power, let's play a minute with Newton's law of universal gravitation.

First, we might ask, does Eq. (4.1) imply an attraction between any two objects whatsoever, say between you and your cat? Newton would say, "yes, indeed," although he'd admit that it doesn't explain why the cat is often found on your lap. The gravitational constant G is exceedingly small, so the attraction between ordinary objects, while detectable in laboratory experiments, is usually negligible. Thus, you don't have to avoid cats or even large boulders for fear of getting stuck to them.

When one mass is as large as the Earth, however, the gravitational force can be considerable, enough to put a slam dunk beyond the reach of ordinary mortals. As soon as Michael Jordan's feet leave the ground, Galileo tells us that the force of gravity will immediately begin to reduce his speed with an acceleration g and eventually return him to Earth. Newton agrees. If M_J is the mass of Jordan and M_E is the mass of the

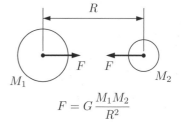

Fig. 4.5 According to Newton, the gravitational force F between to spherical masses, M_1 and M_2, is proportional to the product of the masses and inversely proportional to the square of the distance between their centers.

Earth, then applying $F = MA$ and Eq. (4.1) yields,

$$M_J A = G \frac{M_J M_E}{R^2}, \tag{4.2}$$

where A is Jordan's downward acceleration. (Of course, the Earth also accelerates toward Jordan, but its mass is so large that its acceleration is negligible.) According to Newton, Jordan's acceleration must equal Galileo's g, and, cancelling M_J from both sides of Eq. (4.2), leads to,[2]

$$g = G \frac{M_E}{R^2}. \tag{4.3}$$

Magically, we find that g doesn't depend on the mass M_J of the falling body, just like Galileo said. Newton's formula has passed its first test.

But wait, what value should we use for R in Eq. (4.3)? Every speck of mass in the Earth attracts Jordan, but some bits are nearby while others are on the far side of the Earth. Shouldn't we be adding up the force contributed by each speck according to how far away it is? Newton would agree: the nearby specks contribute much more force than the distant ones, and we really need to add up everything properly. This addition problem presented a formidable obstacle when Newton first thought about gravity, but by the time the *Principia* was published he had the answer: the gravitational attraction of a spherical body is the same as if its entire mass were located at its center. Thus, the R to be used in Eq. (4.3) to calculate the gravitational acceleration of a body on the surface of the Earth is simply the radius R_E of the Earth. That is,

$$g = G \frac{M_E}{R_E^2}. \tag{4.4}$$

Newton probably guessed this result almost immediately, although he didn't have a mathematical proof in hand until years later.

4.5   Circular orbits

Initially, Newton's law of gravitation was simply an hypothesis, a guess waiting to be confirmed by observation and experiment. Newton didn't derive Eq. (4.1) or even suggest a mechanism by which the gravitational force might arise, he simply guessed a force law and began to compare its consequences against the natural world. When Newton turned his attention to celestial motion, he began by considering circular orbits.

An early success in the celestial realm came when Newton compared the acceleration of the Moon predicted by Eq. (4.1) with its actual motion. The easy part of this comparison is evaluating the force law to find the predicted acceleration A_M of the Moon due to the gravitational attraction of the Earth. Using the same steps as for Michael Jordan, we find

$$A_M = G \frac{M_E}{R_M^2}, \tag{4.5}$$

[2] In cancelling M_J from both sides of Eq. (4.2), we acknowledge an equivalence between the mass on the right side, which is called the inertial mass and determines the acceleration in response to a force, and that on the left side, which is called the gravitational mass and determines the gravitational force. This equivalence later became the cornerstone of Einstein's theory of gravity, the general theory of relativity.

where R_M is the distance between the centers of the Earth and the Moon. Just like a falling body on Earth, the acceleration of the Moon doesn't depend on the mass of the Moon. However, A_M is a lot less than g because the Moon is much farther away. In fact, dividing Eq. (4.5) by Eq. (4.4), we obtain,

$$\frac{A_M}{g} = \frac{R_E^2}{R_M^2}, \tag{4.6}$$

so the gravitational acceleration of the Moon is smaller than that of Michael Jordan by the square of the ratio of their distances from the center of the Earth. Using the modern values of $g = 9.81$ m/s^2, $R_E = 6.37 \times 10^6$ m, and $R_M = 3.84 \times 10^8$ m, we obtain $A_M = 2.70 \times 10^{-3}$ m/s^2. If the acceleration of gravity were only A_M at the Earth's surface, almost anyone could jump a kilometer high.

How can we calculate the Moon's acceleration from its motion? Assuming that the Moon orbits the Earth in a circle at constant speed, as is roughly true, its acceleration is due to a change in the direction of its velocity rather than a change in its speed. The situation is illustrated in Fig. 4.6(a). Here we assume that the Moon is initially crossing the X axis of our coordinate system and has a velocity vector V_{old} in the Y direction. After a short time ΔT, the Moon has moved a distance ΔD along its orbit, and its new velocity vector V_{new} has a small component in the X direction. The magnitudes of V_{new} and V_{old} are exactly the same—let's say $|V_{\text{old}}| = |V_{\text{new}}| = V_M$—but the change in direction implies an acceleration. It's this acceleration that we'd like to calculate.

As shown in Fig. 4.6(b), the difference between the new and old velocity vectors is the small vector ΔU, which is basically the X component of V_{new}. Because ΔU is directed toward the center of the circular orbit, the acceleration A_C that it represents is called centripetal acceleration. As always, acceleration is the change in velocity per unit time, so in the present case,

$$A_C = \frac{|\Delta U|}{\Delta T}. \tag{4.7}$$

With a little geometry, we can turn this formula into something more useful. Because the velocity vectors in Fig. 4.6(a) are perpendicular to the radius vectors, the indicated angle between the velocity vectors in

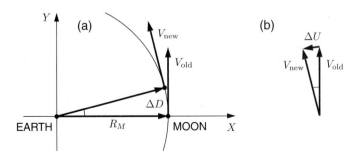

Fig. 4.6 (a) As the Moon circles the Earth it experiences a centripetal acceleration that changes the direction of its velocity vector without changing its speed. (b) The change ΔU in the Moon's velocity is directed toward the Earth.

Fig. 4.6(b) must be the same as that between the radius vectors in Fig. 4.6(a), and the triangles formed by these pairs of vectors are similar. Equating the ratios of corresponding sides in these similar triangles, we have

$$\frac{|\Delta U|}{V_M} = \frac{\Delta D}{R_M}, \tag{4.8}$$

with the slight approximation that results from substituting the distance ΔD along the Moon's orbit for the direct distance between the old and new positions of the Moon. But, remembering that the distance traveled is the velocity times the time, we have

$$\Delta D = V_M \Delta T, \tag{4.9}$$

and combining Eqs. (4.7)–(4.9), we obtain our grand result for the centripetal acceleration of the Moon,

$$A_C = \frac{V_M^2}{R_M}. \tag{4.10}$$

Whew, that took some thinking! I hope it didn't make your brain hurt.

When you derive a new formula like Eq. (4.10), it's a good idea to check whether it makes sense. Although you might not have much intuition about the centripetal acceleration of the Moon, you experience the same type of acceleration every time you go around a curve in a car. Suppose for a moment that V_M is the speed of your car and R_M is the radius of a curve in the road. Is the acceleration greater if you take the curve at a higher speed? Yes, to avoid being thrown against the door, you always slow down before entering a curve, reducing V_M and also A_C. Is the acceleration greater if the curve has a smaller radius? Yes, you need to slow even more if the road bends sharply. Thus, our formula $A_C = V_M^2/R_M$ seems to agree with experience.

Now comes the punch line. Is the centripetal acceleration A_C of the Moon anything like the acceleration A_M calculated from Newton's law of gravity? To calculate A_C we need the Moon's velocity V_M, but that's just the circumference of its orbit $2\pi R_M$ divided by its orbital period $T_M = 27.3$ days or 2.36×10^6 seconds, giving $V_M = 2\pi R_M/T_M = 1.02 \times 10^3$ m/s. Thus, the Moon's centripetal acceleration is $A_C = V_M^2/R_M = 2.71 \times 10^{-3}$ m/s^2, which is almost exactly the same as the gravitational acceleration $A_M = 2.70 \times 10^{-3}$ m/s^2 computed earlier. The gravitational force is just the size needed to explain the centripetal acceleration observed in the motion of the Moon! When the 23 year old Newton made this comparison, he didn't know the lunar distance as accurately as we do today, but the agreement between A_M and A_C was convincing even so. From that moment on, Newton knew that the Moon was very likely held in orbit by the same gravitational force that causes an apple to fall to the ground.

Indeed, it suddenly appeared probable that the entire Solar System, with planets orbiting the Sun and moons orbiting planets, is held together by the force of gravity. As another test of his force law, Newton supposed that the planets follow approximate circular orbits around the Sun. Because the mass M_S of the Sun is much greater than that of any planet, planetary motion is principally determined by the Sun, and we can apply Eqs. (4.5) and (4.10) with appropriate substitutions. Equating the gravitational and centripetal accelerations of a planet yields

$$G\frac{M_S}{R_P^2} = \frac{V_P^2}{R_P},$$
(4.11)

where R_P is the radius of the orbit and V_P is the planet's velocity. But if T_P is the period of the planet then $V_P = 2\pi R_P/T_P$ and Eq. (4.11) can be rewritten as

$$T_P^2 = \frac{4\pi^2}{GM_S}R_P^3.$$
(4.12)

Newton immediately recognized this equation as a statement of Kepler's third law: the square of a planet's period is proportional to the cube of its distance from the Sun. Because Kepler's law can be derived in this way only if the force of gravity falls off as the inverse square of the distance ($F \propto 1/R^2$), Newton now had excellent evidence for his hypothesis. Although he would not complete the full proof of his gravitational theory for another 20 years, the young Newton knew that he had made a great discovery as he worked at home during the plague years.

4.6 Elliptical orbits

When Newton returned to Cambridge in 1667, after the plague had abated, his academic career immediately began to take off. Later that year he was elected a fellow of Trinity College, the following year he received his Master's degree, and in 1669 he was appointed the Lucasian Professor of Mathematics. No ordinary appointment, the Lucasian chair had been magnificently endowed by Henry Lucas five years earlier and provided a generous stipend for the lifetime of the holder.[3] Thus, Newton suddenly gained the financial security that would allow him to devote his life to mathematics and physics.

As a professor, Newton worked intensely, almost without respite, on the problems that intrigued him. Newton's scientific output was both prodigious and of the highest quality, yet he remained relatively unknown within the greater scientific community. Why? One of Newton's early publications on optics led to a bitter dispute, and for years thereafter he was satisfied to solve problems for his own amusement. Newton simply couldn't be troubled to publish his research.

One of the questions that Newton pursued during his years of self-imposed silence concerned planetary motion. While his early results for circular orbits showed promise, a circle is a special case of his theory, and he knew from Kepler that planetary orbits are elliptical. Thus, Newton was naturally led to ask whether a force proportional to $1/R^2$ more generally yields elliptical orbits. As it happened, the astronomer Edmond

[3]The Lucasian chair was recently held by Stephen Hawking, the famed explorer of black holes, and is now held by Michael Green, a string theorist.

Halley, famed discoverer of "Halley's Comet," put this very question to Newton in 1684, and received an immediate answer in the affirmative—a $1/R^2$ force mathematically implies elliptical orbits. At this, Halley was "struck with joy and amazement," astonished that Newton knew the answer but hadn't bothered to publish. Newton was probably also taken aback to realize that others were beginning to think along the lines he had taken up almost 20 years earlier.

Halley's question about the gravitational force law was prompted by a publication of Christian Huygens (1629–1695), a Dutch scientist of international renown, in which he gave the formula for centripetal acceleration. Like Newton, Halley understood that combining centripetal acceleration with a $1/R^2$ gravitational force leads directly to Kepler's third law. Thus, thanks to Huygens and Halley, the cat was out of the bag on gravitation and planetary orbits. If Newton wanted to salvage credit for his theory of gravity, the time had come for publication. With Halley's support and encouragement, Newton thus sat down to write his masterpiece on mechanics and gravitation, the *Principia*. Two years in the writing, the *Principia* was mathematically rigorous and monumental in scope, setting forth the tenets of dynamics in the style of Euclid (*circa* 300 B.C.), whose classic text on geometry had been idolized for two millennia. Newton doubtless intended the *Principia* to be overwhelming in its breadth and rigor, leaving no question as to who invented the science of dynamics. And indeed, with its publication in 1687, the *Principia* suddenly made Newton the acknowledged master of mechanics and the foremost physicist of his day.

In the *Principia*, Newton considered the gravitational interaction of two spherical bodies in meticulous detail, proving that all of Kepler's laws follow from a $1/R^2$ force law. These mathematical demonstrations might be considered the essence of the *Principia*, establishing once and for all that the Solar System is held together by the same mundane force that returns a cannonball to the ground. After this unification of celestial and earthly forces, the awe with which philosophers regard the heavens would never be quite the same as that of the ancients.

We won't trace Newton's proof of Kepler's laws, but it's instructive to see how the problem of orbital motion is formulated. To be specific, let's consider an artificial satellite circling the Earth. Just as with a cannonball, we want to follow the satellite's motion in the X–Y plane, keeping track of its position coordinates, X and Y, and also the X and Y components of its velocity, U and V. From the outset, we know that the satellite's motion is predicted by the equations,

$$T_n = T_o + \Delta T, \tag{4.13}$$

$$X_n = X_o + U_o \Delta T, \tag{4.14}$$

$$Y_n = Y_o + V_o \Delta T, \tag{4.15}$$

$$U_n = U_o + A_X \Delta T, \tag{4.16}$$

$$V_n = V_o + A_Y \Delta T, \tag{4.17}$$

where A_X and A_Y are the X and Y components of its acceleration, and we've used "n" and "o" as abbreviations for "new" and "old." Thanks to Newton, we'll be able to find formulas for A_X and A_Y in terms of X and Y, making Eqs. (4.13)–(4.17) self-contained and amenable to iteration.

According to Newton, A_X and A_Y are related to the respective components of the force, F_X and F_Y, such that

$$A_X = \frac{F_X}{M_S}, \tag{4.18}$$

$$A_Y = \frac{F_Y}{M_S}, \tag{4.19}$$

where M_S is the mass of the satellite. Thus, to evaluate A_X and A_Y, we must break the gravitational force F into its X and Y components, as shown in Fig. 4.7. Because the force triangle with sides F_X, F_Y, and F is similar to the coordinate triangle with sides X, Y, and R_S, we have,

$$F_X = \frac{X}{R_S} F, \tag{4.20}$$

$$F_Y = \frac{Y}{R_S} F, \tag{4.21}$$

where R_S is the distance from the center of the Earth to the satellite. Newton's law of gravity implies

$$F = -G \frac{M_E M_S}{R_S^2}, \tag{4.22}$$

and the Pythagorean theorem tells us that

$$R_S^2 = X^2 + Y^2, \tag{4.23}$$

so, combining Eqs. (4.18)–(4.23), we end up with

$$A_X = -\frac{GM_E X}{(X^2 + Y^2)^{3/2}}, \tag{4.24}$$

$$A_Y = -\frac{GM_E Y}{(X^2 + Y^2)^{3/2}}, \tag{4.25}$$

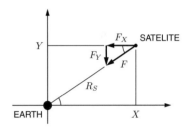

Fig. 4.7 A satellite orbiting the Earth experiences a gravitational force F directed toward the center of the Earth. F can be broken into X and Y components.

which express the acceleration components in terms of the satellite's coordinates. No matter how the satellite arrives at the point (X, Y) or what its velocity is, Newton assures us that its acceleration is given by Eqs. (4.24) and (4.25). Finally, substituting these equations into Eqs. (4.16) and (4.17), we find

$$U_n = U_o - \frac{GM_E X_o}{(X_o^2 + Y_o^2)^{3/2}} \Delta T, \tag{4.26}$$

$$V_n = V_o - \frac{GM_E Y_o}{(X_o^2 + Y_o^2)^{3/2}} \Delta T. \tag{4.27}$$

At last, we know how to compute the trajectory of a satellite. Given the satellite's initial position and velocity in the form of X, Y, U, and V, you can plug these dynamic variables into the right-hand sides of

Eqs. (4.14), (4.15), (4.26), and (4.27) to find their new values after a time ΔT. Moreover, if you are willing to repeat this process many times, you can predict the location and velocity of the satellite at any future time. The equations are a little messy but still simple enough that the computation can be made on a programmable calculator. Just use $GM_E = 3.99 \times 10^{14} \text{ m}^3/\text{s}^2$ and express distances in meters and velocities in meters/second, and everything will work out right. On the other hand, you can save yourself some trouble and see the same result by clicking on Experiment 6 of the Dynamics Lab.

4.7   Clockwork universe

Newton's proof that Kepler's laws of planetary motion follow from a $1/R^2$ law of gravity was a spectacular *tour de force* (pun intended). Could there be any doubt that celestial motion is governed by gravitational attractions? In science there is always room for doubt, and Newton sought to strengthen his case by considering more complex situations. His proof of Kepler's laws considered the interaction of just two bodies, the Sun and a planet, but certain anomalies suggested that lunar motion involved more than just the Moon and the Earth, and Kepler had guessed that the anomalies were caused by the Sun. Thus, a full understanding of the Moon's motion appeared to involve three bodies, not two, and Newton couldn't resist trying to explain lunar motion by considering the mutual gravitational attractions of the Earth, Sun, and Moon.

Having obtained a completely general solution for the two-body problem, Newton might have supposed at first that the motion of three bodies could also be readily solved. Indeed, the equations of motion for three bodies are easily formulated from Newton's laws—the forces depend only on the masses and the distances between the bodies—and today the equations can be solved on a computer without difficulty. In Newton's day, however, obtaining a solution could only mean finding a formula for the motion, a task that proved to be exceedingly difficult. In the end, Newton admitted that "his head never ached but with his studies on the moon."

In his attack on the Moon, Newton began with an elliptical orbit computed in the absence of the Sun, then introduced the Sun as a perturbing force and calculated its effect on the original ellipse. For example, Newton showed that the gravitational pull of the Sun causes the Moon's apogee, or point of closest approach to the Earth, to rotate slowly over the years, in agreement with observations. Nevertheless, Newton wasn't satisfied with the account of lunar motion that originally appeared in the *Principia*, and he returned to the problem in 1694. After a year of intense labor, Newton had refined his theory until it predicted the Moon's position with a maximum error of about 10 arc minutes. Still, an error of this size greatly exceeded any possible observational error, and Newton published his revised theory in 1702 with some reluctance. He was unable to claim total victory.

Newton's revised lunar calculations, completed in 1695 at age 52, brought to a close his life of intense scientific study. While he lived another 32 years, Newton's career became that of public servant and senior statesman of science. In the public sector, Newton was appointed Warden of the Mint in 1696 and promoted to Master of the Mint in 1699. In these posts, which took him to London, Newton oversaw the recoinage of English currency with the precision and efficiency that might be expected of the world's foremost scientist. In the science sector, Newton was elected President of the Royal Society in 1703, the most powerful position in England's science establishment, and produced second editions of his two great works, the *Principia* and *Opticks*. A man of unquestioned eminence, Newton was knighted by Queen Anne in 1705. Newton remained Master of the Mint and President of the Royal Society until his death in 1727, contributing significantly and retaining his mental faculties to the end.

In Newton's wake, lunar motion became a particular fascination of astronomers and mathematicians, and the difficulty of the three-body problem became notorious. Given the significance of lunar tables for navigation, governments and scientific societies offered prizes for improved formulas, attracting the best and brightest to the problem. Initial difficulties led some to speculate that Newton's law of gravity was inadequate to explain the motion of the Moon, but the matter was eventually resolved in Newton's favor. In the early 1750s Alexis Claude Clairaut (1713–1765) and Jean Le Rond d'Alembert (1717–1783) independently demonstrated that, when additional terms were included, Newton's formula substantially agreed with observations. High accuracy would further elude astronomers for almost 200 years, but the Moon had finally become prime evidence supporting Newton's theory of universal gravitation.

As astronomers honed their observational and mathematical skills, the agreement between Newton's theory and observation only improved. After the planet Uranus was discovered in 1781, astronomers were disturbed by the fact that over the years it deviated from its predicted position by as much as an arc minute, even when calculations included the gravitational pull of Jupiter and Saturn. Such was their faith in Newton's laws, however, that some astronomers speculated that an undiscovered planet was responsible for the wayward motion of Uranus. And indeed, all discrepancies were resolved with the discovery of Neptune in 1846.

By the nineteenth century, it was clear that the planets and moons of the Solar System move like clockwork, orchestrated by Newton's laws of motion. Every nuance of planetary motion, no matter how small, appeared to be explained by Newtonian gravity.[4] Was there anything that Newtonian mechanics couldn't predict? Overwhelmed by the success of celestial mechanics, scientists of the nineteenth century began to believe that, at least in principle, all events in the universe might be predetermined and calculable. Pierre Simon de Laplace (1749–1847), one of the world's greatest celestial mechanics, put the idea in a single sentence.

[4]The slow rotation of Mercury's perihelion is an exception. Calculating the influence of the outer planets, in 1843 Urbain Jean Joseph Le Verrier (1811–1877) obtained a value for the perihelion advance of Mercury that was 38 arc seconds per century less than the observed value. This discrepancy was resolved in 1915, when Einstein applied his theory of general relativity. Accounting for Mercury's perihelion advance was the first triumph of general relativity over Newtonian gravity.

An intellect which at any given moment knew all the forces that animate Nature and the mutual position of the beings that comprise it, if this intellect were vast enough to submit its data to analysis, could condense into a single formula the movement of the greatest bodies of the universe and that of the lightest atom: for such an intellect nothing could be uncertain; and the future just like the past would be present before its eyes.

The scenario is now familiar. As the examples in this chapter demonstrate, given a force law and the initial position and velocity of a body, we can easily predict its subsequent motion. In principle, predicting the motion of many bodies is no more difficult: it's just a matter of keeping track of all the coordinates, elementary bookkeeping. Thus, within the context of Newtonian mechanics, Laplace is undoubtedly correct. Science had reached a grand vision in which all can be known and understood.

As Laplace knew well, however, the complexity of any calculation increases rapidly with the number of bodies involved, and beyond a certain point prediction becomes impossible in practice. Newton's famous headache, induced by the three-body problem, is ample evidence of the difficulty. The universe may run like clockwork, but the clock is so complicated that our predictions will always be limited to a small domain. Today, however, we know of another practical limitation to prediction of which Laplace was unaware. In particular, the motion of just a single body subject to Newton's laws may, under certain circumstances, be practically unpredictable over the long term. I refer, of course, to chaotic motion. Chaos is another fly in the ointment of Laplace's clockwork universe.

But chaos is more than just an obstacle to prediction. Having developed his laws to explain the regular dynamics of planets and pendulums, Newton would have been astonished to learn that in some cases his laws also predict essentially random behavior. Sometimes the Newtonian universe runs like clockwork, and sometimes it goes berserk. How could this be? Where is the gremlin lurking in Newton's seemingly innocent equations? These are the questions that we must now attempt to answer.

Further reading

Brahe
- Christianson, J., "The celestial palace of Tycho Brahe", *Scientific American* **204**(2), 118–128 (December 1961).

Celestial mechanics
- Bodenmann, S., "The 18th-century battle over lunar motion", *Physics Today* **63**(1), 27–32 (January 2010).

- Gamow, G., *Gravity* (Anchor Doubleday, Garden City, New York, 1962).

- Gingerich, O., "Ptolemy, Copernicus, Kepler", in *The Eye of Heaven* (American Institute of Physics, New York, 1993) chapter 1.

- Gondhalekar, P., *The Grip of Gravity: The Quest to Understand the Laws of Motion and Gravi-*

tation (Cambridge University Press, Cambridge, 2001).

o Hirshfeld, A. W., *Parallax: The Race to Measure the Cosmos* (W. H. Freeman, New York, 2001).

o Marchant, J., *Decoding the Heavens: A 2,000-Year-Old Computer—And the Century-Long Search to Discover Its Secrets* (Da Capo Press, Cambridge, Massachusetts, 2009).

Copernicus

o Goldoni, G., "Copernicus decoded", *The Mathematical Intelligencer* **27**(3), 12–30 (Summer 2005).

o Ravetz, J. R., "The origins of the Copernican revolution", *Scientific American* **215**(4), 88–98 (October 1966).

Kepler

o Ferguson, K., *Tycho and Kepler: The Unlikely Partnership That Forever Changed Our Understanding of the Heavens* (Walker, New York, 2002).

o Koestler, A., *The Watershed: A Biography of Johannes Kepler* (Anchor Doubleday, Garden City, New York, 1960).

o Wilson, C., "How did Kepler discover his first two laws?", *Scientific American* **226**(3), 92–106 (March 1972).

5 The pendulum—Linear and nonlinear

As the phrase "clockwork universe" implies, clocks are symbolic of perfectly regular motion. The quiet ticktock of a grandfather clock assured generations that all was right with the world, marking the progress of time by the regular swing of a pendulum. Ironically, in this book the pendulum serves as our prime example of a chaotic mechanism. There's no contradiction here: the same Newtonian equations that predict clocklike regularity under some conditions yield random motion under others. In large measure, these contrasting outcomes trace to the difference between what mathematicians call "linear" and "nonlinear" behavior. When a pendulum moves no more than a few degrees from its downward position, its behavior is approximately linear, and chaotic motion is ruled out. On the other hand, if a pendulum is forced to swing full circle, then its behavior is profoundly nonlinear, and chaos is likely.

As an engineering student in the 1960s, I learned the mathematical difference between linear and nonlinear systems, but the curriculum never ventured far into the nonlinear realm. Although the distinction didn't appear to be of great consequence, students were told that nonlinearity was a phenomenon to be avoided. Like unexplored lands on an ancient map, the nonlinear realm was posted, "Abandon all hope ye who enter here." Why the fuss? Most of the advanced techniques of mathematical analysis taught to engineers and scientists, from Fourier and Laplace transforms to Green's functions, apply exclusively to linear systems. The ease with which such systems can be analyzed makes them the mainstay of engineering. In contrast, only a relative handful of nonlinear systems can be solved by mathematical analysis, so scientists are circumspect when they encounter nonlinear behavior. Whatever the mysterious distinction between linear and nonlinear might be, it marks the mathematical boundary between easy and difficult.

More recently, the advent of efficient computing has radically changed our outlook on nonlinearity. While nonlinear equations still can't usually be solved by mathematical analysis, numerical solutions are easily obtained by computer. If we are satisfied with numbers rather than general formulas, then nonlinear equations are often as easy to solve as linear ones. As it happens, we met both linear and nonlinear systems in Chapters 3 and 4 without realizing the difference. In particular, the equations for a mass on a spring are linear while those for a satellite are nonlinear, yet the Dynamics Lab program simulates both with equal ease. In this chapter, we'll rely on the computer to explore the dynamics of a simple pendulum and learn more about the difference between linear and nonlinear behavior.

5.1   Rotational motion

The motion of a pendulum is best described as a rotation. Although we could keep track of the X and Y coordinates of the pendulum bob, it's much easier to specify a pendulum's position as the distance that the bob has traveled or, better yet, the pendulum's angle from the downward direction. As shown in Fig. 5.1, the distance D and the angle θ (Greek theta) amount to the same thing. Indeed, if we measure the pendulum's angle in radians, θ and D are related by

$$\theta = D/R, \tag{5.1}$$

where R is the length of the pendulum. Because angles are natural to rotation, we'll describe the motion of the pendulum in terms of θ rather than D.

If we measure the pendulum's angle in radians (rad), then in exploring its motion we'll encounter angular velocity v measured in radians per second (rad/s), and angular acceleration α (Greek alpha) measured in radians per second per second (rad/s^2). With radians popping up everywhere, we'd better take time out to understand what a radian is. Equation (5.1) actually defines an angle measured in radians: it's the arc length D subtended by the angle at a radius R divided by that radius. Since 360° represents a full circle, its equivalent in radians is the circumference of the circle $2\pi R$ divided by R or 2π rad. Similarly, a half circle is 180° or π rad and a quarter circle is 90° or $\pi/2$ rad. Thus, it's not hard to convert degrees into radians (multiply the angle by $\pi/180$) or convert radians into degrees (multiply by $180/\pi$). Accordingly, a radian is approximately 57.3°.

On the other hand, why do we want to use radians? Aren't degrees just as good? Perhaps surprisingly, radians are much simpler than degrees from a mathematician's point of view. Consider, for example, the trig function $\sin(\theta)$. Scientific calculators allow you to work with θ in either radians or degrees, but, whichever you choose, $\sin(\theta)$ is always calculated according to the infinite series (another product of the calculus),

$$\sin(\theta) = \theta - \frac{\theta^3}{3!} + \frac{\theta^5}{5!} - \frac{\theta^7}{7!} + \frac{\theta^9}{9!} - \cdots, \tag{5.2}$$

which works only for angles in radians. Here the quantity 3!, read "three factorial," is shorthand for $3 \times 2 \times 1 = 6$, and similarly $5! = 5 \times 4 \times 3 \times 2 \times 1 = 120$. If you set your calculator to work in degrees, then it converts θ to radians before Eq. (5.2) is applied. The simplicity of Eq. (5.2) suggests that radians really are the natural way to measure angles.[1]

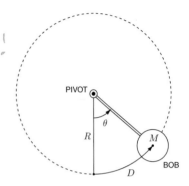

Fig. 5.1 The pendulum consists of a bob of mass M suspended by a rigid shaft from a fixed pivot point.

[1] If you haven't seen Eq. (5.2) before, try it out. You may remember that $\sin(30°) = 1/2$. Since 30° is equivalent to $\pi/6$ radians, plugging $\theta = 0.523598776$ into the right side of Eq. (5.2) should yield $\sin(\theta) = 0.500000000$. The successive terms in Eq. (5.2) get smaller rapidly, so you won't need to add up many before the result is accurate. Of course, Eq. (5.2) is completely general, which allows your calculator to evaluate the sine of any angle.

In rotational motion, the quantities θ, v, and α (angle, angular velocity, and angular acceleration) take the place of D, V, and A (distance, velocity, and acceleration) that we used to describe rectilinear motion. While an angle is a familiar concept, angular velocities and accelerations may seem mysterious. They're actually simple. For example, the second hand of a clock goes around once a minute, so its angular velocity is $v = 2\pi/60 = 0.105$ rad/s, while the hour hand completes a circle in 12 hours or 43,200 s, yielding $v = 1.45 \times 10^{-4}$ rad/s. At the other extreme, an ultra-centrifuge used to purify biological samples might rotate at 100,000 revolutions per minute (rpm),[2] so its angular velocity is $v = 100,000(2\pi)/60 = 1.05 \times 10^4$ rad/s. Because an ultra-centrifuge requires about a minute to achieve operating speed after it's switched on, it experiences an average angular acceleration of $\alpha = 1.05 \times 10^4/60 = 175$ rad/s^2 during the startup period. That is, the angular velocity increases by 175 rad/s during each second of acceleration.

The mathematics of rotational motion is clarified if we suppose that D, V, and A are the distance traveled by the pendulum bob and its velocity and acceleration along the bob's circular path. In this case we not only have $\theta = D/R$ but also

$$v = V/R, \tag{5.3}$$

$$\alpha = A/R. \tag{5.4}$$

In each case, dividing by R converts the distance traveled (per second or per second squared) into the angle traveled (per second or per second squared). Given this correspondence, it follows that the incremental equations for rotational motion are,

$$T_n = T_o + \Delta T, \tag{5.5}$$

$$\theta_n = \theta_o + v_o \Delta T, \tag{5.6}$$

$$v_n = v_o + \alpha_o \Delta T, \tag{5.7}$$

where the subscripts "n" and "o" stand for "new" and "old" as usual. That is, after a time step ΔT we must increment the angle by the angular velocity times the time step and increment the angular velocity by the angular acceleration times the time step. Our program for predicting the motion of a pendulum is thus basically the same as for rectilinear motion. However, the program won't be complete until we relate the angular acceleration to θ, v, and T, so that α is always known and Eqs. (5.5)–(5.7) can be iterated indefinitely into the future.

5.2   Torque

For rotational motion, the equation that corresponds to $F = MA$ is

$$\tau = I\alpha, \tag{5.8}$$

[2] To achieve this speed, the centrifuge must be operated in high vacuum, and the samples experience a centripetal acceleration of $800,000\,g$, which is sufficient to fractionate a solution of macromolecules according to density.

where τ is the torque or twisting force, I is the moment of inertia, and α is the now familiar angular acceleration. Thus, an applied torque produces an angular acceleration that depends on the moment of inertia of the rotating body. The greater the moment of inertia, the larger the torque required to produce the same angular acceleration. Because the torque can often be expressed in terms of θ, v, and T, Eq. (5.8) is the key to evaluating α.

Although $\tau = I\alpha$ may be unfamiliar, it is just $F = MA$ expressed in a form suitable to rotation. Thus, we can guess from the outset that τ and I are simply related to F and M, just as α is related to A. The definition of torque is illustrated in Fig. 5.2 in terms of the force applied to a wrench. Here the torque is given by

$$\tau = RF, \tag{5.9}$$

where R is the distance from the pivot to the point at which the force is applied and F is the component of force perpendicular to the wrench. Why not include forces directed parallel to the wrench? Such forces push the wrench toward or away from the bolt but provide no twist. Why does the radius enter the definition of torque? As you might guess, a wrench of twice the length requires only half as much force to provide the same twist. Thus, our definition of torque makes basic sense.

I first encountered the idea of torque as a kid, watching my dad assemble an engine after reboring the cylinders. When it was time to tighten the head bolts, he got out a special wrench, calibrated in units of distance times force, to apply the torque specified by the manual: enough to keep the head in place, but not enough to snap a bolt. If you've ever used a torque wrench, you know exactly what torque is.

What's a pendulum's moment of inertia I? To get an idea, suppose that the entire mass of the pendulum is concentrated in the bob at radius R. In this case, $F = MA$ can be applied directly to the bob, with the force and acceleration understood to act perpendicular to the shaft of the pendulum. Multiplying $F = MA$ by the radius yields

$$RF = (R^2 M)(A/R). \tag{5.10}$$

Since $RF = \tau$ and $A/R = \alpha$, we see that Eq. (5.10) is equivalent to $\tau = I\alpha$, provided the moment of inertia is

$$I = R^2 M. \tag{5.11}$$

As expected, the moment of inertia is related to the mass, but it also involves the radius. In the general case, I must be calculated by adding up all the bits of mass in the rotating body, with each bit multiplied by the square of its distance from the axis of rotation. However, Eq. (5.11) suffices for our analysis of the pendulum.

Mechanical engineers use a flywheel to provide a large moment of inertia. Once you get a flywheel rotating, it's hard to stop, so connecting a flywheel to an engine keeps the crankshaft turning between the power strokes of the pistons. To make the moment of inertia large, flywheels

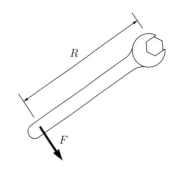

Fig. 5.2 A torque $\tau = RF$ is produced by a force F acting perpendicular to the handle of a wrench at radius R.

must be massive, but, according to Eq. (5.11), it helps to put all the mass at the largest possible radius. Thus, flywheels often consist of a heavy rim connected to a hub with spokes.

5.3   Pendulum dynamics

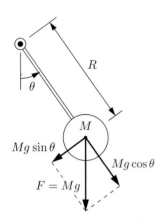

Fig. 5.3 The force of gravity acting on a pendulum bob can be resolved into components that act parallel and perpendicular to the shaft. Only the perpendicular component contributes to the torque.

Now that we've introduced all the players, θ, v, α, τ, and I, writing down the incremental equations for the motion of a pendulum is straightforward. For the moment, we consider just two torques acting on the pendulum: the torque due to the weight of the bob and a torque due to friction. Thus, the total torque is

$$\tau = \tau_G + \tau_D, \tag{5.12}$$

where τ_G is the gravitational torque and τ_D is a frictional torque that damps the motion. Using τ_G and τ_D, we'll be able to explain the natural motion of a pendulum.

As shown in Fig. 5.3, gravity exerts a downward force $F = Mg$ on the pendulum bob. However, only a part of this force, $Mg\sin\theta$, acts perpendicular to the shaft of the pendulum, so the gravitational torque is

$$\tau_G = -RMg\sin\theta, \tag{5.13}$$

where we've multiplied the perpendicular force times the radius at which it acts. The minus sign accounts for the fact that gravity always acts to accelerate the bob toward the downward vertical position ($\theta = 0$). That is, τ_G is negative for $0 < \theta < \pi$ and positive for $-\pi < \theta < 0$.

To fully appreciate equation Eq. (5.13), recall that $\sin\theta$ is a periodic function that oscillates between -1 and 1 over a period of 2π in θ, as shown in Fig 5.4.[3] At $\theta = 0$, the pendulum hangs straight down, and according to Eq. (5.13) there is no gravitational torque. Gravity hasn't been turned off, of course, but when $\theta = 0$ it's pulling exactly against the pivot point, and there's no component of force to cause rotation. The same thing happens at $\theta = \pi$, where the pendulum is straight up: the force of gravity produces no torque, and the pendulum can balance indefinitely. At $\theta = \pm\pi/2$, on the other hand, the pendulum is horizontal, the force of gravity acts perpendicular to the shaft, and $|\tau_G|$ is at its

[3]Strangely, the series expansion of $\sin\theta$, Eq. (5.2), gives no hint of a 2π periodicity, but if you try $\theta = \pi/6 + 2\pi$, you'll find that the series converges to $1/2$ almost as quickly as for $\theta = \pi/6$. Indeed, no matter how many times you add 2π to a given angle, Eq. (5.2) gives the same answer for $\sin\theta$. Amazing!

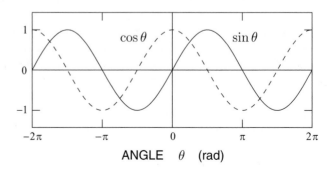

Fig. 5.4 The sine and cosine are oscillatory functions with a period of 2π and a maximum magnitude of 1.

maximum, $\tau_{\max} = RMg$. No wonder those old-time pilots rotated the prop to a horizontal position before trying to start their engine—they wanted to get maximum torque when they pulled down on the prop.

Because the torque is zero at $\theta = 0$ and $\theta = \pi$, the pendulum can remain stationary at either of these positions, pointed either straight down or straight up. With respect to stability, these positions are at opposite extremes. The downward position is stable: if you give the pendulum a nudge, the gravitational torque immediately pulls the bob back towards its original position, producing oscillations about $\theta = 0$. This stability is essential to the pendulum clock. In contrast, the upward position is highly unstable, and the slightest nudge will start the pendulum accelerating away from its balance point. Not surprisingly, chaotic motion in the pendulum can be traced directly to this instability.

Regarding the frictional torque on the pendulum, we'll assume that it's proportional to velocity, just as in Chapter 3. Thus, we have

$$\tau_D = -Cv, \tag{5.14}$$

where the constant C is the damping coefficient. In this equation, the minus sign indicates that the frictional torque always acts to reduce the magnitude of the angular velocity, whether v is positive or negative.

Adding up the gravitational and frictional torques, we obtain the total torque,

$$\tau = -RMg\sin\theta - Cv, \tag{5.15}$$

which completes our picture of the pendulum. We can now solve for the angular acceleration using $\alpha = \tau/I$ and $I = R^2M$ with the result,

$$\alpha = -\frac{g}{R}\sin\theta - \frac{C}{R^2M}v. \tag{5.16}$$

This formula for α in terms of θ and v provides everything we need to iterate Eqs. (5.5)–(5.7) and predict the pendulum's motion as far into the future as desired.

At this point, we could simply proceed with our investigation of the pendulum, but instead we pause to tidy up our equations. We'll use these equations throughout the remainder of the book, so our effort will be rewarded later.

A physicist may find Eq. (5.16) highly satisfactory because the physical constants g, M, R, and C all have significance, but a mathematician usually prefers a simpler form. To achieve this simplicity, we replace all of the dimensioned quantities by new dimensionless quantities that are natural to the problem. For example, instead of using the time T measured in seconds, we adopt a dimensionless time t defined by

$$t = \omega_P T, \tag{5.17}$$

where ω_P is the parameter,

$$\omega_P = \sqrt{g/R}. \tag{5.18}$$

Remembering that g is in units of m/s^2 and R is in m, its easy to verify that ω_P has units of $1/s$ and that t is dimensionless. If we ever need to restore dimensions, we simply divide t by the parameter ω_P.

We can also remove the dimensions of $1/s$ and $1/s^2$ from v and α by adopting the new variables,

$$\tilde{v} = v/\omega_P, \tag{5.19}$$

$$\tilde{\alpha} = \alpha/\omega_P^2. \tag{5.20}$$

Here a tilde $(\sim)$ is used to remind us that $\tilde{v}$ and $\tilde{\alpha}$ are like v and α but with time measured in natural units rather than seconds. Finally, if we introduce a new dimensionless damping constant,

$$\rho = \frac{C}{\omega_P R^2 M}, \tag{5.21}$$

then the incremental equations for the pendulum take the form,

$$t_n = t_o + \Delta t, \tag{5.22}$$

$$\theta_n = \theta_o + \tilde{v}_o \Delta t, \tag{5.23}$$

$$\tilde{v}_n = \tilde{v}_o - (\sin\theta_o + \rho\tilde{v}_o)\Delta t, \tag{5.24}$$

where $\Delta t = \omega_P \Delta T$ and we have used the relation

$$\tilde{\alpha} = -\sin\theta - \rho\tilde{v}, \tag{5.25}$$

to replace the angular acceleration in Eq. (5.24). With a little algebra, you can verify that Eqs. (5.22)–(5.24) are equivalent to Eqs. (5.5)–(5.7) combined with Eq. (5.16). Although dimensioned quantities may be comforting, Eqs. (5.22)–(5.24) still completely describe the behavior of a damped pendulum. Remarkably, our new equations imply that, if we're willing to measure time in units of $1/\omega_P = \sqrt{R/g}$, the pendulum is characterized by a single dimensionless parameter, the damping coefficient ρ.

To make it easier to keep track of the pendulum variables, we list both the dimensioned and dimensionless quantities in Table 5.1. Note that a dimensionless torque $\tilde{\tau}$ can be defined as the ratio of the torque τ to

Table 5.1 You can't tell the players without a program. We list the various constants and variables associated with a pendulum of mass M and radius R, giving both the dimensioned and dimensionless quantities. Here $\omega_P = \sqrt{g/R}$.

Quantity	Dimensioned	Dimensionless
Time	T	$t = \omega_P T$
Angular velocity	$v = V/R$	$\tilde{v} = v/\omega_P$
Angular acceleration	$\alpha = A/R$	$\tilde{\alpha} = \alpha/\omega_P^2$
Torque	$\tau = R^2 M\alpha$	$\tilde{\tau} = \tau/(RMg) = \tilde{\alpha}$
Damping constant	C	$\rho = C/(\omega_P R^2 M)$

the maximum gravitational torque $\tau_{\max} = RMg$. In this case, $\tilde{\alpha} = \tilde{\tau}$, so the dimensionless acceleration and torque are not just proportional but equal.

5.4   Quality factor

Back in the days when the transistor was new, I often studied electronics catalogs to see how I might spend my allowance. Sometimes electronics jargon left me mystified. What was a "SPDT" switch? I puzzled over that one for a long time before somehow guessing that it was a "single-pole double-throw" switch. Another item that bugged me was a "high-Q" ferrite antenna. Presumably high Q is better than low Q, but what is "Q" in the first place? This puzzle was resolved only when I went to college and learned that Q stands for "quality factor" and specifies the degree to which damping is absent in anything that oscillates, from a tuned circuit to a pendulum. In our dimensionless pendulum, the quality factor is precisely $Q = 1/\rho$, so the Q is large when the damping is small.

The easiest way to understand the quality factor Q or damping coefficient ρ is to iterate Eqs. (5.22)–(5.24) and watch what happens. Experiment 7 of the Dynamics Lab lets you do just that by animating the pendulum's motion for whatever damping coefficient you choose. Alternatively, consider Fig. 5.5 which plots the motion for $Q = 5$. Here, the pendulum is given a swift kick in the bob at $t = 0$, so it begins at $\theta = 0$ with the sizeable angular velocity of $\tilde{v} = 4$. As shown in Fig. 5.5(a), the kick is enough to send the pendulum over the top (past $\theta = \pi$), and it continues to rotate, completing one and then two complete revolutions as it passes $\theta = 2\pi$ and 4π. As seen in Fig. 5.5(b), however, friction gradually reduces the pendulum's velocity, and after reaching $\theta = 4\pi$ its speed is no longer sufficient to reach the top. Thus, the pendulum settles into oscillations about $\theta = 4\pi$, and these oscillations gradually decay from the continued action of friction. Altogether, the motion shown in Fig. 5.5 matches our intuition and confirms that Eqs. (5.22)–(5.24) adequately describe a pendulum. These equations of motion are simple but effective and include much more than you'd guess at first sight.

In Fig. 5.5(a), we allow the angle θ to increase indefinitely as the pendulum rotates around and around. Of course, since 2π represents a full circle, the position of the pendulum is the same at $\theta = 2\pi$ and 4π as at $\theta = 0$. Sometimes it's simpler to plot the angle on a single 2π interval, say from $-\pi$ to π, rather than keep track of the total rotation. We adopt this strategy in Fig. 5.6, which compares the pendulum's motion for $Q = 5$ and 10. The case shown in Fig. 5.6(a) is identical to that in Fig. 5.5, but now, when θ first reaches π at about $t = 1$, the plot is continued at $\theta = -\pi$. This jump or discontinuity looks unreal, but it makes perfect sense because $\theta = \pi$ and $-\pi$ correspond to the same position of the pendulum. The effect is like a video game in which your spacecraft exits the top of the screen then reappears at the bottom. However, Fig. 5.6(a) would make even more sense if you were to cut it

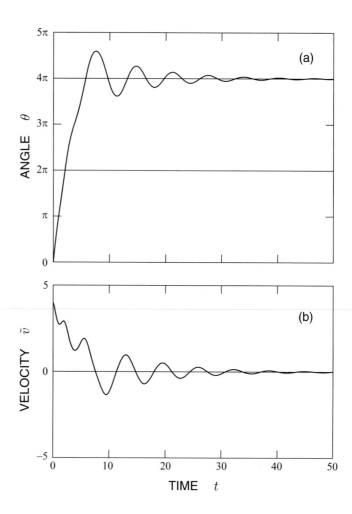

Fig. 5.5 The angle θ (a) and angular velocity $\tilde{v}$ (b) of a damped pendulum as a function of time, assuming that initially $\theta = 0$ and $\tilde{v} = 4$. The quality factor is $Q = 5$ ($\rho = 0.2$).

out and roll it into a cylinder by gluing the top edge to the bottom edge. In cylindrical form, $\theta = \pi$ and $-\pi$ really are the same, and the angle plot is completely continuous.

Comparing Figs. 5.6(a) and (b), we see that when the quality factor increases from 5 to 10, the damping is reduced and friction takes longer to slow the pendulum. Thus, for $Q = 10$ the pendulum completes four rotations rather than two before oscillatory motion begins, and the oscillations decay more slowly. To be quantitative, the amplitude of an oscillator decays by a factor of $\frac{1}{2}$ after approximately $0.22Q$ complete oscillations. This rule of thumb is exemplified in Fig. 5.6, where the oscillation amplitude is reduced by about half after one oscillation for $Q = 5$ and after two oscillations for $Q = 10$. The higher the Q, the more slowly the oscillations decay. Thus, the quality factor tells us quantitatively how long an oscillator will ring. Among mechanical oscillators, tuning forks have very high Q: they ring for several seconds with hundreds of vibrations each second.

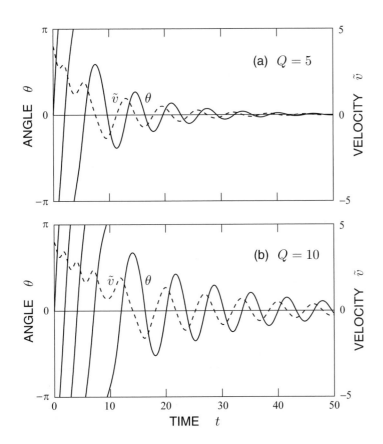

Fig. 5.6 Combined plots of the angle θ (solid curve) and angular velocity $\tilde{v}$ (dashed curve) of a damped pendulum as a function of time, assuming that initially $\theta = 0$ and $\tilde{v} = 4$. In (a) the quality factor is $Q = 5$ ($\rho = 0.2$), and in (b) $Q = 10$ ($\rho = 0.1$). Angles are plotted on the range $-\pi$ to π.

5.5 Pendulum clock

If friction could be eliminated, a pendulum would oscillate forever with a fixed period and make an ideal clock. As noted previously, Galileo suggested that, even with some friction, a pendulum holds promise as a clock because its period is nearly independent of the oscillation amplitude. As he wrote to Admiral Reael in 1636, building an accurate clock is simply a matter of combining a pendulum with a counter to keep track of the oscillations. Galileo began work on his clock only in 1641, however, and failed to complete the task before his death the following year.

Galileo's vision of a clock was realized in 1657 by the Dutch scientist Christian Huygens (1629–1695). Huygens' career was sandwiched between Galileo and Newton in both time and accomplishment: he continued the work of Galileo and anticipated that of Newton. In astronomy, Huygens further developed the telescope and used his improved instrument to discover Saturn's largest moon Titan and to identify the "appendages" of Saturn as an encircling ring. In mechanics, Huygens developed a quantitative theory of the pendulum and was the first to obtain the formula for centripetal acceleration. While this formula was published only in 1673, Huygens had obtained it in 1659, several years before Newton began to study motion. Clearly, Huygens was a creative and accomplished mathematician and scientist in his own right. Indeed, Huygens' wave theory of light would eventually eclipse Newton's particle theory.

For Huygens, the pendulum clock became a lifetime affair. Huygens was 27 in 1657 when he built the first pendulum clock and obtained a patent. The following year he described his invention in a publication entitled *Horologium* or *The Clock*. Motivated by the possibility of solving the problem of longitude, Huygens spent many years perfecting the pendulum clock. In Huygens' day ships were often lost at sea because no adequate means of navigation was available. An accurate clock set to Greenwich time would allow a mariner to determine his longitude anywhere on Earth from a simple observation of the local time. To improve his clock's accuracy, Huygens developed an ingenious "escapement" mechanism which gives the pendulum a kick at just the moment when it least affects its period of oscillation. The escapement is crucial to Huygens' clock: it both supplies the energy needed to overcome friction and counts the pendulum's oscillations. Huygens described his further advances in clock design in publications of 1673 and 1693, the latter just two years before he died. Despite his diligence, however, Huygens failed to devise a useful marine chronometer. This feat was accomplished only in 1760 by an English inventor, John Harrison (1693–1776), with a clock based on a balance wheel rather than a pendulum, an innovation introduced by Huygens in 1675.

Huygens' analysis of the pendulum was remarkable in that it preceded the *Principia*, and used neither the calculus nor the concept of force. Instead, Huygens began with Galileo's rudimentary dynamics, adding only the hypothesis that a pendulum initially at rest can never rise higher than its original position. Combining these meager assumptions with geometric arguments, in 1659 Huygens was able to derive a formula for the period T_{P0} of the small-angle oscillations of a frictionless pendulum. In modern notation, Huygens discovered that

$$T_{P0} = 2\pi\sqrt{R/g}. \tag{5.26}$$

This formula confirms Galileo's empirical observation that the period of a pendulum is independent of its mass and proportional to the square root of its length R. According to Eq. (5.26), if you want to build a clock that ticks off one second each time the pendulum passes $\theta = 0$, then its period will be $T_{P0} = 2$ s (1 s to swing to the left plus 1 s to swing to the right), and the length of the pendulum must be $R = gT_{P0}^2/(4\pi^2) = 0.994$ m. Now you know why a grandfather clock is so tall: its pendulum is nearly a meter long.

Given Huygens' formula for T_{P0}, we can now understand the physical significance of the parameter $\omega_P = \sqrt{g/R}$ that we introduced to create a dimensionless form for the pendulum equations. According to Eq. (5.26), $\omega_P = 2\pi/T_{P0}$, and in our dimensionless system the small-amplitude period is

$$t_{P0} = \omega_P T_{P0} = 2\pi. \qquad (5.27)$$

That is, in Eqs. (5.22)–(5.24) time is referenced to the natural oscillations of the pendulum, such that the small-amplitude period is exactly 2π time units. In fact, if we go beyond Huygens by using calculus, we can show that the small-amplitude oscillations of a frictionless pendulum are described accurately by a sinusoid,

$$\theta(t) = \theta_{\max} \sin(t) \qquad (\rho = 0, \ \theta_{\max} \ll 1), \qquad (5.28)$$

where $\theta_{\max}$ is the maximum displacement of the pendulum. From Eq. (5.2), we again see that the period of oscillation is 2π time units, but we also learn the functional form of the motion. Everything considered, Eq. (5.28) is a remarkably simple formula.

From his derivation of T_{P0}, Huygens knew that Eq. (5.26) is valid only in the limit of small oscillations and that Galileo was not strictly correct in asserting that the period of a pendulum's swing is independent of the amplitude. What happens for a large swing? We can find out by iterating Eqs. (5.22)–(5.24) for $\rho = 0$ and various values of initial velocity. As shown in Fig. 5.7, when the initial kick leads to a swing with $\theta_{\max} = \pi/4$, the oscillations appear to be sinusoidal, but the period is slightly longer than $t_{P0} = 2\pi$. If the initial kick is increased until $\theta_{\max} = \pi/2$ or $90°$, then the period is 18% longer than t_{P0}. In his 1673 publication, Huygens estimated the period for a $90°$ swing to be 17% longer than for small swings but later improved his calculation to obtain 18%. As $\theta_{\max}$ approaches π or $180°$, we see from Fig. 5.7 that the pendulum spends an extended period of time near $\theta_{\max}$, and its motion is far from sinusoidal. This behavior makes sense because, for $\theta_{\max} = \pi$, the pendulum would ideally approach its balance point and never return to smaller angles.

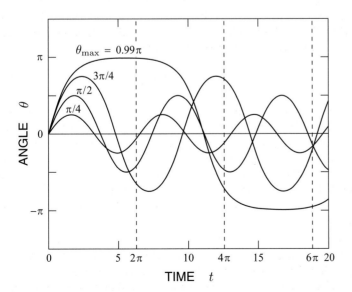

Fig. 5.7 Angle as a function of time for oscillations of a frictionless pendulum $(\rho = 0)$ with various maximum amplitudes $\theta_{\max}$. The oscillation period (2π) in the limit of small amplitudes is indicated by dashed vertical lines.

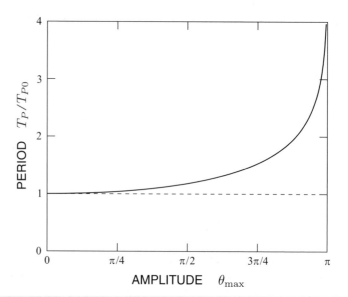

Fig. 5.8 The period T_P of a friction-less pendulum $(\rho = 0)$ as a function of the amplitude $\theta_{\max}$ of the oscillation. The period is normalized to the period $T_{P0} = 2\pi\sqrt{g/R}$ for small-amplitude oscillations.

Our results for a frictionless pendulum are summarized in Fig. 5.8, where we plot the oscillation period T_P as a function of the amplitude $\theta_{\max}$. As anticipated, T_P goes to infinity in the limit as $\theta_{\max}$ approaches π, while T_P is nearly constant for small $\theta_{\max}$. Thus, Galileo was infinitely wrong at large amplitudes but nearly correct for small ones.

While the amplitude dependence of a pendulum's period presents a practical difficulty, the ultimate accuracy of any clock is determined by the quality factor of its oscillator. Higher Q's allow more accurate clocks. For an ordinary pendulum clock the Q is of order 10^3, and only modest accuracy is possible. On the other hand, a slab of crystalline quartz can vibrate with a Q as high as 10^6, giving modern watches a distinct advantage over grandfather clocks. Still, for the highest accuracy we rely on atomic oscillators with quality factors as high as 10^9. No wonder that navigation with the Global Positioning System, based on atomic clocks, has become the ultimate solution to the longitude problem.

5.6   Frequency

Because periodic motion is the antithesis of chaos, we'll be seeing it often, and we'll find the notion of frequency useful. The frequency of an oscillation is precisely the inverse of its period. Thus, if a pendulum's period is T_{P0}, then its frequency f_0 is,

$$f_0 = 1/T_{P0}. \tag{5.29}$$

For example, a grandfather clock with period of $T_{P0} = 2$ seconds per cycle oscillates with a frequency of $f_0 = 0.5$ cycles per second. However, scientists have given "cycles per second" the name "hertz" after an early radio pioneer, so we can more properly say that a grandfather clock oscillates at 0.5 Hz.

The notion of frequency can be applied to anything periodic, from musical tones to radio waves. Piano tuners know that the string for middle C must be tuned to a frequency of 261.63 Hz, the quartz crystal in my digital watch vibrates at 32,768 Hz or 32.768 kHz (kilohertz), my favorite radio station can be found at 91,500,000 Hz or 91.5 MHz (megahertz) on the FM dial, and by definition the cesium atoms in an atomic clock oscillate at 9,192,631,770 Hz or about 9 GHz (gigahertz). Oscillations are all around us, and sometimes it's more convenient to specify their frequencies than their periods.

5.7   Nonlinearity

As noted earlier, the pendulum of a clock never moves far from $\theta = 0$, a regime where its behavior is simple and, as a mathematician would note, linear. However, when a pendulum is allowed to reach large angles, it may rotate rather than oscillate. To an expert, the fact that a pendulum can display two distinct types of motion, rotation and oscillation, is a sure sign of nonlinearity, a mathematical condition necessary to a range of complex behavior, including chaotic motion.

Is nonlinearity an exotic phenomenon? Far from it! Stanislaw Ulam, a Polish-American mathematician, once remarked that using the term "nonlinear science" is like calling the bulk of zoology the study of non-elephant animals. Although nonlinearity is notoriously difficult to treat mathematically, in the physical world, nonlinear systems are the rule rather than the exception. Thus, by undertaking a study of chaos, we are surveying common, if mathematically treacherous, terrain. Complexity arising from nonlinearity is everywhere—it's just more difficult to understand than the occasional linear system that every engineer is taught to analyze.

You can't have chaos without nonlinearity, but understanding chaos doesn't require knowing the technical definition of a nonlinear system. To get an idea of the distinction, however, let's compare the pendulum, a nonlinear system, with a mass on a spring, a linear system. As noted previously, the spring force (Eq. (3.6)) is proportional to the displacement of the mass. In this case, graphing the force versus position yields a straight line, a reflection of the system's linearity. In contrast, according to Eq. (5.13), the gravitational torque on a pendulum is proportional to $\sin \theta$, not θ itself. Thus, for a pendulum the relation between torque and position is highly nonlinear, and the $\sin \theta$ term is the culprit that makes the pendulum prone to chaos.

Why is the pendulum linear for small angles? If ($|\theta| \ll 1$), then θ^2, θ^3, etc. are even smaller than θ, so the only significant contribution to the series formula for $\sin \theta$ (Eq. (5.2)) is the θ term. Thus, in the limit of small angles, we have $\sin \theta \approx \theta$, the pendulum is nearly linear, and a pendulum clock should be perfectly regular. But when the pendulum is allowed to swing full circle, its motion is highly nonlinear, and chaos can be expected.

5.8   Where's the chaos?

With any damping at all, a pendulum gradually slows and comes to a stop, no matter how large an initial kick we give it. In spite of the pendulum's nonlinearity, there is no chance of observing persistent chaotic motion in the absence of a persistent driving force. All the same, the natural, unforced motion of the pendulum gives an important clue to the origin of chaos.

In particular, the unforced motion of a pendulum can exhibit sensitivity to initial conditions, a hallmark of chaotic motion. Consider, for example, the two trajectories plotted in Fig. 5.9. The initial angle is $\theta(0) = 0$ in both cases, but the initial velocities differ slightly: $\tilde{v}(0) = 2.20$ in one case and 2.21 in the other. This small difference in the initial velocity makes a big difference in the final trajectories. As Fig. 5.9 shows, the trajectories are nearly identical at first, but, when the pendulum pauses near $\theta = \pi$, the instability of the balance point takes over and the trajectories diverge radically. With the smaller initial velocity, the pendulum doesn't quite reach the balance point and falls back to oscillate about $\theta = 0$. With the larger initial velocity, the pendulum passes over the top and ends up oscillating about $\theta = 2\pi$. The resulting divergence of trajectories is entirely understandable, even trivial, but lies at the heart of chaotic motion.

The sort of divergence shown in Fig. 5.9 represents exceptional behavior for an unforced pendulum. In this system, most trajectories with similar initial conditions do not radically diverge. Thus, for example, if

Fig. 5.9 The natural motion of a pendulum can exhibit sensitivity to initial conditions, even if it doesn't display persistent random behavior. Here we plot two trajectories that diverge radically although their initial conditions are nearly identical. For $\theta(0) = 0$ and $\tilde{v}(0) = 2.20$, the pendulum approaches the balance point but falls back to oscillate about $\theta = 0$. On the other hand, for $\theta(0) = 0$ and $\tilde{v}(0) = 2.21$, the initial velocity is just enough to put the pendulum over the top, and it ends up oscillating about $\theta = 2\pi$. The damping parameter is $\rho = 0.1$.

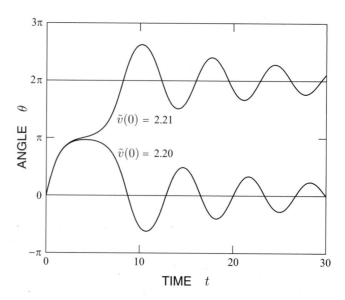

we had plotted the motion for $\tilde{v}(0) = 2.19$ in Fig. 5.9, it would be close to that for $\tilde{v}(0) = 2.20$, and the motion for $\tilde{v}(0) = 2.22$ would be close to that for $\tilde{v}(0) = 2.21$. The instability evident in Fig. 5.9 is an isolated instance.

When a pendulum moves chaotically, by contrast, all of the trajectories specified by initial conditions within a broad range of angle and velocity diverge rapidly from one another. Moreover, the divergence is not limited to a single event but persists indefinitely. Such wholesale instability boggles the mind. No wonder scientists were reluctant to admit the possibility of chaos. How could anything be unstable everywhere all of the time, especially when it's described by the same sort of equations that predict perfectly regular motion in other situations?

In partial answer to this question, we note that chaos arises in a pendulum when it's driven by a periodic torque. The applied torque keeps the pendulum in motion, in spite of any friction, and forces it to pass through the balance point over and over again. While the driven pendulum may settle into regular periodic motion, chaotic motion is also possible. What causes the chaos? Consider the following simplistic picture as an hypothesis. Each time the pendulum approaches the balance point, some neighboring trajectories diverge rapidly. Because the opportunity for divergence is repeated *ad infinitum*, any two trajectories, no matter how close they were initially, are eventually separated. Thus, sensitivity to initial conditions on the scale needed to explain chaos may result because the balance-point instability is encountered repeatedly.

In the next chapter, we take a look at the periodic motion of a driven pendulum, a phenomenon that's actually useful and not as simple as you might suppose. In the chapter after that, we enter the crazy world of chaotic motion.

Further reading

Clocks

○ Andrewes, W. J. H., "A chronicle of time-keeping", *Scientific American* **287**(3), 76–85 (September 2002).

• Baker, G. L. and Blackburn, J. A., "The pendulum clock", in *The Pendulum: A Case Study in Physics* (Oxford University Press, Oxford, 2005) chapter 10.

○ Ekeland, I., "Keeping the beat", in *The Best of All Possible Worlds: Mathematics and Destiny* (University of Chicago Press, Chicago, 2006) chapter 1.

○ Jespersen, J. and Fitz-Randolph, J., *From Sundials to Atomic Clocks*, 2nd edition (Dover, Mineola, New York, 1999).

○ Newton, R. G., *Galileo's Pendulum: From the Rhythm of Time to the Making of Matter* (Harvard University Press, Cambridge, 2004).

○ Sobel, D., *Longitude: The True Story of the Lone Genius Who Solved the Greatest Scientific Problem of His Time* (Walker, New York, 1995).

Huygens

- Gindikin, S. G., "Christiaan Huygens, pendulum clocks, and a curve 'not at all considered by the ancients'", in *Tales of Physicists and Mathematicians* (Birkhäuser, Boston, 1988), pp. 75–94.

- Huygens, C., *The Pendulum Clock or Geometrical Demonstrations Concerning the Motion of Pendula as Applied to Clocks* (Iowa State University Press, Ames, 1986).

Sychronization—The Josephson effect

6

In February of 1665, Christian Huygens made a remarkable observation while confined to bed with an illness. Idling away the hours, he noticed that two of his famous pendulum clocks, hanging next to each other on the wall, were moving in exact synchrony. When one pendulum moved to the right, the other moved to the left. Hour after hour, the pendulums moved apart and then back together, dancing in exact step with one another. Huygens quickly realized that this synchrony was unlikely to persist by chance. Sure enough, when he moved one clock to another wall, the clocks broke step, and a day later had drifted apart by 5 seconds. Returning the clocks to their original positions, Huygens found that synchrony was restored within half an hour.

Further experiments by Huygens confirmed what you might guess: the clocks had synchronized because they were linked by vibrations transmitted through the wall. However improbable it sounds, the vibrations caused one clock to speed up or the other to slow down, until they moved in lock step. Today we recognize this behavior as an example of entrainment, a phenomenon that requires nonlinearity. Huygen's observations, duly published by the Royal Society of London, initiated the study of spontaneous synchronization or entrainment.

Synchronization is occasionally observed in nature. The rotation of the Moon, for example, is synchronized with its orbital motion, so that one rotation is completed in the same time as one orbit. As a consequence, the same side of the Moon always faces Earth, and the far side remained a mystery until space probes circled the Moon and sent back pictures. Entrainment is also found in biological systems. Our sleep cycle is synchronized with the rising and setting of the Sun, a fact well known to modern travelers who struggle to reset their internal clocks when they fly to distant time zones. Curiously, some species of fireflies spontaneously synchronize their flashes, as if orchestrated by an unseen conductor. An entire meadow may light up when the fireflies all flash at once. Of course, there is no maestro, just a visual interaction that mutually coordinates the fireflies.

In this chapter, we begin to explore the nonlinear nature of the driven pendulum, examining four phenomena that require nonlinearity: hysteresis, multistability, synchronization, and symmetry breaking. We also introduce the Josephson junction, a quantum device that can be

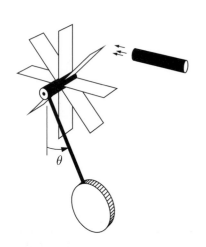

Fig. 6.1 With an applied torque, a pendulum can keep on rotating. Here, air blown through a straw pushes on the vanes of a pinwheel to produce a torque.

entrained with a microwave oscillator to create an incredibly precise standard of voltage: its equation of motion is exactly like that of a driven pendulum.

6.1   Hysteresis

In Chapter 5, we calculated the natural motion of a pendulum in the absence of a driving torque. Without an external torque, the pendulum may oscillate or even rotate for a while, but it eventually comes to a stop, and that's that. How can we apply a torque without otherwise constraining the pendulum's motion? One way is to attach a pinwheel to the pendulum and blow on the pinwheel through a straw, as in Fig. 6.1. While not very practical, the pinwheel and straw give a good idea of how a torque can be applied.

The equations of motion for the pendulum are easily modified to include a driving torque. In our dimensionless system, we simply add a term to the angular acceleration (Eq. 5.25) to get

$$\tilde{\alpha} = \tilde{\tau}_a(t) - \sin\theta - \rho\tilde{v}, \tag{6.1}$$

where $\tilde{\tau}_a$ is the applied torque in units of the maximum gravitational torque (RMg), and we allow $\tilde{\tau}_a$ to depend on time. In this case, the equations of motion become,

$$t_{\mathrm{n}} = t_{\mathrm{o}} + \Delta t, \tag{6.2}$$

$$\theta_{\mathrm{n}} = \theta_{\mathrm{o}} + \tilde{v}_{\mathrm{o}}\Delta t, \tag{6.3}$$

$$\tilde{v}_{\mathrm{n}} = \tilde{v}_{\mathrm{o}} + (\tilde{\tau}_a(t_{\mathrm{o}}) - \sin\theta_{\mathrm{o}} - \rho\tilde{v}_{\mathrm{o}})\Delta t. \tag{6.4}$$

Once $\tilde{\tau}_a(t)$ is specified, we can iterate Eqs. (6.2)–(6.4) to calculate new values of t, θ and $\tilde{v}$, using, as always, a computer program like that diagrammed in Fig. 3.2. The future of a driven pendulum isn't hard to predict!

First, let's use Eqs. (6.2)–(6.4) to see what happens when the applied torque is a constant, $\tilde{\tau}_a(t) = \tilde{\tau}_0$. You can imagine the result. If you blow softly through the straw $(\tilde{\tau}_0 < 1)$, the pendulum will rise until the gravitational torque balances the applied torque $(\sin\theta = \tilde{\tau}_0)$, and hang there as long as you keep blowing. On the other hand, if you blow hard enough $(\tilde{\tau}_0 > 1)$, the applied torque will exceed the maximum gravitational torque, and the pendulum begins to rotate, eventually settling into a regular pattern of rotation. For a constant torque, the pendulum's steady-state motion couldn't be simpler: the pendulum is displaced from $\theta = 0$ but remains stationary for $\tilde{\tau}_0 < 1$, and it settles into periodic rotation for $\tilde{\tau}_0 > 1$.

Well, almost. Given the pendulum's nonlinearity, we should always be prepared for a surprise. In Fig. 6.2, we plot the average velocity $\tilde{v}_a$ of the steady-state motion as a function of the applied torque $\tilde{\tau}_0$. In this case, the damping is light, with $\rho = 0.1$. As $\tilde{\tau}_0$ increases from 0, the velocity at first remains 0 as well, reflecting the fact that the gravitational torque

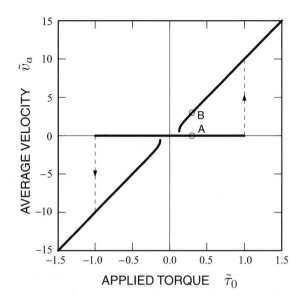

Fig. 6.2 Average angular velocity $\tilde{v}_a$ for the steady-state motion of the pendulum as a function of the applied torque $\tilde{\tau}_0$. The damping constant is $\rho = 0.1$.

balances the applied torque and the pendulum is stationary at any given $\tilde{\tau}_0$. The pendulum's angle increases as $\tilde{\tau}_0$ increases, but it otherwise remains still. When $\tilde{\tau}_0$ exceeds 1, however, the pendulum begins to rotate and $\tilde{v}_a$ suddenly jumps from 0 to about 10. It takes some time for the pendulum to build up speed, but, because Fig. 6.2 displays only the steady-state average, we see a jump in $\tilde{v}_a$. Such a jump is never seen in a linear system, where increasing the applied force always leads to a proportionate response.

For the pendulum, however, this jump isn't too surprising. Once the maximum gravitational torque is exceeded and the pendulum begins to rotate, the gravitational torque roughly averages to zero, because it's negative as the pendulum climbs to its highest point and positive as it comes down again. When rotating, the pendulum thus accelerates until the damping torque ($-\rho\tilde{v}$ in Eq. (6.1)) balances the applied torque on average. Indeed, equating these average torques yields $\rho\tilde{v}_a = \tilde{\tau}_0$, so that $\tilde{v}_a$ increases in proportion to $\tilde{\tau}_0$, as seen in Fig. 6.2 for $\tilde{\tau}_0 > 1$.

Perhaps a bigger surprise results if the applied torque is reduced again while the pendulum is rotating. For $\rho = 0.1$, instead of coming to a stop as soon as $\tilde{\tau}_0 < 1$, the pendulum keeps on rotating until $\tilde{\tau}_0$ drops to 0.127, as shown in Fig. 6.2. This behavior is an example of hysteresis: the persistence of an effect after the cause is removed. In the case of a lightly damped pendulum, rotation is induced by an applied torque in excess of 1 but persists when the torque is reduced below 1. In everyday life, hysteresis is not particularly remarkable—a copper wire doesn't spring back to its original shape after it's bent and an electromagnet doesn't lose all its magnetism when the current is switched off. On the other hand, hysteresis is a sign of nonlinearity, and it is often associated with chaos in dynamical systems.

Why does the pendulum keep going after the torque is reduced below that required to start it rotating? For the pendulum, hysteresis is due to a combination of inertia and light damping. As with an engine's flywheel, the pendulum's inertia helps keep it rotating even during parts of the cycle when the net acceleration acts to slow it down. For $\rho = 0.1$ the slowing due to friction isn't too large, and, as long as $\tilde{\tau}_0 > 0.127$, the positive acceleration produced by blowing on the pinwheel is enough to keep the pendulum going. On the other hand, when $\tilde{\tau}_0$ drops to 0.127, the pendulum can no longer reach the upward vertical and it settles back to the stationary state in which the applied torque is balanced by the gravitational torque.

This return to zero velocity completes the "hysteresis loop" shown in Fig. 6.2, created as the applied torque is raised from 0 to something greater than 1, then reduced again below 0.127. For applied torque in the range $0.127 < \tilde{\tau}_0 < 1$, we can't say for sure whether the pendulum is going to be rotating or stationary unless we know the history of how torque was applied: if $\tilde{\tau}_0$ was slowly increased from 0, then the pendulum will be stationary, but if $\tilde{\tau}_0$ was reduced from above 1, then the pendulum will be rotating. Thus, thanks to its nonlinearity, a pendulum has a memory of what has happened to it in the past. A pendulum isn't particularly intelligent, but it can learn.

To see exactly how hysteresis works in the pendulum, try Experiment 8 of the Dynamics Lab.

6.2   Multistability

Looked at from a slightly different perspective, the hysteresis of a pendulum is an example of multistability. As the name implies, a system exhibits multistability if there are two or more possibilities for the steady-state motion in a given situation. In particular, for $\rho = 0.1$ and $\tilde{\tau}_0 = 0.3$, the pendulum can either be stationary or rotating, as shown by the points labeled A and B in Fig. 6.2. Both states are entirely stable and can persist indefinitely as long as the applied torque is maintained. Multistability is a big word for a simple idea, but it's a nonlinear phenomenon that we'll see more of as we investigate chaotic motion.

Before examining the action at points A and B in greater detail, I want to introduce another way of looking at the pendulum equations that we'll use in the future. While you might suppose that Eqs. (6.2)–(6.4) are unique to the pendulum, they actually describe several different physical systems, and we can sometimes gain additional insight by thinking in terms of a particular realization. Thus, instead of a pendulum, we can think of Eqs. (6.2)–(6.4) as specifying the motion of a BB (a small round pellet) on a corrugated surface like that of an old-fashioned washboard, as shown in Fig. 6.3. Physicists like to say that a BB on a washboard is an "analog" of the pendulum.

To make the washboard an exact analog of the pendulum, we assume that the corrugations are sinusoidal, that the BB experiences a friction

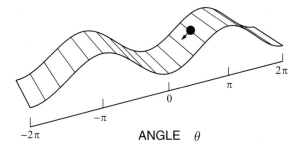

-2π $-\pi$ 0 π 2π

ANGLE θ

Fig. 6.3 A BB rolling on an infinite washboard is analogous to a pendulum.

force proportional to velocity, and that it experiences an applied force $\tilde{\tau}_0$, perhaps delivered by blowing through a straw. If the BB's position is measured by the variable θ, which increases by 2π as the BB moves from the bottom of one corrugation to the bottom of the next, then its motion is predicted by Eqs. (6.2)–(6.4). In this analogy, the pendulum hanging down corresponds to the BB resting at the bottom of one of the valleys. Similarly, the balance point of the pendulum corresponds to the BB sitting still at the top of a hill. You can easily imagine that a pendulum falling from its balance point is mimicked by a BB rolling down a corrugation of the washboard. On the other hand, when the pendulum advances by 2π, it returns to its starting position, while a similar advance for the BB moves it to an adjacent corrugation. Thus, the washboard analog is useful when we want to keep track of the accumulated increase in θ. A pendulum that has advanced by 8π has returned to its starting position, but a BB is four corrugations farther along the washboard.

Now let's take a closer look at the stationary and rotating motion corresponding to points A and B in Fig. 6.2. In Fig. 6.4, we plot both trajectories in a way that shows how the BB moves across the washboard. The washboard itself is plotted in (b), and just above in (a) we see how the position of the BB changes in time (plotted vertically). For trajectory A, the BB simply sits near the bottom of a valley in the washboard, displaced slightly from the low point by the applied force $\tilde{\tau}_0 = 0.3$ For trajectory B, the force keeps the BB moving across the washboard, slowing slightly as it goes uphill and gaining speed again as it falls into a valley, but relentlessly rolling along. This trajectory is exactly periodic in the sense that the BB goes through the same undulation in velocity each time it goes over a corrugation, and the analogous motion of the pendulum brings it back to the same angle after each period.

Both types of motion, A and B, can persist indefinitely, and they illustrate multistability in a nonlinear system. Indeed, if we consider the washboard analog, multistability takes on new dimensions. Thus, while we chose to show trajectory A in Fig. 6.4 as a BB stuck near the bottom of the valley near $\theta = 0$, we could have chosen to put the BB in the valley at $\theta = 2\pi$ or $\theta = -2\pi$. Clearly, there are an infinite number of choices and an infinite number of trajectories on the washboard that represent solution A. Similarly, the periodic solution shown in Fig. 6.4

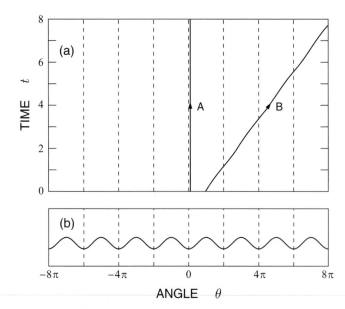

Fig. 6.4 The trajectories corresponding to points A and B in Fig. 6.2 are plotted as a function of time in (a), while the washboard is plotted in (b). The applied force is $\tilde{\tau}_0 = 0.3$ and the damping is $\rho = 0.1$.

can be offset in θ by any multiple of 2π and still be a valid steady-state trajectory. Rather than just two simultaneously stable solutions, for the washboard, we have two infinite sets of stable solutions. Multistability has gone berserk.

You might ask, however, which of the infinite number of stable trajectories will result if we simply begin calculating the motion with $\rho = 0.1$ and $\tilde{\tau}_0 = 0.3$? In this case, the outcome depends entirely on the initial values chosen for the position $\theta(0)$ and velocity $\tilde{v}(0)$. If the chosen $\theta(0)$ places the BB near the bottom of a valley and the magnitude of $\tilde{v}(0)$ is not too large, we expect the BB to settle into a fixed position like A in Fig 6.4. If $\tilde{v}(0)$ is large, then the BB is likely to continue moving and eventually adopt regular periodic motion. Unless we happen to pick exactly the right initial angle and velocity, there is always an initial transient in which the motion is not of type A or B, but the BB eventually approaches one type of motion or the other.

6.3   Synchronization

As you might have guessed by now, driving a pendulum with a constant torque is not sufficient to produce chaotic motion: the steady state is either stationary or periodic, and those are the only possibilities. However, the pendulum can go chaotic if we apply an additional torque that is periodic in time. In this section, we take the plunge into the realm of chaos by considering an applied torque that includes both a constant and a periodic component. That is, we consider a torque of the form,

$$\tilde{\tau}_a(t) = \tilde{\tau}_0 + \tilde{\tau}_1 \sin(\omega t), \qquad (6.5)$$

where the sine function is used to create a periodic time dependence. Although we'll postpone looking at chaos until the next chapter, with Eqs. (6.2)–(6.5), everything is in place for taking the pendulum beyond regular motion. First, however, let's look at the kinds of motion that Galileo and Newton would have expected from these equations. We'll find that the pendulum often sychronizes with the periodic torque, moving with the same periodicity as the drive, a phenomenon that is sometimes technologically important.

With the addition of a periodic torque, we add two new parameters to the equation of motion: $\tilde{\tau}_1$ and ω. Because $\sin(\omega t)$ in Eq. (6.5) oscillates between $+1$ and -1, the parameter $\tilde{\tau}_1$ sets the maximum amplitude of the periodic torque in units of the maximum gravitational torque. Thus, setting $\tilde{\tau}_1 = 12$, as in the example to be discussed, allows the applied periodic torque to reach 12 times the gravitational torque exerted on the pendulum when $\theta = \pi/2$ or $90°$. The parameter ω, on the other hand, fixes the period of the applied torque. Because the sine function has a period of 2π in angle, the time periodicity t_{P1} of $\tilde{\tau}_a$ is given by $\omega t_{P1} = 2\pi$ or,

$$t_{P1} = 2\pi/\omega. \tag{6.6}$$

Recalling that the period of the pendulum's natural oscillation (Eq. (5.27)) is $t_{P0} = 2\pi$, we see that

$$t_{P1} = t_{P0}/\omega. \tag{6.7}$$

Thus, setting $\omega = 2.5$, as in our first example, makes the period of the applied torque 40% of the pendulum's natural oscillation period.

Alternatively, we can think of ω in terms of frequency. If we define $\tilde{f}_0 = 1/t_{P0}$ and $\tilde{f}_1 = 1/t_{P1}$ as the dimensionless frequencies of the pendulum's natural oscillation and the applied oscillatory torque, then Eq. (6.7) becomes,

$$\omega = \tilde{f}_1/\tilde{f}_0. \tag{6.8}$$

Thus, ω is ratio of the drive frequency to the natural frequency of the pendulum, and we'll refer to ω as the frequency parameter. When we confront chaotic motion in the next chapter, this parameter will occupy center stage.

What's a pendulum's response to a periodic torque? Of course, everyone has had the experience of pushing a swing and watching the oscillations build until friction sets a limit. Surely, something similar will result if we compute the motion predicted by Eqs. (6.2)–(6.5). On the other hand, when we push a swing, the force we apply is timed to match the natural oscillation period of the swing. In contrast, the applied torque given by Eq. (6.5) may have either a longer ($\omega < 1$) or a shorter ($\omega > 1$) period than the pendulum. With this in mind, can you still guess what the motion will be? There are all kinds of possibilities, especially when the applied torque is large enough that the pendulum can rotate full circle. Perhaps the swing will settle into periodic motion with the same period as the applied torque. Or, maybe the conflict between the pendulum's natural period and that of the applied torque will never be resolved, and the pendulum will play an infinite a game of cat and mouse with the applied torque—always managing to do the unexpected. Then again, rotation might dominate and impose a period set by the time required to journey full circle. In fact, all three of these possibilities can be realized, depending on the parameters of the system, ρ, $\tilde{\tau}_0$, $\tilde{\tau}_1$, and ω.

To get started, let's repeat the calculation of average velocity as a function of the constant torque component $\tilde{\tau}_0$, but now with the oscillatory torque added in. Results are shown in Fig. 6.5 for $\rho = 0.1$, $\tilde{\tau}_1 = 12$, and $\omega = 2.5$. Comparing Fig. 6.5 to the case without an oscillatory torque (Fig. 6.2), we see that some features are familiar, including a range of $\tilde{\tau}_0$ near 0 where the average velocity is 0 and a range $\tilde{\tau}_0 > 1$ where $\tilde{v}_a$ increases in proportion to $\tilde{\tau}_0$. These portions of Fig. 6.5 are almost identical to Fig. 6.2, as if the oscillatory torque had no effect at all. However, we also find a series of horizontal lines or "steps" at non-zero velocities, a strange result given our expectation that increasing the constant component of torque will increase the angular velocity. These constant-velocity steps are the product of synchronization.

For example, the step at $\tilde{v}_a = 2.5$ corresponds to motion in which the pendulum advances by exactly one revolution during each cycle of the oscillatory drive. In this case, the average velocity is a simple consequence of synchronization. That is, because the pendulum advances by one revolution or 2π during one drive period $t_{P1} = 2\pi/\omega$, the average angular velocity is $\tilde{v}_a = 2\pi/t_{P1} = \omega = 2.5$. Given this result, the step at $\tilde{v}_a = 5$ must correspond to motion in which the pendulum completes two full revolutions during each drive cycle. And, more generally, the step at

$$\tilde{v}_a = n\omega, \tag{6.9}$$

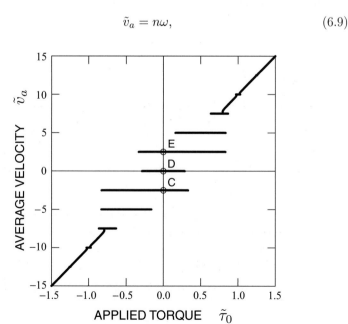

Fig. 6.5 Average angular velocity $\tilde{v}_a$ for the steady-state motion of a pendulum as a function of the constant component of applied torque $\tilde{\tau}_0$, in the presence of an oscillatory component $\tilde{\tau}_1 \sin(\omega t)$. In this computation, the frequency parameter and amplitude of the oscillatory torque are fixed at $\omega = 2.5$ and $\tilde{\tau}_1 = 12$, while the damping is $\rho = 0.1$.

represents motion in which the pendulum advances by exactly n revolutions during the drive period. In Fig. 6.5, we find steps of constant velocity for integer values of n between -4 and $+4$, so the pendulum can advance by up to four full revolutions per drive cycle. Thus, not only do we observe synchronization between the pendulum and the oscillatory drive, but we find several types of synchronization. Seemingly nothing is simple with nonlinear systems.

The most remarkable aspect of synchronization is that it fixes the average angular velocity over a range of $\tilde{\tau}_0$. For example, the $n = 1$ step in Fig. 6.5 extends from $\tilde{\tau}_0 = -0.33$ to $\tilde{\tau}_0 = 0.83$, and the average velocity is exactly constant over this range. Moreover, the damping ρ and the amplitude $\tilde{\tau}_1$ of the oscillatory torque can also be varied within limits without upsetting synchronization. Engineers like to say that the pendulum rotation is "locked" to the oscillatory drive and refer to the parameter range $-0.33 < \tilde{\tau}_0 < 0.83$ as the "locking range" of the $n = 1$ step. Of course, the synchronization observed here is basically the same as that discovered by Christian Huygens more than 400 years ago when he saw two pendulum clocks swinging in unison.

6.4   Symmetry breaking

Let's take a closer look at the some of the synchronized trajectories that lead to constant-velocity steps in Fig. 6.5. In Fig. 6.6, we plot the steady-state motion associated with points C, D, and E in Fig. 6.5. These trajectories represent synchronized motion on the $n = -1$, 0, and $+1$ steps in the absence of a constant component of torque ($\tilde{\tau}_0 = 0$).

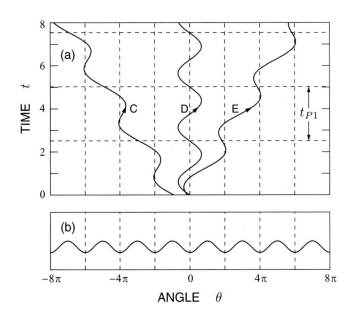

Fig. 6.6 The trajectories corresponding to points C, D, and E in Fig. 6.5 are plotted as a function of time in (a), while the washboard is plotted in (b). The constant component of applied torque is 0, the periodic component has amplitude $\tilde{\tau}_1 = 12$ and frequency parameter $\omega = 2.5$, and the damping is $\rho = 0.1$.

Trajectory D is the simplest, showing the sort of motion we anticipate from pushing a swing: the pendulum oscillates back and forth about its equilibrium position at $\theta = 0$. The amplitude of the oscillation, about $2\pi/3$ or $120°$, is a little larger than the average child on a swing might like, and the period of the oscillation, $t_{P1} = 2.513$, is just 40% of the swing's natural period. But otherwise, this is the sort of response anyone would expect from a swing driven by a periodic torque. Viewed in terms of the washboard analog, the BB oscillates back and forth within the confines of the valley at $\theta = 0$.

Trajectories C and E of Fig. 6.6 show how, under the same conditions as trajectory D, the pendulum can rotate either backward or forward by exactly one revolution during each drive cycle. The oscillatory component of torque sets the pace, and the pendulum follows along. In Fig. 6.6, the periodicity is made clear by the rectangular grid of lines spaced by 2π in angle and by t_{P1} in time. Thus, the "S" made by trajectory E during one interval of t_{P1} is repeated in the next interval but shifted forward in angle by 2π. Similarly, trajectory C repeats during each drive period but with the angle shifted backwards by 2π.

The periodic trajectories shown in Fig. 6.6 are another example of multistability. In this case, we find that three different steady-state solutions are possible for the same parameter values. Or, because we can add any multiple of 2π to any of these trajectories, we really have three infinite sets of stable solutions. As before, exactly which solution is realized in a computation depends on the values chosen for the initial angle and velocity. Experiment 9 of the Dynamics Lab allows you to try different values of $\theta(0)$ and $\tilde{v}(0)$ and observe how the pendulum settles into one of the three types of synchronized motion. Try it!

While trajectories C and D seem reasonable enough, they are peculiar from one point of view. Thinking physically for a moment, we are led to ask why the pendulum should move steadily either backward or forward. Because the constant component of torque is 0 and the oscillatory component does not favor motion in one direction or the other, it is strange to find trajectories that progress steadily in one direction or the other. As it happens, trajectories C and D are examples of a phenomenon physicists call "symmetry breaking." These trajectories are solutions that violate the symmetry of the equations from which they derive. As you might guess, symmetry breaking is another manifestation of nonlinearity—the world would be truly boring if it were linear.

Symmetry breaking is as common in our everyday world as an ordinary light switch. The typical switch can toggle between on and off and is mechanically symmetric, but we always find the switch in one of these two positions, never in the symmetric center position. A toggle switch breaks its own symmetry. Our pendulum example is similar, but there are three stable states, $n = -1$, 0, and $+1$, and the pendulum can be found in any of the three, even though the $n = -1$ and $+1$ states break symmetry.

In recent years, symmetry breaking has achieved a certain fame among physicists because a fundamental property of matter seems to result from a broken symmetry. In 1964, Peter Higgs at Edinburgh University noted that, while the equations of quantum field theory have a symmetry implying that all elementary particles should be massless, when this symmetry is broken, mass is a consequence. Higgs's idea is now taken seriously, and the highest energy particle accelerators are currently searching for the "Higgs boson," a particle predicted by his theory of the origin of mass. Little did Newton suspect that mass might one day be explained in such terms.

Even when a symmetry is broken, however, it still has consequences. In the case of the driven pendulum, the $n = +1$ trajectory relentlessly advances in angle even though the equations do not favor one direction or the other. On the other hand, it is balanced by the $n = -1$ trajectory that moves in the opposite direction. Moreover, these two trajectories are intimately related. If trajectory C in Fig. 6.6 is reflected about the $\theta = 0$ axis and shifted backward in time by $t_{P1}/2$ then it exactly coincides with trajectory E. Thus, the symmetry between positive and negative angles remains; it's just hidden when the pendulum settles into one of the trajectories C or E.

6.5   Josephson voltage standard

Synchronization of a pendulum may seem interesting but of no practical consequence. However, the discussion in this chapter is directly relevant to a device used around the world in the precise measurement of voltage. Called a Josephson junction, the device appears to have no connection with a pendulum, but, like a BB on a washboard, it's dynamically analogous. If you understand the pendulum, you'll understand the Josephson junction and why sychronization is crucial to making a voltage standard.

Because I helped develop improved voltage standards while working at the National Institute of Standards and Technology (NIST), I think Josephson junctions are cool. Actually, they're downright cold.

Physically, a Josephson junction is no more complicated than a capacitor, one of the simplest of all electronic components. As shown in Fig. 6.7, it consists of an insulating film "I" sandwiched between two metal electrodes "S," forming an "SIS" junction between the electrodes. If the insulator is relatively thick, the device is just a capacitor. However, if the insulator amounts to only a few atomic layers, then electrons can quantum mechanically tunnel through it, and a current can flow from one electrode to the other. At room temperature the device then acts like a capacitor with a resistor wired in parallel—still nothing very special. But something magic happens when the junction is cooled to within a few degrees of absolute zero (-273.15 degree Celsius). Provided the metal electrodes are made of the right material, say lead or niobium, they become superconducting, and the device performs weird electrical tricks just like those of a pendulum. To get a better idea of how Josephson junctions work, let's take a quick look at superconductivity.

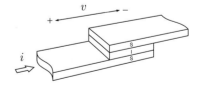

Fig. 6.7 A Josephson junction consists of an insulating layer I sandwiched between two metal films S. It's just like a capacitor except that the metal films are superconducting and the insulator is thin enough that electrons can quantum mechanically tunnel between the superconductors.

Superconductivity made a big splash toward the end of the 1980s when materials were found exhibiting zero resistance at higher temperatures. However, the phenomenon itself was discovered by Heike Kamerlingh-Onnes (1853–1926), a Dutch physicist, way back in 1911. Actually, superconductivity was discovered by an assistant named Gilles Holst, but he worked under the direction of Kamerlingh-Onnes, and it was Kamerlingh-Onnes who received the Nobel Prize in 1913. What Holst observed was that the electrical resistance of mercury dropped suddenly to zero when he cooled a sample below about −269 degree Celsius. Was the resistance exactly zero? Later experiments confirmed that a current flowing around a ring of superconducting wire persists without decay as long as the ring is kept cold. Superconductivity allows a kind of perpetual motion, much like that of electrons orbiting an atom's nucleus, but a supercurrent is macroscopic rather than microscopic.

In 1911, the unique properties of the superconducting state were not entirely appreciated, but, as experimental results accumulated, superconductivity became more and more mysterious. How could electrical resistance disappear so completely, and why did superconductors expel magnetic fields? Despite the best efforts of theoreticians, these riddles were not solved until 1957. In that year John Bardeen, Leon Cooper, and Robert Schrieffer published a detailed microscopic theory of superconductivity. According to this theory, called the BCS theory after the authors' initials, superconductivity results when the conduction electrons of a metal form bound pairs, called Cooper pairs, that condense into a macroscopic quantum state. It's a highly sophisticated mathematical theory, but its predictions were confirmed by experiment, and Bardeen, Cooper, Schrieffer received the Nobel Prize for their work in 1972.

In 1962, not long after the BCS theory appeared, Brian Josephson, a graduate student at Cambridge University (Newton's alma mater), found himself studying the intricacies of the new theory and wondering if there was an experiment that might reveal the underlying quantum nature of the superconducting state. In this quest, he considered what would happen if Cooper pairs were to tunnel from one superconductor to another. In short order he published a paper predicting that several strange phenomena, now called the Josephson effects, would be observed in a superconducting tunnel junction.

Not everyone was convinced by Josephson's paper. Bardeen in particular argued that if the probability of one electron tunneling was small, then the probability of two electrons tunneling simultaneously as a Cooper pair would be utterly negligible. Josephson and Bardeen debated the issue at a conference held in London in September of 1962, a David and Goliath affair that pitted a graduate student against a Nobel laureate. (Bardeen had already received a Nobel Prize in 1956 for the invention of the transistor—the prize for the BCS theory would be his second.) However, as Galileo knew so well, the outcome of a scientific debate is determined by observation and experiment, not force of argument. As in the biblical battle, the unlikely contestant proved to be the victor. Josephson's outrageous theory was vindicated by experiments made the following year, and superconducting tunnel junctions have been called Josephson junctions ever since. To top it off, Josephson received a Nobel Prize of his own in 1973.

Exactly how is a Josephson junction like a pendulum? Although the full theory is extraordinarily complicated, Josephson realized that a simple approximation would account for the basic dynamics of a superconducting junction. In this approximation, the voltage v across a junction is related to the current i flowing through it (see Fig. 6.7) in the same way as the angular velocity of a pendulum is related to the applied torque. That is, the voltage v is analogous to the velocity $\tilde{v}$, and the current i is analogous to the applied torque $\tilde{\tau}_a$. Thus, a physicist concerned with Josephson junctions would find Figs. 6.2 and 6.5 completely familiar, except that she would see them as voltage plotted as a function of current rather than velocity as a function of torque. These are typical oscilloscope traces for a Josephson junction, well known to engineers and physicists in the business.

Figure 6.5 is particularly relevant here. It corresponds to a situation in which a Josephson junction is subjected to an oscillatory current in addition to a constant current. Typically, the oscillatory component has a frequency in the microwave region, say between 10^{10} and 10^{11} Hz, that is, between 10 and 100 GHz. In this case, the steps of constant average velocity in the pendulum become steps of constant average voltage in the junction. Josephson predicted such steps in his first paper on superconducting tunnel junctions and gave the average voltages on the nth step as,

$$v_a = n\frac{hf}{2e},\tag{6.10}$$

where h is Planck's constant, f is the frequency of the microwave current, and e is the charge of the electron.

Equation (6.10) is remarkable for several reasons. Firstly, it is exactly analogous to Eq. (6.9) and is the result of synchronization between an angle variable θ internal to the junction and the applied microwave current. It's not important to understand the exact nature of the angle θ in a junction—suffice it to say that θ is the quantum variable Josephson wanted to reveal when he began to think about superconductive tunneling. Secondly, although Eq. (6.10) relates two macroscopic variables, voltage and frequency, it includes Planck's constant h, which is normally associated only with the microscopic domain. Clearly, Josephson had succeeded in revealing the quantum nature of the superconducting state. Thirdly, the relation between voltage and frequency specified by Eq. (6.10) involves only the fundamental constants h and e. This fact opened the door to building a better standard of voltage, because frequency is the quantity known to the highest precision of all, presently to 1 part in 10^{15}. If voltage is fundamentally related to frequency, then voltage might be measured with the same high precision.

Josephson's paper immediately set metrologists (measurement scientists, not weathermen) thinking about the possibility of a new, highly precise standard of voltage. The principle question was whether there are corrections to Eq. (6.10) that might be needed to account for the effect of temperature, the superconducting material, or any other variable. After numerous tests, metrologists were satisfied that Eq. (6.10) is highly accurate, and in 1972, the Josephson volt was adopted as a worldwide standard. To date, no corrections to Eq. (6.10) have been discovered, and a comparison between two junctions driven by the same microwave source proved that their voltages were the same to better than 3 parts in 10^{19}, an incredible level of precision. As long as synchronization is not disrupted by noise, a Josephson voltage standard is believed to be as accurate as the frequency of the microwave source that drives it.

Synchronization is key to the Josephson voltage standard, because it makes the average voltage entirely insensitive to parameters like the constant component of current (as illustrated in Fig. 6.5), the amplitude of the microwave current, and the damping. On the other hand, synchronization requires nonlinearity, and nonlinearity raises the specter of chaotic behavior. In the next chapter, we'll see how the presence of chaos can displace synchronization and has an impact on the design of voltage standards.

Further reading

Josephson effect

- Anderson, P. W., "How Josephson discovered his effect", *Physics Today* **23**(11), 23–29 (November 1970).
- Baker, G. L. and Blackburn, J. A., "Superconductvity and the pendulum", in *The Pendulum: A Case Study in Physics* (Oxford University Press, Oxford, 2005) chapter 9.
- Langenberg, D. N., Scalapino, D. J., and Taylor, B. N., "The Josephson effects", *Scientific American* **214**(5), 30–39 (May 1966).
- McDonald, D. G., "The Nobel laureate versus the graduate student", *Physics Today* **54**(7), 46–51 (July 2001).

Superconductivity

- Dahl, P. F., *Superconductivity: Its Historical Roots and Development from Mercury to the Ceramic Oxides* (American Institute of Physics, New York, 1992).

- Bruyn Ouboter, R. de, "Heike Kamerlingh Onnes's discovery of superconductivity", *Scientific American* **276**(3), 98–103 (March 1997).
- Nobel, J. de, "The discovery of superconductivity", *Physics Today* **49**(9), 40–42 (September 1996).
- MacDonald, D. K. C., *Near Zero: The Physics of Low Temperature*, (Anchor Doubleday, Garden City, New York, 1961).
- Matthias, B. T., "Superconductivity", *Scientific American* **197**(5), 92–103 (November 1957).
- Simon, R. and Smith, A., *Superconductors: Conquering Technology's New Frontier* (Plenum, New York, 1998).

Symmetry breaking

- Close, F., *Lucifer's Legacy: The Meaning of Asymmetry* (Oxford University Press, Oxford, 2000).

○ Kane, G., "The mysteries of mass", *Scientific American* **293**(1), 40–48 (July 2005).

○ Lederman, L. M. and Hill, C. T., "Broken symmetry", in *Symmetry and the Beautiful Universe* (Prometheus, Amherst, New York, 2004) chapter 9.

○ Quinn, H. R. and Nir, Y., *The Mystery of the Missing Antimatter* (Princeton University Press, Princeton, 2008).

○ Randall, L., "The origin of elementary particle masses: Spontaneous symmetry breaking and the Higgs mechanism", in *Warped Passages: Unraveling the Mysteries of the Universe's Hidden Dimensions* (Harper Collins, New York, 2005) chapter 10.

○ Sharon, E., Marder, M., and Swinney, H. L., "Leaves, flowers and garbage bags: Making waves", *American Scientist* **92**, 254–261 (2004).

Synchronization

○ Bak, P., "The devil's staircase", *Physics Today* **39**(12), 38–45 (December 1986).

○ Buck, J. and Buck, E., "Synchronous fireflies", *Scientific American* **234**(5), 74–85 (May 1976).

○ Dermott, S. F., "How Mercury got its spin", *Nature* **429**, 814–815 (2004).

○ Klarreich, E., "Huygen's clocks revisited", *American Scientist* **90**, 322–323 (2002).

○ Peterson, I., "Call of the firefly", in *The Jungles of Randomness: A Mathematical Safari* (Wiley, New York, 1998) chapter 4.

○ Strogatz, S. H., *Sync: The Emerging Science of Spontaneous Order* (Hyperion Press, New York, 2003).

○ Strogatz, S. H. and Stewart, I., "Coupled oscillators and biological synchronization", *Scientific American* **269**(6), 102–110 (December 1993).

Part III
Random motion

Chaos forgets the past

7

Early Josephson standards were important because they established the first reference voltage that was stable over long periods. These standards were so difficult to use, however, that they were seldom seen outside of national laboratories, the main problem being that they produced only a small voltage. Because improving measurement standards is a primary mission of NIST, my colleagues and I naturally asked how we could increase the voltage of a Josephson standard. In trying to answer this question, I personally encountered chaos for the first time.

As we saw in the last chapter, a Josephson junction produces an accurate voltage when its angle advances in synchrony with a microwave drive. According to Eq. (6.10), the voltage is proportional to the step number n and the microwave frequency f, so higher voltages can be obtained by increasing either n or f. Unfortunately, the constant of proportionality, $h/2e$, is very small, amounting to just $2.067834\,\mu V/GHz$ (microvolt per gigahertz). Thus, even the relatively high frequency of 10 GHz produces a first step at only 20 μV. Clearly, it is advantageous to use a high-order step.

In early standards, the junction was driven with enough 10 GHz microwave power to generate constant-voltage steps up to about $n = 250$, yielding accurate voltages near 5 mV (millivolt). However, because 5 mV is still relatively small compared to that of the electrochemical cells and zener diodes used as secondary standards, typically 1 to 10 V, accurate calibrations based on the Josephson effect required the specialized expertise found in a national laboratory.

Moreover, there appeared to be no simple way to push the voltage of a junction much beyond 5 mV. Working with microwaves substantially above 10 GHz is difficult, and Josephson steps above about $n = 250$ are too small to be useful. By small, I mean that the constant-voltage step extends over a small range of current, just as the $n = 4$ constant-velocity step at $\tilde{v}_a = 10$ in Fig. 6.5 extends over a small range of torque. Even at $n = 250$, the current must be carefully adjusted to insure that the junction is biased within the range of synchronization. Thus, obtaining more than 5 mV from a single junction is problematic.

What about connecting several Josephson junctions in series? As you may know, the 12 volt battery in your car is made by wiring together six wet cells, each generating 2 volts. (That's why there are six reservoirs to be filled when you top up the battery fluid.) Josephson junctions can be connected in the same way to add their voltages, so the total voltage can be as large as you want, at least in principle. The problem with this scheme is that it requires a separate, carefully adjusted bias current for each junction. Thus, a series array of junctions was a possible solution to our problem, but not an appealing one.

This impasse was well understood by the summer of 1977 when my NIST colleague Donald Sullivan hosted a visit from Mogens Levinsen, a Danish physicist from the University of Copenhagen. Enjoying a beer in Sullivan's backyard while looking over Levinsen's recent Josephson simulations, the two were struck by an interesting idea. The idea was a possible solution to our voltage-standard problem, and it appeared to be technically sweet. The graph that fired their imaginations was much like that in Fig. 6.5, showing the somewhat unusual phenomenon in which steps of constant velocity (voltage) extend across the point of zero torque (current). Sullivan and Levinsen asked themselves, why not operate a voltage standard at a point like E in Fig. 6.5 where no bias current is required? You wouldn't get much voltage from a single junction because the step order would be small, but you wouldn't need a separate bias for each junction in a series array. Because all the junctions would be at zero current, you could stack up as many as needed to get a large voltage, and the array would be as easy to operate as a single junction. Nifty!

Excited by this idea, Sullivan immediately invited me to join him in the lab to try it out. As it happened, other colleagues had recently fabricated microcircuits including large arrays of junctions connected in series, so we set up the experiment in almost no time. However, our experiment turned out to be a case of little ventured, little gained. Mysteriously, we didn't see the large steps spanning zero current that were predicted by Levinsen's simulation. What we found instead were miserable little steps that were barely stable and entirely useless for a voltage standard. Three years later we would understand that the unstable mess we had observed was an example of chaotic motion intrinsic to the sinusoidally driven junction.

Our initial experiment was so disappointing that we set aside series arrays for more than a year, returning to other, more promising work. When Sullivan suggested that I give the idea another try, my thought was that further experiments would be pointless if we didn't understand our initial failure. In the meantime, Levinsen had published a paper on his simulations, laying out all the various conditions required for a voltage standard using steps at zero bias. Although our first experiment met these conditions, I thought through the problem again and soon began to understand what might have gone wrong.

In his paper, Levinsen hadn't expressly considered the frequency parameter $\omega = \tilde{f}_1/\tilde{f}_0$, but I realized that it might be crucial to obtaining the steps we wanted. A simple heuristic argument suggested that large steps would certainly result if ω was much greater than 1, but all bets were off if ω was less than 1. While I couldn't prove the case (remember, nonlinear systems are notoriously intractable), it was easy to check a few cases. First, what about Levinsen's simulations? Sure enough, in that work ω was 2.24, large enough that large steps could be expected.

And what about the experiment I had performed with Sullivan? In that case, ω was 0.49, small enough that large steps might not be observed, as indeed they weren't. It began to look as if large steps really did require $\omega \gg 1$.

Shortly after arriving at this conclusion, I fabricated a new batch of junctions with a natural frequency f_0 well below the frequency f_1 of my microwave generator. I immediately observed exactly the kind of large, zero-bias steps that Levinsen had found in his simulations. At last, the array voltage standard was back on track, and development could proceed. But a question remained: what exactly was the problem with operating at ω less than 1? At the time (1979), we had no idea that chaotic behavior was our nemesis, and, like most physicists, we hadn't even heard of chaos. Nevertheless, as a side project, I set out to simulate pendulum dynamics for $\omega < 1$, just to see if synchronization was in fact unstable in this realm.

In this chapter, we follow the trail I first explored in 1979, but with the full benefit of hindsight. We'll look first at a precursor to chaos called period doubling at $\omega = 1.4$, then lower ω to 0.8, where chaos achieves full flower, and begin our long look at chaos itself.

7.1   Period doubling

In the last chapter, we computed the response of a pendulum to a periodic driving torque and, as seems natural, found that the pendulum often settles into synchronized motion with the same period as the drive. If the pendulum were linear, this would be the only possible result. However, by lowering the drive frequency from $\omega = 2.5$ to 1.4, we find a curious new type of motion in which the period is two drive cycles rather than one.

In this chapter and several that follow, our primary concern is with a pendulum driven by an applied torque $\tilde{\tau}_a(t) = \tilde{\tau}_1 \sin(\omega t)$ that is purely sinusoidal and has no constant component. In this case, Eqs. (6.2)–(6.4) can be rewritten as,

$$t_n = t_o + \Delta t, \tag{7.1}$$

$$\theta_n = \theta_o + \tilde{v}_o \Delta t, \tag{7.2}$$

$$\tilde{v}_n = \tilde{v}_o + (\tilde{\tau}_1 \sin(\omega t_o) - \sin \theta_o - \rho \tilde{v}_o) \Delta t. \tag{7.3}$$

These are the master equations that allow us to update the pendulum's angle and velocity, θ and $\tilde{v}$, as time advances, Δt by Δt. In principle, Eqs. (7.1)–(7.3) are simple enough that they could be iterated on a programmable calculator, but, like the proverbial rabbit drawn from a hat, the resulting motion is often unexpected.

As a case in point, consider Fig. 7.1. Here we plot the pendulum angle versus time for several steady-state trajectories at $\omega = 1.4$, $\tilde{\tau}_1 = 4.9$, and $\rho = 0.1$. To facilitate thinking in terms of the washboard analog, angle is plotted on the horizontal axis, and dashed vertical lines mark the valleys

Fig. 7.1 Steady-state motion of a sinusoidally driven pendulum with $\omega = 1.4$, $\tilde{\tau}_1 = 4.9$, and $\rho = 0.1$. All of the eight trajectories shown have a period twice that of the applied torque. Dashed vertical lines mark the valleys of the washboard analog, and dashed horizontal lines mark the period of the applied torque. Dotted trajectories are equivalent to solid trajectories shifted in time by one drive period. Circles mark the beginning of each drive cycle for the solid $n = 1$ trajectory.

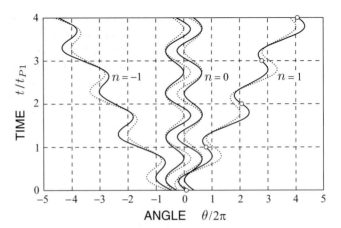

of the washboard. Similarly, dashed horizontal lines mark the beginning of each drive cycle. Each trajectory begins with the BB near the central valley at $\theta = 0$, and, buffeted by the applied force, it scurries first left and then right as time proceeds, much as we saw in Fig. 6.6 for $\omega = 2.5$. At $\omega = 1.4$, however, we find eight distinct steady-state solutions rather than three, so multistability has dramatically increased.

Nevertheless, we can still identify each trajectory by a step order n according to its average velocity, $\tilde{v}_a = n\omega$. That is, each trajectory in Fig. 7.1 advances in angle on average by $n2\pi$ per drive cycle, with $n = -1, 0$, or 1. Even so, the $n = 1$ trajectories don't advance by exactly 2π on every cycle. Careful examination of Fig. 7.1 reveals that the solid $n = 1$ trajectory advances by less than 2π on the first cycle and by more that 2π on the second. This pattern is repeated with each succeeding pair of drive cycles, so the trajectory is periodic, but its period is $2t_{P1}$. In Fig. 7.1, this alternating pattern is made clear by circles that mark the beginning of each drive cycle.

As it happens, all of the trajectories in Fig. 7.1 have twice the period of the drive and are said to be period doubled. If you think about it, period doubling is a strange phenomenon. Even though the only driving force in the system has period t_{P1}, the pendulum chooses to respond with motion having a period of $2t_{P1}$. The situation is reminiscent of our earlier observation of motion that persistently advances in a particular direction when the force doesn't favor one direction or the other. In that case, the pendulum broke a spatial symmetry of the problem, creating the $n = -1$ and 1 steps at $\tilde{\tau}_0 = 0$. With period doubling, we have a broken time symmetry. The drive force during the time interval $t = t_{P1}$ to $2t_{P1}$ is the same as during the interval $t = 0$ to t_{P1}, but the pendulum chooses to move differently on the second interval.

Of course, there really isn't any difference between the first drive cycle and the second, and this symmetry is represented in Fig. 7.1 by the solid and dotted trajectories. For each solid trajectory, there is a corresponding dotted trajectory that is identical but offset in time by one drive cycle. Thus, the $n = 1$ dotted trajectory advances by more than 2π on the first drive cycle and less than 2π on the second, just the opposite of its solid companion. Taken together, the possible states of motion represented by a pair of solid and dotted trajectories reflect the periodicity of the drive, but the actual motion is one or the other. To see an animation of period doubled motion in the pendulum, check out Experiment 10 of the Dynamics Lab.

Occasionally, period doubling is even observed in nature. You may remember from the last chapter that the rotation of our moon is synchronized with its orbital motion, such that it completes one rotation during each orbit. The rotation of the planet Mercury is also synchronized with its orbital motion, but Mercury revolves three times during two orbits. Like our period-doubled pendulum, Mercury breaks the time symmetry of its orbital motion, and its rotation is different on alternate orbits.

While interesting in itself, period doubling also foreshadows the onset of chaos. With period doubling, time symmetry is broken in a modest way, with periodicity simply extended to two drive cycles. For chaos, on the other hand, time symmetry is broken on a grand scale: a chaotic trajectory never repeats itself—the period is infinite. Let's have a look.

7.2   Random rotation

Figure 7.2 shows five typical trajectories of a sinusoidally driven pendulum for $\omega = 0.8$, $\tilde{\tau}_1 = 1.6$, and $\rho = 0.1$. Clearly, none of these trajectories is periodic on a short time scale. Are we seeing irregular initial transients that will disappear as soon as the pendulum has time to settle down? No, each trajectory was allowed time to approach a steady state before beginning the plot. In fact, these trajectories are completely characteristic of the pendulum's long-term motion, and they illustrate predictable random behavior or chaos.

After exercising our neurons so diligently to learn the science of dynamics, anticipating chaos at every turn, Fig. 7.2 may seem a little

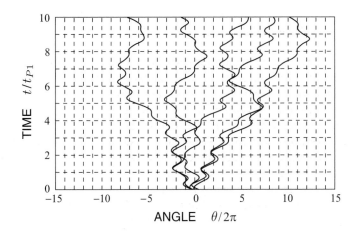

Fig. 7.2 Chaotic motion of a sinusoidally driven pendulum is illustrated by five trajectories for $\omega = 0.8$, $\tilde{\tau}_1 = 1.6$, and $\rho = 0.1$. To insure that these trajectories represent steady-state motion, they were initiated at $t = -30t_{P1}$ with $\tilde{v} = 0$ and angles θ distributed uniformly between $-\pi$ and π. The initial transients between $t = -30t_{P1}$ and 0 were computed but not plotted. At $t = 0$, the angle of each trajectory was shifted by a multiple of 2π, as needed to return it to the central valley of the washboard, and the subsequent trajectory was plotted from $t = 0$ to $10t_{P1}$.

disappointing. Our inner kid is apt to say, "What's the big deal? I could make up something more chaotic than that." Even so, there are many surprises to be found in chaotic motion. Some of these surprises were introduced in Chapter 1, but we begin now to open up the Pandora's box of chaos and take a good look at what's inside. In this chapter and the two that follow, we examine the random nature of chaotic motion, checking how it compares with conventional ideas of randomness. In Chapters 10 through 14, we take a closer look at the butterfly effect and its consequences for predictability. Then, having surveyed chaos as a phenomenon, in Chapter 15 we begin to examine the mathematical underpinnings of chaos and search for its ultimate cause.

To get started, let's return to Fig. 7.2 and consider the motion in terms of the washboard analog. The computer experiment presented here actually began at $t = -30t_{P1}$ with five BBs spaced uniformly between $-\pi$ and π in the central valley of the washboard, each with an initial velocity of zero. Using Eqs. (7.1)–(7.3), the motion of each BB was independently computed from $t = -30t_{P1}$ to 0, allowing the BB to settle into its steady-state behavior. At $t = 0$ each BB was then shifted in angle by a multiple of 2π to return it to the central valley of the washboard, without changing its velocity or relative position within the valley. Finally, the continuation of each trajectory was plotted from $t = 0$ to $10t_{P1}$. The plotted motion is different for each BB because they all started with different initial angles. Also, because initial transients have been removed, each trajectory presents a sample of long-term motion on the washboard.

Although all five BBs begin at $t = 0$ in the central valley, they quickly fan out, some moving to the left and some to the right. None of the BBs advances in a regular fashion, unlike the $n = -1$ or 1 trajectories in Fig. 7.1. Instead, at odd intervals a given BB reverses its general direction of motion, and wanders haphazardly forwards and backwards across the washboard. The irregular intervals between reversals suggests that there is a random element in the motion. At the same time, the trajectories are smooth and continuous, following as always from Newton's equations of motion, Eqs. (7.1)–(7.3). Thus, in spite of their irregularity, the trajectories are perfectly predictable. If we begin with the same values of initial angle and velocity, we'll compute the same curves.

The fact that irregular motion can derive from equations lacking an explicitly random term is, of course, the central paradox of chaos. Although a kid asks "So what?" when shown Fig 7.2, a scientist is apt to say, "Wow, I didn't know you could get something so complicated from such simple equations—is this really just a pendulum driven by a periodic torque?" Since we see ordinary kinds of chaos all around us, irregularity is never surprising. On the other hand, when Galileo and Newton developed the science of dynamics, their goal was to understand regular motion, from the swing of a pendulum to planetary orbits, without a thought about irregularity. They would surely have been surprised by Fig. 7.2, and these fathers of dynamics would no doubt have asked lots of questions. Does the irregular motion persist forever without repeating? Is it as random as it appears or is there hidden order? Are you sure there's no source of noise in your calculation? How can anything be predictable and random at the same time? Indeed, these are just the kinds of questions scientists began asking themselves in the 1960s and 1970s when chaotic motion first began to receive wide attention. They're also the questions we'll be trying to answer in the next chapters.

To get a better idea of the long-term behavior of this motion, let's abstract from each drive cycle just the net advance of the BB, or equivalently the net rotation of the pendulum. During the ith drive cycle the net rotation is the angle $\theta(t)$ evaluated at the end of the cycle $(t = it_{P1})$ minus the angle at the beginning $(t = (i-1)t_{P1})$. Thus, we define,

$$R_i = \frac{\theta(it_{P1}) - \theta((i-1)t_{P1})}{2\pi}, \qquad (7.4)$$

where we have divided the change in angle by 2π, so that $R = 1$ corresponds to a net advance of exactly one revolution of the pendulum or one ripple of the washboard.

In Fig. 7.3 we compare the net rotation R_i as a function of time i for steady-state motion corresponding to $\omega = 2.5$, 1.4, and 0.8. At $\omega = 2.5$ the pendulum advances by exactly one revolution during each drive cycle, so R is always 1, and the plot in Fig. 7.3(a) is totally boring. At $\omega = 1.4$, we find period doubling, so the rotation alternates between values slightly less than 1 and slightly greater than 1, as in Fig. 7.3(b). Finally, at $\omega = 0.8$, the rotation is entirely inconsistent from cycle to cycle, varying between roughly -2 and 2 in an irregular way, as in Fig. 7.3(c). Here we see more clearly what chaos is all about.

Even a cursory look at Fig. 7.3(c) reveals that chaotic motion is not completely random. There is some irregularity here, but the rotation

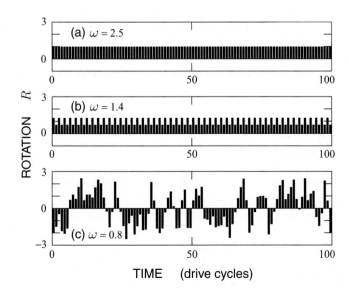

TIME   (drive cycles)

Fig. 7.3 The net rotation R per drive cycle of a pendulum is plotted as a function of time for three cases: (a) $\omega = 2.5$ and $\tilde{\tau}_1 = 12$, (b) $\omega = 1.4$ and $\tilde{\tau}_1 = 4.9$, and (c) $\omega = 0.8$ and $\tilde{\tau} = 1.6$. In each case the damping is $\rho = 0.1$.

tends to stay positive (or negative) for several drive cycles before reversing direction. It's not as if the next rotation is drawn out of a hat without regard for what happened during the last drive cycle. And the physical origin of the tendency to keep moving in the same direction is clear. Just as Galileo argued so long ago, objects in motion tend to keep going. That is, the inertia of the pendulum or BB helps to keep it moving in the same direction.

What about the reversals of direction? Why doesn't the motion just keep going one direction or the other? A convincing answer to this question is difficult to find. Because chaotic motion occurs only when $\omega = \tilde{f}_1/\tilde{f}_0 = t_{P0}/t_{P1}$ is less than about 1, the cause may be linked to the fact that the period t_{P1} of the driving force is greater than the period t_{P0} of the natural oscillation of the BB in a valley. Perhaps a long drive period allows the BB time to begin a natural oscillation that conflicts with motion that would otherwise be imposed by the drive. But this is pure speculation. In the end, it's difficult to give any simple physical reason for the irregular reversal of motion apparent in Fig. 7.3(c), except to say that it results from Eqs. (7.1)–(7.3) and occurs as well in a real pendulum with similar parameters. At this point, we're no smarter than the kid who says, "Chaos happens."

One of the best ways to get a feel for chaotic motion is to watch it happen, and Experiment 11 of the Dynamics Lab lets you do just that. Even if you try no other experiment, you should give this one a spin. The animated pendulum always moves smoothly and precisely in response to gravity and the applied periodic torque, yet it never settles into a routine.

The entire sequence in Fig. 7.3, from synchronized motion ($\omega = 2.5$) to period doubling ($\omega = 1.4$) to chaos ($\omega = 0.8$), is mimicked by the rotation of orbiting bodies in our Solar System. We previously noted the analogy between the Moon and Fig. 7.3(a) and the less strict analogy between Mercury and Fig. 7.3(b). (Mercury's rotation is period doubled but not in the same way as shown here.) To complete the sequence, we note that Hyperion, a potato-shaped moon of Saturn, is observed to rotate randomly, and its motion has been proved chaotic by detailed simulations. Our pendulum isn't alone in its antics, and, as far as nature is concerned, chaos isn't new.

7.3   Statistics

A longer look at chaos in the driven pendulum is given by Fig. 7.4, which extends Fig. 7.3(c) to include 500 drive cycles. Inspecting this log of rotations in detail, we see that the pendulum hasn't begun to repeat itself after 500 cycles. We find sequences of three or four rotations that are seemingly repeated after some long interval, but agreement never goes beyond a few rotations in a row. Indeed, even when the record of rotations is extended to a million drive cycles, we find no repetition. The motion of the chaotic pendulum truly appears to be without periodicity.

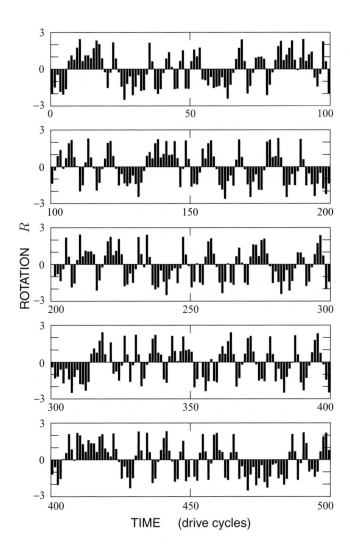

ROTATION R

TIME (drive cycles)

Fig. 7.4 The net rotation of a chaotic pendulum as a function of drive cycle for $\omega = 0.8$, $\tilde{\tau}_1 = 1.6$, and $\rho = 0.1$.

All the same, there is some monotony in the motion shown in Fig. 7.4. After looking at the first 400 drive cycles, we aren't surprised by the next 100 cycles. Individual rotations are never more than about 2.6 revolutions, and they appear to be well mixed between forward and backward. This monotony suggests that chaotic motion can be characterized statistically in the long run even if it's irregular.

The basic statistical tests that we apply to chaos are no different than those that a teacher uses to make sense of exam scores. Just as a teacher computes the average score, we can calculate the average rotation R_a over N drive cycles,

$$R_a = \frac{1}{N}(R_1 + R_2 + \cdots + R_N)$$

$$= \frac{1}{N}\sum_{i=1}^{N} R_i, \tag{7.5}$$

where we have again used a capital sigma (Σ) to denote a sum. As always, the average of N quantities is the sum of the quantities divided by N. Considering $N = 10^6$ cycles of the chaotic series begun in Fig. 7.4, we find that the average rotation is $R_a = -0.0019$. While this average doesn't make sense for a set of test scores, it makes perfect sense for the rotations of a chaotic pendulum. Because the sinusoidal torque doesn't favor either forward or backward rotation, we can expect the average to be near 0, as observed.

Another statistic that a teacher might compute is the standard deviation of a set of test scores. Denoted σ, the standard deviation gives an idea of how much typical scores deviate from the average. If the teacher posts both the average and the standard deviation, then students can get a good idea of how well they performed. A score more than a standard deviation above average is excellent, while a score more than σ below average reveals ample room for improvement.

The square of the standard deviation σ^2 is called the variance, and for our collection of a set of N rotations it's given by,

$$\sigma_R^2 = \frac{1}{N} \sum_{i=1}^{N} (R_i - R_a)^2. \tag{7.6}$$

In other words, σ_R^2 is the average of the square of the departure of the rotation from average. For the chaotic trajectory in Fig. 7.4, we find a standard deviation of $\sigma_R = 1.4$ for $N = 10^6$, indicating that rotations typically deviate from 0 by roughly ± 1.4. Given the data in Fig. 7.4, this σ_R makes perfect sense.

A teacher who wants to make absolutely sure that students understand how well they performed might go a step beyond the standard deviation by plotting the distribution of test scores. A plot of this kind often takes the form of a bar chart, with the height of each bar representing the number of students receiving grades between say 80 and 85 or between 85 and 90. Similarly, in Fig. 7.5, we plot the distribution of rotations for a chaotic pendulum. Here, each bar shows the fraction of 10^6 drive cycles in which the rotation fell within a particular interval of rotations 0.1 revolutions wide. For example, the bar at $R = 1$ tells us that about 3.4% of the rotations were between $R = 0.95$ and 1.05.

Unlike the average R_a and standard deviation σ_R, the distribution plotted in Fig. 7.5 reveals something about chaos that we might not have guessed from a brief inspection of the rotation sequence in Fig. 7.4. In particular, some values of rotation occur much more frequently than others. There are peaks in the distribution at $R = -1.5, -0.2, 0.7$, and 2.1 that we might not have expected. These peaks probably reflect the tendency of the pendulum to rotate an integral number of revolutions per drive cycle. More importantly, the peaks demonstrate that chaotic motion retains a good deal of structure peculiar to the system in which it arises.[1] Chaos isn't completely random, even if it includes a random element.

[1] Because the drive doesn't favor forward or backward rotations, the observant reader might wonder why the distribution in Fig. 7.5 is not symmetric about $R = 0$. The asymmetry results because rotations are computed over a drive period extending from $t = 0$ to t_{P1} or equivalent. If the period is chosen instead as $t = -t_{P1}/2$ to $t_{P1}/2$ or equivalent, then the distribution of rotations has an asymmetry exactly opposite to Fig. 7.5.

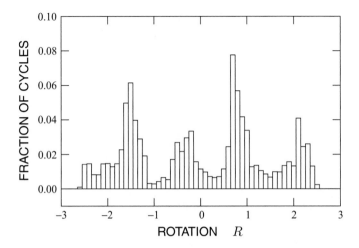

Fig. 7.5 The distribution of rotations for 10^6 drive cycles of a chaotic pendulum with $\omega = 0.8$, $\tilde{\tau}_1 = 1.6$, and $\rho = 0.1$. Each bar shows the fraction of cycles in which the rotation falls within an interval 0.1 revolution in width.

7.4 Correlation

So, is the rotation of a chaotic pendulum random or not? To answer this question, we turn now to a more complicated statistical measure called a correlation function. Ordinarily, it wouldn't make sense for a teacher to compute the correlation function for a set of test scores, but let's suppose that the students taking the test are arranged in one long row. In that case, the correlation function might reveal whether students sitting together copied each other. The correlation function is almost magic.

For a sequence of rotations R_i, the correlation function $C_R(m)$ tells us something about the relation between a given rotation and a rotation the occurs m drive cycles later. By definition, the correlation function is

$$C_R(m) = \frac{1}{\sigma_R^2 N} \sum_{i=1}^{N} (R_i - R_a)(R_{i+m} - R_a). \qquad (7.7)$$

Although the meaning of this function isn't at all obvious, an example will help make sense of it.

For our example, we compute the correlation function for the digits of π, a case in which randomness is well established. To do this, we'll assume that R_i in Eq. (7.7) refers to the ith digit of the decimal expansion of π rather than the ith rotation of a driven pendulum. Using the first 10,000 digits of π, we first calculate the average $R_a = 4.49$ and the standard deviation $\sigma_R = 2.86$. These numbers make sense because the average 4.49 is near the middle of the range of digits (0 to 9), and the standard deviation of 2.86 seems typical of how far a digit is likely to differ from the average.

When the sum defined by Eq. (7.7) is computed for values of m between -20 and 20, we obtain the correlation function shown in Fig. 7.6 for the digits of π. Here, each $C_R(m)$ is plotted as a bar which displays the correlation between two digits separated by m decimal places. How

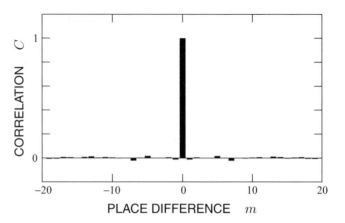

Fig. 7.6 Correlation C of the digits of π as a function of the place difference m, computed for the first 10,000 digits. (To accommodate the fact that R_{i+m} is required by Eq. (7.7) for m between -20 and 20, the sum was computed over the range $i = 21$ to 9,980, so that $i + m$ ranged from 1 to 10,000. In this case, N was taken to be 9,960.)

can we understand this strange plot? First, consider the tall bar at the center of Fig. 7.6, indicating that $C_R(0) = 1$. Looking back at Eq. (7.7), we see that when $m = 0$ the terms in the sum reduce to $(R_i - R_a)^2$, just like the terms in Eq. (7.6) for the variance σ_R^2. As a result, $C_R(0)$ is exactly 1, no matter what the R_i are. This makes sense because with $m = 0$ we are comparing each R_i with itself, and perfect correlation is assured.

What about the very small values of $C_R(m)$ in Fig. 7.6 obtained for $m \neq 0$? To understand this result, it is useful to rewrite Eq. (7.7) as an equivalent but even more daunting equation,

$$C_R(m) = \frac{1}{\sigma_R^2 N} \sum_{D=0}^{9} \left[(D - R_a) \sum_{\substack{i=1 \\ R_i=D}}^{N} (R_{i+m} - R_a) \right]. \qquad (7.8)$$

This expression results directly from Eq. (7.7) by the following argument. In the problem being considered, R_i is always one of the digits 0 to 9. Suppose that we gather together all the terms in Eq. (7.7) for which R_i is a particular digit D and factor out the quantity $(D - R_a)$ that appears in each of these terms. We are then left with the interior sum in Eq. (7.8) over the terms $(R_{i+m} - R_a)$, where R_{i+m} is the digit occurring m decimal places after an instance of the digit D. Thus, the bracketed quantity $[\cdots]$ in Eq. (7.8) includes all the terms in Eq. (7.7) for which $R_i = D$, and by summing over D we obtain all the terms of the original sum.

The inner sum in Eq. (7.8),

$$S_D = \sum_{\substack{i=1 \\ R_i=D}}^{N} (R_{i+m} - R_a), \qquad (7.9)$$

is key to understanding the correlation function. This sum is a little strange because it includes only those digits R_{i+m} which are positioned m decimal places after an instance of the digit D. But this is exactly why we've introduced the correlation function: it tells us whether or not the digit D is followed more frequently by some digits than others. For example, if $D = 5$, $m = 1$, and the very next digit after a 5 is almost always a 7, then S_5 will include many terms in which $R_{i+1} = 7$ is greater than the average $R_a = 4.49$, and S_5 will be positive. In this case, S_5 gives a positive contribution to $C_R(1)$, corresponding to a correlation between successive digits. If, on the other hand, all the digits that follow a 5 occur in nearly the same proportion as in the entire N-digit sample of π, then some of the terms R_{i+1} will be greater than the average R_a while others will be less, and the average contribution to S_5 will be very nearly 0. In this case, S_5 does not contribute to $C_R(1)$, indicating a lack of correlation between successive digits when the first digit is 5.

When applied to the digits of π, we are tempted to conclude if $C_R(m)$ is near 0 then digits separated by m decimal places are uncorrelated. In fact, however, $C_R(m)$ may be zero because the contributions S_D from various digits happen to cancel. Thus, although a non-zero $C_R(m)$ is a sure sign of correlations, when $C_R(m)$ is 0, we can only say that correlations haven't been detected. Nevertheless, the fact that $C_R(m)$ is near zero in Fig. 7.6 for all m except 1, is good evidence that we are dealing with a random collection of digits.

Now let's apply the correlation function to the rotations of a driven pendulum. In Fig. 7.7 we compare $C_R(m)$ computed for the three types of steady-state motion shown in Fig. 7.3: (a) periodic motion in which the pendulum advances by 1 revolution in each drive cycle, (b) period-doubled motion in which the pendulum advances by two revolutions in two drive cycles, and (c) chaotic motion. In Fig. 7.7(a), the correlation C is uniformly 1 for all values of time difference m. This makes sense because all the R_i are 1 and they are all perfectly correlated. Actually, this result is a swindle because for $R_i = 1$ we have $R_a = 1$, $\sigma_R = 0$, and $C_R(m) = 0/0$. Since $0/0$ is not a well defined number, the assignment $C_R(m) = 1$ was simply chosen to satisfy our intuition.

In Fig. 7.7(b), we see that the correlation of the period-doubled motion oscillates between 1 and -1 as a function of m. In this case, R_i itself oscillates between $R_1 = 1.284$ and $R_2 = 0.716$, so the average rotation is $R_a = 1$, and the standard deviation is $\sigma_R = 0.284$. When m is even, all the terms in Eq. (7.7) are of the form $(R_1 - R_a)^2$ or $(R_2 - R_a)^2$, and the correlation is exactly 1 because we are always comparing cycles with equal rotations. When m is odd, on the other had, the terms in the correlation sum are always of the form $(R_1 - R_a)(R_2 - R_a)$, and we are comparing cycles with different rotations. Because R_1 is greater than R_a and R_2 is less than R_a, the product $(R_1 - R_a)((R_2 - R_a)$ is negative, and the correlation proves to be exactly -1. Thus, rotations that differ by an odd number of cycles are precisely anticorrelated: one rotation is as much greater than the average as the other is less.

Now we are in a position to understand the correlation function for chaotic motion in Fig. 7.7(c). As usual, $C_R(0) = 1$, indicating that every rotation is perfectly correlated with itself. At $m = \pm 1$, we find

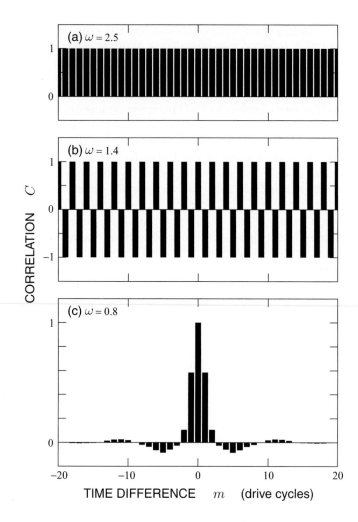

Fig. 7.7 Correlation C of the rotation per drive cycle of a pendulum as a function of the time difference m in drive cycles. Correlation functions are shown for the three cases for which rotations are plotted in Fig. 7.3: (a) $\omega = 2.5$ and $\tilde{\tau}_1 = 12$, (b) $\omega = 1.4$ and $\tilde{\tau}_1 = 4.9$, and (c) $\omega = 0.8$ and $\tilde{\tau} = 1.6$. In each case the damping is $\rho = 0.1$. In (c) the correlation function was computed for $N = 10^6$ drive cycles.

$C = 0.584$, which confirms our suspicion that there are strong correlations between successive rotations. If a teacher found this kind of correlation in test scores, he might conclude that some students were cheating. Of course, it might also be that students of similar ability like to sit together. In the case of the pendulum, however, the large correlation for $m = 1$ can be attributed to inertia. Once the pendulum gets started in one direction, it's likely to keep rotating in the same direction.

Surprisingly, the positive correlation in Fig. 7.7(c) does not extend beyond $m = \pm 2$. For larger time differences, between $m = \pm 3$ and ± 8, the correlation is slightly negative. This tells us that if the pendulum is rotating in one direction now, there's a better than even chance that it will be rotating in the opposite direction five drive cycles from now. I'm glad I was never asked to bet on that proposition—I wouldn't have guessed such anticorrelation from the sequence of rotations in Fig. 7.4.

Most important, Fig. 7.7(c) reveals that for time differences greater than about 13 drive cycles, the correlation between rotations is negligible. Barring an improbable cancellation of terms in Eq. (7.7), we can conclude that there is no significant correlation between rotations well separated in time. This is an amazing conclusion. Beginning with equations completely lacking a random element, we have precisely predicted the motion of a pendulum and yet discovered that the motion is random for all practical purposes. What would Galileo or Newton have thought?

To put this result in perspective, suppose that we build a box with a chaotic pendulum hidden inside and arrange a mechanism that periodically prints out rotations at intervals of 14 or more drive cycles. If we didn't know what was in the box, we could tabulate the values of rotation and determine their distribution, recreating Fig. 7.5, but nothing more could be learned. There might as well be someone in the box drawing numbers from a hat well stocked in proportion to the tabulated distribution. Thus, the wonderfully precise dynamics of Galileo and Newton is reduced to predicting randomness.

7.5   Voltage-standard redux

You might think that, in pursuing the driven pendulum into the realm of $\omega < 1$, I would have discovered chaotic motion for myself, but I never did. Instead, because my main interest was in synchronization, I investigated the stability of periodic motion. I learned that the stable synchronized motion found for $\omega \gg 1$ doesn't completely disappear for $\omega < 1$ but is largely replaced by similar motion that is entirely unstable.

This conclusion is illustrated by Figs. 6.5 and 7.8. Both figures plot the pendulum's average velocity $\tilde{v}_a$ as a function of the constant component of torque $\tilde{\tau}_0$, and both reveal wide ranges of torque where $\tilde{v}_a$ is exactly constant at $n\omega$. In Fig. 6.5 for $\omega = 2.5$, these constant-velocity steps are all stable. In Fig. 7.8 for $\omega = 0.8$, only the solid portions of the curves are stable. The open lines are the ghosts of departed trajectories, marking where synchronized motion exists as a possibility but is unstable and can't be observed in practice. Thus, according to Fig. 7.8, broad steps are expected in principle for $n = 0$ through 4 at $\omega = 0.8$, but only a small fragment of the $n = 1$ step is actually stable. Reducing ω below 1 has decimated stable synchronization.

Before leaving Fig. 7.8, it is interesting to point out the stable step at $\tilde{v}_a = 1.2$, midway between the $n = 1$ and $n = 2$ steps. On this step, as you might guess, the pendulum advances by three revolutions during two drive cycles. This motion is exactly analogous to that of Mercury which, as you'll recall, completes three revolutions during two orbits.

Having completed my stability analysis, by 1980 I understood why the early voltage-standard experiment with Sullivan had been a failure. The fragmentary steps that we observed in 1977 were not due to some unknown defect in our junctions, nor were they due to external noise plaguing our apparatus. The small steps were the remnants of stable

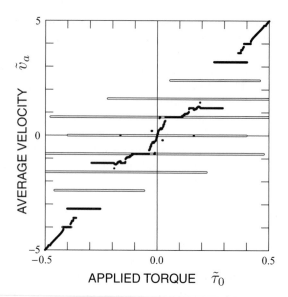

Fig. 7.8 Average velocity $\tilde{v}_a$ for the steady-state motion of a pendulum as a function of the constant component of applied torque $\tilde{\tau}_0$, in the presence of an oscillatory component $\tilde{\tau}_1 \sin(\omega t)$. In this computation, the frequency parameter and amplitude of the oscillatory torque are fixed at $\omega = 0.8$ and $\tilde{\tau}_1 = 1.6$, while the damping is $\rho = 0.1$. Stable motion is plotted by solid lines and unstable motion by open lines.

synchronization that persist for $\omega < 1$, and the instability in this regime is entirely explained by the pendulum model of the Josephson junction. Although this result confirmed my hunch, I was still surprised to learn something new from the simple pendulum model, 17 years after constant-voltage steps were first observed.

I was even more surprised by a preprint that I received just as my paper on the stability of synchronization was being submitted for publication. A preprint is a prepublication copy of a paper circulated by the authors to publicize work that's about to appear in print. The one I received in 1980 had spread quickly through the community of Josephson physicists, although the names of its three authors, Bernardo Huberman, James Crutchfield, and Norman Packard, were not familiar. At first the paper seemed completely crazy. Based on analog simulations of a driven pendulum, the authors reported high levels of noise arising from "chaotic behavior" intrinsic to the Josephson junction in the parameter range $\omega < 1$.

I was flabbergasted. Certainly the Huberman paper jibed with my discovery of instabilities, but what was "chaotic behavior," and how could a completely noise-free system generate noise? To me, like most physicists in 1980, the concept of chaos was completely unknown. Our naiveté, however, was about to end. Chaos would soon become the latest buzz word in science, and, within a few years, most scientists would gain some familiarity with chaos. By 1990, even nonscientists were generally aware of the hoopla surrounding it.

The hoopla was well deserved. For hundreds of years, we had used the dynamics of Galileo and Newton to predict the future and tested our predictions to the nth degree, without realizing that randomness and apparent unpredictability were latent within the science of dynamics itself. Now, dynamics had a new and unanticipated role to play. The new role didn't negate all that had gone before—astronauts could still rely on Newton's equations to get to the Moon—but it opened up new vistas. With the advent of chaos theory, scientists understood that in some cases Newtonian dynamics could explain the origin of noise or account for our inability to accurately predict the future. These new roles for dynamics were surprising, almost contradictory to our old vision, and it took time for scientists to reconcile themselves to the idea of chaos.

Although the seeds of chaos theory date back more than a century, it was only in the 1960s that the theory began to be developed in earnest and gradually make its way into the wider scientific community. When Huberman, Crutchfield, and Packard discovered chaos in the driven junction in 1980, they were relatively new to the field but already expert, part of the gathering tidal wave that was about to flood all of science. At the time, Crutchfield and Packard were still working toward their doctorates at the University of California at Santa Cruz, where they were members of the Dynamical Systems Collective, a group of graduate students interested in chaos. For them, their brief paper on chaos in the Josephson junction was no more than a warm-up exercise for their dissertations. For those of us trying to use junctions in practical circuits, it opened up a whole new perspective on junction dynamics.

Clearly, chaos must be avoided in the voltage standard and other Josephson circuits for proper operation to be assured. However, to understand the full significance of chaos for voltage standards, we need to first understand why the frequency parameter ω is crucial. In a nutshell, ω controls the importance of nonlinearity in the driven pendulum. When $\omega \gg 1$, the periodic torque is largely taken up in accelerating the pendulum, and the response is nearly linear. When $\omega \ll 1$, the periodic torque acts largely in opposition to the gravitational torque, the one nonlinear element in the system, and the response is highly nonlinear. Since nonlinearity is required for chaos, this explains why chaos in prevalent for $\omega < 1$.

On the other hand, you'll recall from the last chapter that nonlinearity is also required for synchronization. Thus, we arrive at the usual engineering pickle. We need nonlinearity to make a voltage standard, but, if the nonlinearity is too strong, we get chaos instead. There must be an optimum value for ω that gives the strongest possible synchronization, but what is it? To answer this question, I considered what happens when the pendulum is buffeted by external noise and, after lengthy calculations, determined that an ω of roughly 3 makes synchronization the least susceptible to disturbances. This simple result solved the design problem posed by chaos in the driven junction.

The Josephson voltage standard was probably the first instance in which chaos was consciously accounted for in the design of a practical device. However, the same sort of situation is likely to be found in many systems that use a nonlinear effect to achieve a desired goal. In nonlinear systems, there is always a chance of finding chaos when the nonlinearity is too strong, and the optimum nonlinearity may locate the system close to a region of chaotic behavior.

Now that chaos has entered the mindset of scientists and engineers, it will never disappear. These days, when we discover noise in a nonlinear system, we're as likely to look for the source in the system's noise-free equations as in its external environment. If the noise is chaotic, we'll want to carefully map the range of parameters evoking chaos, because the optimum operating point may be nearby. Chaos has brought a whole new perspective to nonlinear design.

In the end, the series-array voltage standard proved to be a success. Chaos was just one of many problems to be solved, and scientists at both NIST and its sister institution in Germany, the Physikalisch-Technische Bundesanstalt, contributed innovative ideas to the final design. In 1985, both laboratories fabricated microcircuit arrays of more than 1,000 Josephson junctions that produced accurate voltages in excess of 1 V, and by 1987 arrays with more than 10,000 junctions extended the output to over 10 V. These standards are now routinely used by government and industrial laboratories around the world as the ultimate authority in voltage measurement. The thousands of junctions in these systems all operate near $\omega = 3$, quietly avoiding the nearby chaotic regime.

Further reading

Dynamical Systems Collective

- Bass, T. A., *The Eudaemonic Pie* (Houghton Mifflin, Boston, 1985).
- Gleick, J., "The Dynamical Systems Collective", in *Chaos: Making a New Science* (Viking, New York, 1987), pp. 241–272.

Hyperion

- Killian, A. M., "Playing dice with the solar system", *Sky and Telescope* **78**, 136–140 (1989).
- Parker, B., "The strange case of Hyperion and other mysteries", in *Chaos in the Cosmos* (Plenum, New York, 1996) chapter 10.
- Peterson, I., "Hyperion tumbles", in *Newton's Clock: Chaos in the Solar System* (Freeman, New York, 1993) pp. 199–222.

Pendulum chaos

- Baker, G. L. and Blackburn, J. A., "The chaotic pendulum", in *The Pendulum: A Case Study in Physics* (Oxford University Press, Oxford, 2005) chapter 6.
- Kautz, R. L., "Chaos in a computer-animated pendulum", *American Journal of Physics* **61**, 407–415 (1993).

Statistics

- Stigler, S. M., *The History of Statistics: The Measurement of Uncertainty before 1900* (Belknap Harvard University Press, Cambridge, 1986).

Voltage standard

- Kautz, R. L., "Noise, chaos, and the Josephson voltage standard", *Reports on Progress in Physics* **59**, 935–992 (1996).

Chaos takes a random walk

8

The science of dynamics developed by Galileo and Newton aims at the exact prediction of future events. In contrast, the science of probability gives us predictions for the likelihood of various possible futures. While exact results are the holy grail of science, in complex situations, physicists often find probability theory to be a simple and useful option. Sometimes probabilities are the alternative to making no prediction at all.

Given the random nature of chaotic motion, we are bound to ask whether the general outcome might not be described by probabilities almost as well as by a full dynamical calculation. In this chapter, we find that the net advance of a chaotic pendulum is accurately modeled by what is known in probability theory as a random walk or drunkard's walk. Imagine, if you will, a man so inebriated that he is as likely to take a step backward toward the tavern as a step forward toward home. Without knowing anything more, we can't predict exactly when he'll get to bed, but we can determine where along the street he's likely to be found after a given number of steps.

Although knowing anything at all about a drunkard's walk may seem unimportant, it happens that many physical systems behave just like a drunkard in the way they move. Famously, in the same year (1905) that Albert Einstein (1879–1955) discovered special relativity, he also analyzed the motion of dust motes in water using probability theory. Einstein hypothesized that the motes would be kicked about by fast moving water molecules and were equally likely to move in any direction. His analysis predicted how fast a mote would diffuse through water and proved to be in agreement with previous observations of what is called "Brownian" motion. A mundane piece of science? Not at all. Einstein's work provided the best evidence of the day that matter is made of atoms!

The chaotic walk of a driven pendulum is illustrated in Fig. 8.1 by 20 typical trajectories similar to those shown in Fig. 7.2 but extended to 50 drive cycles. In terms of the washboard analog, the trajectories all begin at $t = 0$ in the central well of the washboard ($-\pi < \theta < \pi$), but with various initial angles and velocities. Examining any single trajectory, we find that the pendulum repeatedly reverses its direction of motion at odd intervals, moving toward positive θ for a few cycles, then heading in the negative direction for a while. Because there is no clear

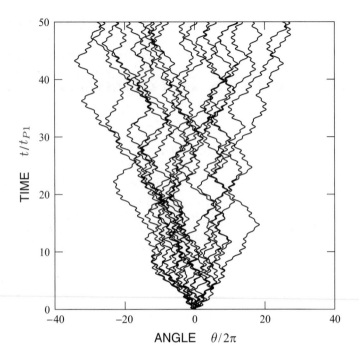

Fig. 8.1 Chaotic motion of a sinusoidally driven pendulum is illustrated by 20 trajectories for $\omega = 0.8$, $\tilde{\tau}_1 = 1.6$, and $\rho = 0.1$. To insure that these trajectories represent steady-state motion, they were initiated at $t = -30t_{P1}$ with $\tilde{v} = 0$ and angles θ distributed uniformly between $-\pi$ and π. The initial transients between $t = -30t_{P1}$ and 0 were computed but not plotted. At $t = 0$, the angle of each trajectory was shifted by a multiple of 2π, as needed to return it to the central valley of the washboard, and the subsequent trajectory was plotted from $t = 0$ to $50t_{P1}$.

pattern in the interval between reversals, it's not hard to imagine that the pendulum executes some form of random walk or Brownian motion. If $\theta = 0$ represents the location of the drunkard's tavern, it would be difficult indeed to predict when the pendulum might arrive home at say $\theta/2\pi = 20$.

Does the pendulum's behavior actually follow the laws of probability? To find out, we now take up the study of chance, the very antithesis of deterministic dynamics. Here we throw exact prediction to the winds and assume that motion is completely random. All the same, we'll find that probabilistic analysis fits predictable chaotic motion to a tee.

8.1   Probability

Like dynamics, probability theory has its roots in the Italian Renaissance. The story begins with Gerolamo Cardano (1501–1576), a physician and mathematician born 63 years before Galileo. Why was a physician interested in probability? Cardano, despite his scholarly nature, was passionately devoted to gambling, a popular pastime among the nobility of his day. Indeed, while a student at the University of Padua, gambling was not only Cardano's favorite diversion, it was also his main source of income.

How could Cardano win consistently? Was it a matter of luck, or did Cardano know something that his adversaries didn't? In a dice game, the numbers thrown are certainly determined more by chance than anything predictable, so an element of luck is important. But as Cardano realized, some numbers are thrown more frequently than others, and, to make an intelligent bet, you need to know the likelihood of throwing each possible combination. This was exactly Cardano's edge: by developing a rudimentary theory of probability, he could calculate the mathematical odds of winning and put his money on favorable outcomes. With the odds on his side, Cardano was almost certain to win.

The concept of probability is simple. Consider, for example, an ordinary die with sides numbered 1 through 6. On a given roll, the die can land with any side facing up, so, if the die is perfectly balanced, there are six possible outcomes, each equally likely. In this case, we say that the probability of rolling any given number is 1/6, because we expect each number to turn up roughly $\frac{1}{6}$th of the time.

In general then, for a situation in which several outcomes are possible, the probability of a given outcome is just the fraction of trials in which the outcome is expected to occur. Probabilities are always between 0 and 1. A probability of 0 corresponds to an outcome that's impossible (such as throwing 7 with a single die), and 1 to an outcome that's certain (such as rolling a positive number).

Of course, knowing the probabilities for a single die wouldn't have given Cardano an edge over anyone. However, he was also able to compute the probabilities for rolling various numbers with two and three dice, and these results were not commonly known in his day. Although Cardano never stated them explicitly, his calculations relied on two elementary rules for combining probabilities.

   I. The probability of observing either of two mutually exclusive events is the sum of the probabilities of the two events.

   II. The probability of observing both of two independent events is the product of the probabilities of the two events.

For our limited investigation of probability, these are the only rules we'll need.

Rule I is almost intuitively obvious. With a single die, rolling a 2 and rolling a 5 are mutually exclusive, since only one face can be up on a given throw. Thus, the probability of rolling either a 2 or a 5 is

$$P(2 \text{ or } 5) = P(2) + P(5) = \frac{1}{6} + \frac{1}{6} = \frac{1}{3}. \tag{8.1}$$

Rule I tells us that the probability of an event goes up when more possible outcomes are included in the event. What's the probability of throwing a 1, 2, 3, 4, 5, or 6 with a single die? According to rule I the answer is $\frac{1}{6} + \frac{1}{6} + \frac{1}{6} + \frac{1}{6} + \frac{1}{6} + \frac{1}{6} = 1$. That is, it's certain we'll roll some number between 1 and 6.

Rule II applies when we throw two dice. In this case, rolling 2 on the first die and rolling 5 on the second are independent events. After all, the dice can be rolled one after the other to eliminate any possible dependence. According to rule II, the probability of observing the combination $(\text{die1}, \text{die2}) = (2, 5)$ is thus

$$P(2, 5) = P(2)P(5) = \frac{1}{6} \times \frac{1}{6} = \frac{1}{36}. \tag{8.2}$$

This result makes sense because die1 will be 2 on $\frac{1}{6}$th of the trials, and of these trials die2 will be 5 only $\frac{1}{6}$th of the time. Rule II tells us that we are being picky by requiring both a 2 and a 5, and the probability is smaller than that for rolling just a 2 or just a 5.

Following in Cardano's footsteps, we can now calculate the probability that the numbers appearing on two dice will sum to N. We already know that any combination $(\text{die1}, \text{die2})$ is rolled with a probability of $1/36$, so it's just a matter of adding up the probabilities of all the combinations for which $\text{die1} + \text{die2} = N$. Our results, listed in Table 8.1, show that a sum of 7 is most likely, resulting from six different combinations, and occurs with probability $1/6$. Similarly, $N = 2$ and $N = 12$ are least likely because they are each represented by just one combination, and they occur with probability $1/36$.

Because Cardano's gambling partners lacked the information in Table 8.1, perhaps having only a vague notion that some sums are more likely than others, they were at a distinct disadvantage when it came to playing dice games. For example, Cardano might have offered 100 *reals* to anyone who rolled snake eyes (double ones), provided they paid him 10 *reals* each time they failed. This offer might seem like a good opportunity to someone unfamiliar with the actual odds. But Cardano knew that snake eyes are thrown on average only once in 36 tries, so he could expect to pay out an average of $(1/36)100 = 2.78$ *reals* per trial, while he would take in an average of $(35/36)10 = 9.72$ *reals*. Although he might occasionally have to make a large payout, on average Cardano would net 6.94 *reals* with every roll. No wonder he was an avid gambler!

Table 8.1 The probability $P(N)$ of throwing the sum N with a pair of dice. The various combinations of numbers appearing on the two dice for a given sum are also listed.

Sum N	Combinations (die1,die2)	Probability P
2	(1,1)	1/36
3	(2,1) (1,2)	2/36
4	(3,1) (1,3) (2,2)	3/36
5	(4,1) (1,4) (3,2) (2,3)	4/36
6	(5,1) (1,5) (4,2) (2,4) (3,3)	5/36
7	(6,1) (1,6) (5,2) (2,5) (4,3) (3,4)	6/36
8	(6,2) (2,6) (5,3) (3,5) (4,4)	5/36
9	(6,3) (3,6) (5,4) (4,5)	4/36
10	(6.4) (4,6) (5,5)	3/36
11	(6,5) (5,6)	2/36
12	(6,6)	1/36

Like Galileo and Newton, both of whom postponed publication of their work from youth to old age, Cardano hesitated to publish his theory of probability. Perhaps Cardano simply wanted to maintain his advantage in gaming, but his treatise entitled *De Ludo Aleae* or *On Games of Chance*, completed around 1564, wasn't published until 1663, nearly a century after his death. In the interim, Galileo published a note on probabilities for three dice around 1600, the French mathematicians Blaise Pascal (1623–1662) and Pierre de Fermat (1601–1665) began their famous correspondence on probability in 1654, and Huygens published his book, *Computations in Games of Chance*, in 1657. Thus, although Cardano's work was eclipsed even before it appeared in print, he was nonetheless the first to understand and use probability.

Curiously, all of the pioneers of probability theory were prompted in one way or another by gaming. Although the casinos of today still rely on knowing the odds, probability is surely used now by scientists more than anyone.

8.2 Quincunx

Perhaps the simplest device that leads to a random walk is the quincunx, an example of which could be found at the dime store in my hometown. The dime store's quincunx, illustrated in Fig. 8.2, consisted of a glass fronted box with just enough space between front and back for a penny to slip in between. Designed to separate pennies from patrons, the quincunx promised a prize for a lucky random walk. To play the game, a patron inserted a penny in a slot at the top, watched it bounce through an array of pegs, and hoped it would land in one of the bins at either corner. The actual machine included several more rows of pegs than that shown in Fig. 8.2, and I don't recall ever hearing of anyone winning. Although none of us kids could calculate probabilities, we quickly guessed that reaching a corner was much less likely than rolling snakes eyes with a pair of dice. With that realization, I found it more satisfying to buy a root beer barrel with my penny than take a chance on the quincunx.

How does the quincunx represent a random walk? When a penny dropped through the slot encounters its first peg, it must bounce either left or right before it can drop to the next row of pegs. After dropping through, the penny is confronted with another peg and the same two possibilities: bounce either left or right. Thus, with each row of pegs, the penny is forced to take a step in one direction or the other, and, if left and right occur with equal probability regardless of previous steps, its horizontal motion executes a random walk. The qunicunx is so simple and efficient, with results automatically tabulated in the bins at the bottom, that the English scientist Francis Galton (1822–1911) used it extensively in studies of heredity. Nowadays, the qunicunx is often featured at science museums.

Our challenge is to analyze the quincunx using our limited knowledge of probability theory. It's actually pretty simple. Suppose that when a peg is encountered, the probability of taking a bounce to the right is p_r and the probability of going left is $p_l = 1 - p_r$. Later we'll be interested in the case $p_r = p_l = \frac{1}{2}$, but let's keep the discussion general for now.

Since the bounce on each successive row is assumed to be independent of previous bounces, we can apply rule II and compute say the probability of falling through six rows of pegs in the sequence right, right, left, right, left, right simply by multiplying the probabilities:

$$P(r, r, l, r, l, r) = p_r p_r p_l p_r p_l p_r = p_r^4 p_l^2. \tag{8.3}$$

Because multiplication doesn't depend on order, we get the same result for any trajectory with four r's and two l's. Thus, more generally, if the quincunx has n rows of pegs, then any path that includes m bounces to the right and $(n - m)$ to the left will occur with probability

$$p_r^m p_l^{n-m}. \tag{8.4}$$

To get our final result, we now apply rule I and add the probabilities for all paths that have the same m to get $P_n(m)$, the probability of netting m bounces to the right. Since all paths with a given m have the same probability, we simply multiply expression (8.4) by the number of paths with m bounces right and $(n - m)$ bounces left, a number conventionally denoted $\binom{n}{m}$,

$$P_n(m) = \binom{n}{m} p_r^m p_l^{n-m}. \tag{8.5}$$

This formula will tell us all we need to know about the quincunx and other random walks, just as soon as we evaluate the curious number $\binom{n}{m}$.

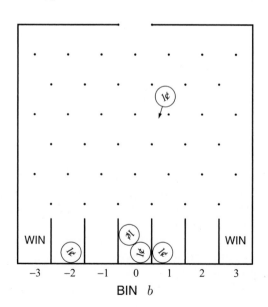

Fig. 8.2 A quincunx for pennies. Falling through an array of pegs, a penny ends up in one of several bins at the bottom.

8.3 Pascal's triangle

When the number of rows in the quincunx is small, calculating $\binom{n}{m}$ is easy. For example, if $n = 2$ then there are two trajectories with one bounce to the right, (r, l) and (l, r), so $\binom{2}{1} = 2$. Similarly if $n = 3$ there are three trajectories with two bounces right, (r, r, l), (r, l, r), and (l, r, r), so $\binom{3}{2} = 3$. As more rows are added, however, listing all of the possibilities becomes a tedious way of evaluating $\binom{n}{m}$. Fortunately, Pascal, one the founders of probability theory mentioned earlier, encountered the same problem and advocated (but did not originate) a neat geometric way to tabulate $\binom{n}{m}$.

Called Pascal's triangle, the method is illustrated in Fig. 8.3 as it applies to the quincunx. Here, each penny is labeled with the value of $\binom{n}{m}$ corresponding to its location in row n on diagonal m. It's easy to see that the diagonals correspond to a given number of bounces m to the right. For example, positions along the left edge of the triangle can only be reached if the penny always bounces to the left at every row, so $m = 0$. Similarly, a position along the $m = 1$ diagonal can be reached only if the penny bounces to the right exactly once during its journey.

In Pascal's triangle, the number $\binom{n}{m}$ labeling each penny is the number of ways the penny might have reached its position. As Pascal realized, the values of $\binom{n}{m}$ for a given row can be calculated easily from the values in the row above. For example, the value of $\binom{3}{1}$ follows from the fact that the position $(n, m) = (3, 1)$ can only be reached from positions $(n, m) = (2, 0)$ and $(2, 1)$ in the row above, the first of which can be reached in only one way and the second of which can be reached in two ways. Thus, there are $1 + 2 = 3$ ways of reaching position $(n, m) = (3, 1)$ and $\binom{3}{1} = 3$. In general then, each number in the interior of the triangle

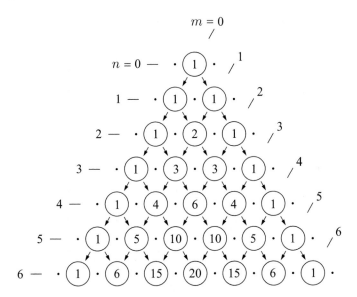

Fig. 8.3 Pascal's triangle as applied to the quincunx. Each penny is labeled with the number of ways in which it might have arrived at its location. Thus, $\binom{n}{m}$ can be evaluated by finding the penny in row n along the diagonal m.

is the sum of the two nearest numbers in the row above. Of course, along the edges of the triangle, where $m = 0$ or $m = n$, there is only one way to reach each position, since the penny must bounce either always left or always right to stay along an edge. So $\binom{n}{0} = \binom{n}{n} = 1$ in every row. Thus, Pascal's triangle is easy to fill in: first put 1's in all the edge positions, then, working from the top down, add pairs of adjacent numbers in one row to fill in the next row. Following these rules, you can quickly verify the numbers in Fig. 8.3.

Now that you understand where $\binom{n}{m}$ comes from, it's time to admit that there's a simple formula for this number, namely

$$\binom{n}{m} = \frac{n!}{m!(n-m)!}. \tag{8.6}$$

Keeping in mind that $n!$ (read n factorial) is defined by $n! = n \times (n-1) \times \cdots \times 2 \times 1$, with the extra, curious proviso that $0! = 1$, you'll be able to quickly verify that Eq. (8.6) reproduces Pascal's triangle. The number $\binom{n}{m}$ is exactly the number of ways in which m things can be chosen among n things. Thus, in the quincunx, $\binom{n}{m}$ is the number of ways m bounces to the right can be chosen among n rows of pegs. For this reason, the notation "$\binom{n}{m}$" is read "n choose m."

Equations (8.5) and (8.6) completely pin down probabilities for the general quincunx. Before putting probabilities to work, however, it's always a good idea to check that, when summed over all possible outcomes, the probabilities add to 1. Is it certain that a penny will end up in one bin or another after falling through the quincunx? The answer to this question happens to fall out of what's known as the binomial expansion. You may remember from algebra that when a binomial such as $(p_r + p_l)$ is taken to the nth power, it can be expanded as a sum of the form,

$$(p_r + p_l)^n = p_l^n + \binom{n}{1} p_r p_l^{n-1} + \binom{n}{2} p_r^2 p_l^{n-2} + \cdots + p_r^n$$

$$= \sum_{m=0}^{n} \binom{n}{m} p_r^m p_l^{n-m}. \tag{8.7}$$

On the last line of Eq. (8.7), we have exactly the sum of the probabilities of all the possible outcomes of the quincunx, a quantity we expect to be exactly 1. On the other hand, because we assume that the penny must bounce either right or left at each peg, we have $p_r + p_l = 1$, so $(p_r + p_l)^n = 1^n = 1$. Thus, the sum is indeed 1, and the probabilities add up perfectly! Because of its close association with the binomial expansion, the probability distribution given by Eq. (8.5) is called the binomial distribution.

Let's apply the binomial distribution to the quincunx shown in Fig. 8.2 with $n = 6$ rows of pegs. Equation (8.5) gives the probability that the penny will bounce to the right exactly m times as it falls from top to bottom. Assuming $p_r = p_l = \frac{1}{2}$, the probability is

$$P_n(m) = \binom{n}{m}\frac{1}{2^n}. \tag{8.8}$$

So what's the probability of winning the prize when you insert a penny? Since m is related to the bin number b according to $m = b + n/2$, the probability of hitting a corner bin is $P_6(0) + P_6(6) = \frac{1}{64} + \frac{1}{64} = \frac{1}{32}$. Thus, you can expect to win once in 32 tries, and the store can't afford to give a prize worth more than 32¢. On the other hand, if the quincunx had 10 rows of pegs then the probability of winning would be $\frac{1}{512}$, and the store would break even with a prize of \$5.12. As the popularity of lotteries suggests, people often like to bet on long shots, so a store might offer a \$2 prize for winning with 10 rows of pegs. If so, would you play the game or buy a root beer barrel with your penny?

The probability distribution given by Eq. (8.8) can be verified by trial and error using the virtual quincunx presented in Experiment 12 of the Dynamics Lab.

8.4 Diffusion

As a penny falls through a quincunx, its horizontal position tends to wander farther away from its initial value, just as a drunkard might wander farther from the tavern. Physicists call this type of motion "diffusion," and we are now ready to investigate it in detail. To get started, let's take a look at Fig. 8.4, which shows the probability of a penny landing in bin b of a quincunx after falling through $n = 6$, 12, or 24 rows of pegs. The probabilities calculated from Eq. (8.8) are plotted as bars, while the continuous curves show an approximation we'll discuss later. In all cases, the penny is most likely to end up in bin 0, but there is a small probability of reaching $b = \pm n/2$. Thus, the further the penny falls through the quincunx, the wider the range of bins in which it can be found. This haphazard spreading of the penny's probable location is characteristic of diffusion, and we now investigate it quantitatively.

In the last chapter we used the standard deviation σ to measure how far random numbers varied from their mean, and we'll use the same measure here to look at diffusion. What's the standard deviation σ_b of the bin number b from its mean value of 0? By definition σ_b^2 is the average of $(b-0)^2$ or $(m-n/2)^2$. Because we know the probability of each m, we can add up all the deviations $(m-n/2)^2$ multiplied by their probabilities $P_n(m)$ to get the desired average:

$$\sigma_b^2 = \sum_{m=0}^{n}(m-n/2)^2 P_n(m)$$

$$= \sum_{m=0}^{n}(m-n/2)^2\binom{n}{m}\frac{1}{2^n}$$

$$= \frac{n}{4}. \tag{8.9}$$

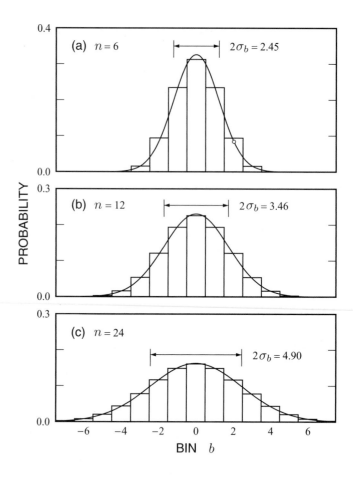

Fig. 8.4 Probability distributions for quincunxes with $n = 6$, 12, and 24 rows of pegs based on Eq. (8.8). Bars record the probability of penny a falling into each bin. Bell shaped curves apply in the limit of large n.

If you're thinking that the last step above is far from obvious, you're absolutely right. It's easy to verify that $\sigma_b^2 = n/4$ for small values of n but difficult to prove the general case. We'll accept the result on faith.

Equation (8.9) tells us something interesting about diffusion. In general, the bulk of a probability distribution can be found within $\pm\sigma$ of its mean. Thus, according to Eq (8.9), most pennies will be found in bins between $b = -\sigma_b = -\frac{1}{2}\sqrt{n}$ and $b = \sigma_b = \frac{1}{2}\sqrt{n}$. As expected, pennies spread over more and more bins the farther they fall through a quincunx. For the cases in Fig. 8.4, with $n = 6$, 12, and 24 rows of pegs, $2\sigma_b$ increases from 2.45 to 3.46 to 4.90 bins. It's perhaps surprising that σ_b increases in proportion to $\sqrt{n}$ rather than n, but the distribution spreads out slowly because the penny is always as likely to bounce to the right as the left. Assuming that the time T required to fall through a quincunx is proportional to the number of rows of pegs, we can also say that σ_b increases in proportion to $\sqrt{T}$. This dependence on the square root of time is a fundamental characteristic of diffusion. Cars move with purpose as $D \propto T$, but drunkards wander aimlessly, and their probability distribution spreads as $\sigma_b \propto \sqrt{T}$.

What about the smooth curves plotted in Fig. 8.4? These exemplify the most famous distribution in probability theory, called the bell curve or normal distribution. The bell curve appears here courtesy of a remarkable theorem proven by Laplace in 1810—the same Laplace who hypothesized that the universe is entirely predictable in principle. (In spite of his bias, Laplace clearly had a foot in the camps of both determinism and chance.) Called the central limit theorem, Laplace's result relates to the sum of many random variables with identical distribution functions. In the quincunx, the final bin number b is the sum of all the steps to the right or left taken by the penny as it falls through n rows of pegs. Regardless of how the individual steps are distributed, as long as they are distributed with the same probability each time a step is taken, Laplace was able to show that for large n the sum b is distributed according to

$$P(b) \approx \frac{1}{\sqrt{2\pi}\,\sigma_b}\, e^{-(b-b_a)^2/2\sigma_b^2} \qquad (n \gg 1), \qquad (8.10)$$

where b_a and σ_b are the average and standard deviation of b. In this formula, $e = 2.718\ldots$ is the base of the natural logarithm, and the function e^x can be evaluated by hitting the appropriate key on your calculator—even if you're not sure exactly what it means. Can you verify that Eq. (8.10) yields $P(2) \approx 0.0856$ for $b_a = 0$ and $\sigma_b = \sqrt{6}/2$, as indicated by the circle in Fig. 8.4(a)? First calculate $-b^2/2\sigma_b^2$, then punch the e^x key, and divide the result by $\sqrt{2\pi}\sigma_b$. Try It! We'll learn more about the exponential function e^x in Chapter 10.

The bell curve specified by Eq. (8.10) is ubiquitous in probability theory because it applies to many situations. Whenever an outcome depends on the sum of many small effects, the normal distribution is likely to result. Thus, supposing that the size of an orange depends on when its blossom matures, the amount of sunlight and water it receives, the age of the tree, etc., we wouldn't be surprised to find that the oranges in a grove are distributed in size according to a bell curve. Indeed, the bell curve is so common that science museums can't resist building a quincunx to show how it happens.

In our analysis of the quincunx, we assumed that the bin number is the sum of independent steps to the right or left, so the distribution of pennies must follow a bell curve for large enough n. This result is completely confirmed by Fig 8.4. To obtain the bell curves shown here, we substitute $b_a = 0$ and $\sigma_b = \sqrt{n}/2$ before plotting Eq. (8.10).[1] In each case, the match between the bell curve and the binomial distribution is remarkably accurate, even for relatively small n. Thus, in accord with the central limit theorem, the position of a diffusing particle is characterized not only by a standard deviation proportional to $\sqrt{T}$, but by a probability distribution in the shape of a bell curve. A random walk may seem like a higgledy-piggledy mess, but its average properties are known with precision.

[1] Equation (8.10) assumes that b is a continuous random variable and specifies that the probability of finding b in the range between $b - \delta/2$ and $b + \delta/2$ is $\delta P(b)$. For the quincunx, however, b is an integer and we want to know the probability that b is between $b - 1/2$ and $b + 1/2$. In this case, $\delta = 1$, and the probability of b is just $P(b)$.

8.5   Chaotic walk

Now that we've learned something about diffusion according to probability theory, it's time to see how the chaotic walk shown in Fig. 8.1 stacks up. Thinking in terms of the washboard analog, we consider trajectories of BBs that begin in the central well of the washboard ($-\pi < \theta < \pi$) and move chaotically across the washboard in response to a sinusoidal periodic force. The system is much like the quincunx, with the valleys of the washboard substituting for the bins of the quincunx, but the BB moves predictably according to Eqs. (7.1)–(7.3), while the penny is assumed to move randomly. Thus, there is no particular reason to believe that a BB's chaotic motion will mimic diffusion, although the possibility is suggested by the trajectories in Fig. 8.1.

To compare a chaotic walk with a random walk, we perform a numerical experiment involving 100,000 BBs. As in Fig. 8.1, each BB begins at $t = 0$ in the central valley of the washboard ($-\pi < \theta < \pi$) with a distinct (not to say random) initial position and velocity. The equations of motion are then applied to calculate each BB's trajectory across the washboard during 100 periods of the sinusoidal force. With this data, we can compute a distribution function for the location of the BBs at any time up to $t = 100t_{P1}$. Suppose that b is an integer designating the valley between $\theta = 2\pi b - \pi$ and $2\pi b + \pi$. Assuming our collection of 100,000 BB trajectories is an unbiased sample, we can estimate the probability of finding a BB in valley b at a given time after it is observed in valley 0, by taking the ratio of the number of sample trajectories in valley b to the total number in the sample. Probabilities estimated in this way are shown as bar charts in Fig. 8.5 for times $t/t_{P1} = 25$, 50, and 100.

As expected, the probability distributions in Fig. 8.5 spread out as time proceeds. Do they have the shape expected for diffusive motion? Yes, bell curves with standard deviations σ_b matching those of the bar charts are in close agreement. Does σ_b increase in proportion to $\sqrt{t}$? Yes, as Table 8.2 shows, the ratio $\sigma_b/\sqrt{t}$ is nearly the same for $t/t_{P1} = 25$, 50, and 100. Thus, it appears that the chaotic motion of a driven pendulum (or particle on a washboard) mimics the diffusive motion of a random walk with remarkable precision. The pendulum advances as if it were driven by chance rather than determinism, with its motion characterized by the bell curve, an icon of probability theory. As we have now begun to expect, the predictable motion of a chaotic pendulum imitates unpredictable motion far better than we might otherwise have thought. It's a paradox that begs explanation.

At one level, we can explain the chaotic walk in terms of the correlation function calculated in Chapter 7. As was noted, the net advance of the BB during one drive cycle is not correlated with the net advance just a few drive cycles earlier or later. Thus, like a drunkard, the BB tends to "forget" what it was doing, so we can't be too surprised that its motion mimics a random walk. On the other hand, the lack of memory remains a mystery given that equations of motion permit perfect knowledge of both past and future. The essential paradox remains unresolved.

Table 8.2   The standard deviation σ_b of the angle of a chaotic driven pendulum as a function of the time t in drive periods, based on a collection of 100,000 trajectories. The ratio $\sigma_b/\sqrt{t}$ is nearly constant, indicating that σ_b increases in proportion to $\sqrt{t}$.

Time	Deviation	Ratio
t	σ_b	$\sigma_b/\sqrt{t}$
25	9.812	1.962
50	13.859	1.960
100	19.680	1.968

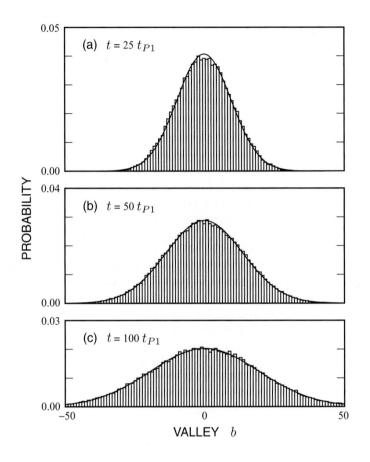

Fig. 8.5 The probability distribution of the angle of a sinusoidally driven chaotic pendulum after 25, 50, and 100 drive periods, based on a numerical experiment including $100,000$ trajectories like those illustrated in Fig. 8.1. To insure that these trajectories represent steady-state motion, they were initiated at $t = -30t_{P1}$ with $\tilde{v} = 0$ and angles θ distributed uniformly between $-\pi$ and π. The initial transients between $t = -30t_{P1}$ and 0 were computed and discarded. At $t = 0$, the angle of each trajectory was shifted by a multiple of 2π, as needed to return it to the central valley of the washboard, and the subsequent trajectory was calculated from $t = 0$ to $100t_{P1}$. The location of each BB was recorded at the end of 25, 50, and 100 drive periods, and this information was used to estimate the probability of occupying each valley. The probability of occupation is taken to be the number of BBs found in a valley divided by the total number of BBs. The parameters of the pendulum are $\omega = 0.8$, $\tilde{\tau}_1 = 1.6$, and $\rho = 0.1$. Bell curves with the same area and standard deviation are plotted for each distribution.

8.6   In search of true randomness

Having introduced probability in this chapter, it seems appropriate to comment on the mathematical ideal of randomness. As noted previously, true randomness is usually taken to imply both "jumbled" in the sense of statistically random and also "unpredictable." Chaotic motion is jumbled, even by statistical tests, but in principle it's completely predictable. On the other hand, mathematical probability is usually tied to ideal randomness, including complete unpredictability. This makes probability something we can manipulate with precision. For example, our derivation of the binomial distribution involved no approximation, following precisely from definitions and explicit assumptions.

Keeping the ideal nature of probability in mind, we now ask a simple question. Can we use a computer to simulate a process that truly follows the binomial distribution?

In Experiment 12 of the Dynamics Lab, the binomial distribution is simulated by applying Newton's laws to a penny falling through a quincunx. Because the penny bounces many times in passing through each row of pegs, it approximates the ideal in which steps to the left and right are of equal probability and independent of previous steps. On the other hand, the penny's motion is actually predictable, and, as we'll see later, the quincunx is another example of a chaotic system. Clearly, Experiment 12 is not based on a truly random process.

What about a more direct approach? All we really need is a source of truly random lefts and rights that will decide which way our drunkard steps at each point in time. Then we'll have a process that follows a true binomial distribution. Most programming languages include a random-number generator that should solve our problem. Typically such generators produce a sequence of numbers evenly distributed on the internal 0 to 1, so our drunkard can be programmed to step left when the next number in the sequence is less than 0.5 and right when it's larger.

The problem is that computer-based random-number generators don't generate truly random numbers. Given a seed number, they apply a simple algorithm to generate a random number that in turn serves as the seed for the next random number. It's a completely predictable process and the numbers generated are only pseudorandom. Indeed, at the heart of a random-number generator one usually finds a simple chaotic process. Computers are simply not made to create truly random events.

We conclude that if truly random numbers are to be found, they'll be found in the real world. Of course, if Laplace were right, everything in the real world is also predictable in principle, in which case sequences of truly random numbers would be scarce indeed. We'll return to this topic in Chapter 14, but suffice it to say for now that true randomness is not easy to find.

Further reading

Brownian motion

○ Bernstein, J., "Do atoms exist?", in *Secrets of the Old One: Einstein, 1905* (Springer Copernicus, 2006) chapter 3.

○ Haw, M., "Einstein's random walk", *Physics World* **18**(1), 19–22 (January 2005).

○ Haw, M., *Middle World: The Restless Heart of Matter and Life* (Macmillan, London, 2007).

○ Hersh, R. and Griego, R. J., "Brownian motion and potential theory", *Scientific American* **220**(3), 67–74, (March 1969).

● Klafter, J., Shlesinger, M. F., and Zumofen, G., "Beyond Brownian motion", *Physics Today* **49**(2), 33–39 (February 1996).

○ Lavenda, B. H., "Brownian motion", *Scientific American* **252**(2), 70–85 (February 1985).

○ Peterson, I., "Trails of a wanderer", in *The Jungles of Randomness: A Mathematical Safari* (Wiley, New York, 1998) chapter 8.

○ Slade, G., "Random walks", *American Scientist* **84**, 146–153 (1996).

Cardano

○ Ekert, A., "The gambling scholar", *Physics World* **22**(5), 36–40 (May 2009).

○ Ore, O., *Cardano: The Gambling Scholar* (Princeton University Press, Princeton, 1953).

Pascal's triangle

○ Edwards, A. W. F., *Pascal's Arithmetical Triangle: The Story of a Mathematical Idea* (Johns Hopkins University Press, Baltimore, 2002).

- Gardner, M., "Pascal's triangle", in *Mathematical Carnival* (Knopf, New York, 1975) chapter 15.
- Schumer, P. D., "Pascal potpourri", in *Mathematical Journeys* (Wiley, Hoboken, 2004) chapter 16.

Probability

- Aczel, A. D., *Chance: A Guide to Gambling, Love, the Stock Market, and Just About Everything Else* (Thunder's Mouth Press, New York, 2004).
- Ayer, A. J., "Chance", *Scientific American* **213**(4), 44–54 (October 1965).
- David, F. N., *Games, Gods, and Gambling: A History of Probability and Statistical Ideas* (Charles Griffin, London, 1962).
- Kac, M., "Probability", *Scientific American* **211**(3), 99–108 (September 1964).
- Packel, E., *The Mathematics of Games and Gambling* (Mathematical Association of America, Washington, 1981).
- Weaver, W., *Lady Luck: The Theory of Probability* (Anchor Doubleday, Garden City, New York, 1963).

Randomness

- Beltrami, E., *What Is Random? Chance and Order in Mathematics and Life* (Copernicus Springer-Verlag, New York, 1999).
- Bennett, D. J., *Randomness* (Harvard University Press, Cambridge, 1998).
- Casti, J., "Truly, madly, randomly", *New Scientist* **155**(2096), 32–35 (23 August 1997).
- Chaitin, G., *Meta Math! The Quest for Omega* (Pantheon, New York, 2005).
- Ford, J., "How random is a coin toss?", *Physics Today* **36**(4), 40–47 (April 1983).
- Peterson, I., *The Jungles of Randomness: A Mathematical Safari* (Wiley, New York, 1998).
- Kac, M., "What is random?", *American Scientist* **71**, 405–406 (1983).
- Poincaré, H., "Chance", in *Science and Method* (Dover, New York, 1952) chapter 4.
- Mackenzie, D., "On a roll", *New Scientist* **164**(2211), 44–47 (6 November 1999).
- Stewart, I., "In the lap of the gods", *New Scientist*, **183**(2466), 28–33 (25 September 2004).

9

Chaos makes noise

In preceding chapters, we established the random character of chaotic motion in a driven pendulum, showing that the pendulum's rotation isn't correlated over more than a few drive cycles and that its net advance mimics a random walk. Now we consider yet another sign of randomness in the chaotic pendulum: the presence of noise. Of course, a well-oiled pendulum is nearly silent, whether it's chaotic or not, so the noise we're interested in isn't a sound as such. But it's so closely related to the kind of sound we call noise that we'll end up listening to it.

It's possible to turn just about any oscillatory function into sound, simply by shifting the oscillations into the audible range of frequencies. Biologists, for example, make the subsonic sounds of whales audible to humans by playing their tapes at a speed faster than they were recorded. For a driven pendulum, the oscillations of interest come from the pendulum's velocity. Plotted as a function of time in Fig. 9.1, we show velocity waveforms for cases in which (a) the motion has the same period as the sinusoidal drive, (b) the period is twice that of the drive, and (c) the motion is chaotic and never repeats. All three of these waveforms have an oscillatory character and can be heard as sound if converted to a pressure wave in the audible range. In this chapter, we'll compare the sound equivalents of these oscillations.

Why do we want to listen to a waveform rather than simply inspect its graph? As it happens, our ears can perceive something that our eyes can't. Our ears naturally interpret sounds in terms of their pitch or frequency. Almost anyone can tell whether a series of musical tones is ascending or descending in frequency, and a lucky few, those with perfect pitch, can even identify the precise notes being played. While frequency analysis of this kind can also be done mathematically, our ears perform the job without effort. This ability is a truly amazing property of human hearing, and one that helps us understand the related mathematical process. Here, we'll look at frequency analysis from two directions, using both mathematics and our ears.

9.1   Beethoven's Fifth

To get started with sound, we consider the first four notes of Beethoven's Fifth Symphony. During the Second World War, these notes attained a certain fame derived from Winston Churchill's habit of holding up his fingers to form a "V" for "Victory." As it happens, the first notes

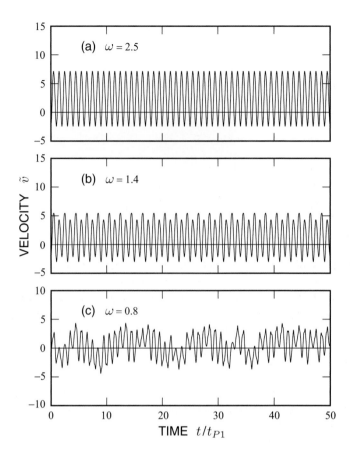

Fig. 9.1 The velocity $\tilde{v}$ of a driven pendulum plotted as a function of time for three cases: (a) $\omega = 2.5$ and $\tilde{\tau}_1 = 12$, (b) $\omega = 1.4$ and $\tilde{\tau}_1 = 4.9$, and (c) $\omega = 0.8$ and $\tilde{\tau} = 1.6$. In each case the damping is $\rho = 0.1$.

of the Fifth Symphony have a rhythm, da-da-da-daaaa, that mimics the Morse code for the letter "V", dot-dot-dot-dash. Throughout the war, the British Broadcasting Corporation thus played Beethoven's four-note motif in each of it's European broadcasts to symbolize inevitable triumph.

Famous as it is, we'll use Beethoven's motif as an introduction to frequency analysis. Figure 9.2 shows both the score of the notes opening the Fifth Symphony and the pressure wave produced when these notes are played by an orchestra. The da-da-da-daaaa motif is three eighth notes, G's, followed by a half note, an E♭. While the notes are simple enough, when played by an orchestra they become a complex sound waveform. Figure 9.2 shows the burst of sound associated with each note. The notes are played in about 1.5 second, but the sound persists as a reverberation that lasts another 1.5 second.

To reveal the true complexity of this sound, the amplitude of the pressure wave is plotted on an expanded time scale in Fig. 9.3. Here, the three second sound wave is broken into 15 traces, each 0.2 second long, and plotted as successive lines in a single graph. As it happens, each eighth note lasts about 0.2 second, so the first three lines of Fig. 9.3 correspond to the three G's of the motif. The E♭ nominally lasts about

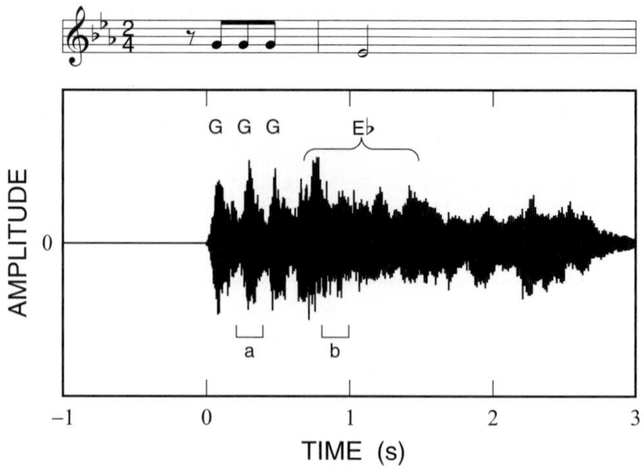

130 *Chaos makes noise*

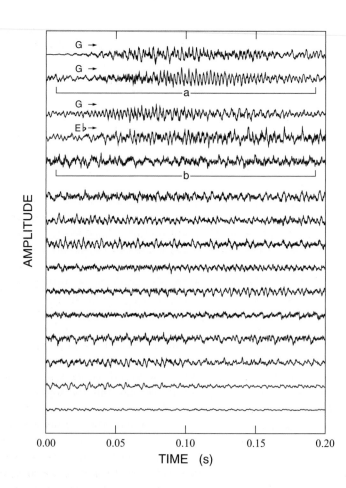

Fig. 9.2 The first four notes of Beethoven's Fifth Symphony as they appear in the score and as the sound wave produced by an orchestra. The intervals a and b were chosen for Fourier analysis.

Fig. 9.3 The opening motif of Beethoven's Fifth on an expanded time scale. Here, three seconds of sound are broken into 15 traces, each 0.2 second long, plotted in order from top to bottom. The intervals a and b were chosen for Fourier analysis.

0.8 second and corresponds to the following four lines. The remaining lines are devoted to reverberation.

There are two striking features about Fig. 9.3. First, we should be impressed by the fact that a sound can be represented by a time function $F(t)$ and plotted as we would any other function. This makes sense, of course, since a microphone merely records the pressure wave at its location, and the pressure has just one value at any given time. On the other hand, we might naively ask, where are all the various instruments—the strings, woodwinds, horns, etc.—to be found in this $F(t)$. Isn't there something more to sound? No, no matter how many musicians are playing, at any given instant the pressure at your ear drum has a single specific value and that is all there is to sound. A typical compact disc includes pressure amplitudes recorded 44,100 times during each second of sound, but that's all—just a single $F(t)$.[1] We can think of the entire Fifth Symphony as a single, super-long, time function.

The second remarkable feature of the function shown in Fig. 9.3 is that it's meaning as sound is completely hidden. No matter how many times you've listened to the Fifth Symphony, you'd be unlikely to guess that Fig. 9.3 represents Beethoven's famous motif. On the other hand, if this function is used to drive a speaker, it's immediately recognized by anyone familiar with the symphony. Indeed, with a good ear, you might even be able to guess the instrumentation. Clearly, Fig. 9.3 means a lot more to the ear than the eye.

Our perception of sound is a complicated phenomenon, involving the propagation of sound waves in the ear's cochlea, nerve impulses generated by hair cells spaced along the cochlea, and massively parallel signal processing in the brain. Nonetheless, one feature is certain: hair cells at different positions in the cochlea respond to different frequencies of sound. Thus, it's no wonder that people are good at distinguishing the frequencies of musical notes, even when the waveforms are as messy as those in Fig. 9.3. The brain receives frequency information directly from the ear.

Visually, the three G's recorded in the first three lines of Fig. 9.3 are not particularly similar, even though our ears hear the same note repeated three times. Clearly, the essential frequency information is not perceived as easily by the eye as the ear. How can we ferret out this information mathematically?

9.2 Fourier

The mathematical procedure for discovering the frequency components hidden in a waveform is known as Fourier analysis, after the French scientist Joseph Fourier (1768–1830). Unlike Galileo and Newton, who made their great discoveries in dynamics while in their early twenties, Fourier was nearly 40 by the time he produced his scientific masterpiece, *The Analytical Theory of Heat*, in 1807. Although you may wonder

[1] Okay, you're right, two functions are required for stereo.

what heat has to do with frequency analysis, let's look first at Fourier's adventures before he turned to thermal physics.

The son of a master tailor, Fourier was born in Auxerre, France but orphaned just before his tenth birthday. Because he was a bright lad, the citizens of Auxerre arranged for Fourier's education at the local Royal Military School, run by Benedictine monks and normally reserved for the children of the nobility. At age 13, Fourier fell in love with mathematics and, while at the military school, mastered all six volumes of Bézout's *Course of Mathematics*. Fourier completed his schooling at the Collège Montaigu in Paris, graduating at age 17.

Two career paths were now open to Fourier: the military and the priesthood. He first applied to enter the artillery or the army engineers. Although his application was supported by Adrien-Marie Legendre (1752–1833), a famous mathematician, it was rejected because he was not of noble birth. Thus, in 1787 Fourier entered the Benedictine abbey at St. Benoit-sur-Loire as a novice.

Two years later, as Fourier prepared to take the vows of priesthood, his life was overtaken by the events of history. In that year, 1789, the French Revolution began and with it the process that replaced the monarchy by a republic. Most important to Fourier, by order of the new Constituent Assembly, religious vows were summarily prohibited. Fourier, now 21, gave up his bid to become a priest and left the abbey to seek a career in mathematics. In December of 1789, he traveled to Paris and presented a paper on algebraic equations at the Royal Academy of Science, his first original contribution. Early the next year, he returned to Auxerre to teach rhetoric and mathematics at his old military school.

While at Auxerre, Fourier was attracted by the ideals of the Revolution—liberty, equality, fraternity—and by the promised renaissance in science and culture. In 1793 he joined the local Revolutionary Committee, but later spoke out against the Reign of Terror, in which thousands of opponents to the Revolution were guillotined without trial. For speaking out, Fourier was imprisoned in July of 1794 and might himself have fallen to the guillotine. Fortunately, our hero was released the following month, after Robespierre, architect of the Terror, was executed.

In December of 1794 Fourier was nominated to study at the newly formed École Normale in Paris. While the École Normale would last only a few months, Fourier so impressed his teachers that, after the school was dissolved, he was appointed a lecturer in mathematics at the École Polytechnique. The same year, 1795, Fourier was again imprisoned, now on charges of abetting the Terror rather than opposing it, but he was soon released. Fourier taught at the École Polytechnique for three years, and in 1797 succeeded Joseph-Louis Lagrange (1736–1813) as professor of analysis and mechanics.

In ordinary times, Fourier might have spent the remainder of his career as a professor at the Polytechnique, but the Revolution continued to interfere. In 1798, Fourier was tapped by Napoléon Bonaparte

(1769–1821) to join a military and scientific expedition to Egypt. The expedition was successful in that it conquered Egypt for the French and collected many artifacts of the ancient Egyptian civilization. Fourier's role was that of scientist and administrator. He proved capable in both realms, participating in archaeological explorations, collating the scientific discoveries, and successfully concluding the many delicate negotiations required to maintain the expedition's tenuous military position. He returned to France three years later in 1801, when Egypt was handed over to the British.

Although Fourier briefly resumed teaching at the Polytechnique, in 1802 Napoléon, then First Consul and soon to be Emperor, appointed Fourier Prefect of the Department of Isère, the administrative district surrounding Grenoble. At 34, Fourier threw himself into the task of governing and again demonstrated his facility for administration and the art of gentle persuasion that kept the prefecture running smoothly for the next 13 years. During his years in Grenoble, Fourier was also heavily involved in the production of a multi-volume work entitled *Description of Egypt* which presented the scientific results of the Egyptian expedition. But wearing the hats of prefect and egyptologist was not enough for Fourier, who still regarded himself as a mathematician. Thus, it was at Grenoble that Fourier also produced his great work on heat flow.

9.3   Frequency analysis

The mathematical problem of heat flow had been attacked earlier by Jean-Baptiste Biot (1774–1862) but in a limited way. Inspired by Biot's paper of 1804, Fourier plunged headlong into the problem, repeating previous experiments and deriving the now classic equation for the general case. As he discovered, heat flow is governed by a type of equation, known as a partial differential equation, then at the forefront of mathematical research. In the following century, partial differential equations would be found to describe everything from fluid flow (the Navier–Stokes equation) to electromagnetic fields (Maxwell's equations) to the curvature of spacetime (Einstein's equation of general relativity) to the quantum-mechanical motion of subatomic particles (the Schrödinger equation). While Fourier's equation for heat flow is simple, even linear, in his day the mathematical solution of simple partial differential equations was far from obvious. Fourier's triumph was to find a general method of solution that applied to many such equations. Later in the century, the physicist James Clerk Maxwell would describe Fourier's work as a "great mathematical poem."

At the heart of Fourier's method is the idea that an arbitrary function $F(t)$, defined on an interval $0 < t < T$, can be written as a constant plus the sum of an infinite number of sinusoidal functions with periods of T, $T/2$, $T/3$, etc. For example, we can write,

$$F(t) = a_0 + a_1 \cos(2\pi t/T + \phi_1) + a_2 \cos(4\pi t/T + \phi_2)$$
$$+ a_3 \cos(6\pi t/T + \phi_3) + \cdots$$
$$= a_0 + \sum_{m=1}^{\infty} a_m \cos(m2\pi t/T + \phi_m), \tag{9.1}$$

where the constants a_0, a_1, a_2, $a_3, \ldots$ and ϕ_1, ϕ_2, $\phi_3, \ldots$ can be selected to duplicate any function $F(t)$. This is an astonishing statement. Somehow, adding together a set of oscillatory functions with amplitudes set by the coefficients a_m and time offsets set by the angles ϕ_m can produce functions of any kind, oscillatory or not. Although the Swiss mathematician Daniel Bernoulli (1700–1782) introduced trigonometric expansions of this type, Fourier was the first to understand their universal nature and devise a general method for calculating the constants a_m and ϕ_m. Appropriately, such an expansion is now known as a Fourier series.

What does all this have to do with frequency analysis? While Fourier introduced his series in the context of solving partial differential equations, it's precisely what we need to determine the frequency components found in anything from music to the motion of a chaotic pendulum. Because $\cos(x)$ has a periodicity of 2π, the mth term in Eq. (9.1) has a time periodicity of T/m and a frequency of $f_m = m/T$. Thus, Eq. (9.1) breaks the function $F(t)$ into a sum of sinusoids having frequencies ranging from $1/T$ to infinity. This decomposition allows us to identify the frequencies that contribute to $F(t)$ through the amplitudes a_m— the larger $|a_m|$, the more prominent the component at frequency f_m in $F(t)$. Given Fourier's method of calculating the a_m, we can analyze the frequency content of any waveform.

As an example, consider the square wave plotted in Fig. 9.4. The square wave is often used as a diagnostic tool by electrical engineers precisely because it includes sinusoidal components spanning a wide range of frequencies. With Fourier analysis, we have a way to determine what those components are. Because the square wave is periodic, we need only apply Fourier's method to one period of the wave, say from 0 to T, to determine the amplitudes and phases of its sinusoidal components. The result is

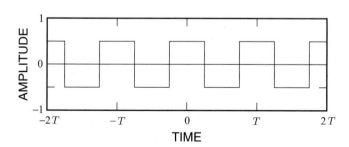

Fig. 9.4 A square wave.

$$a_m = \begin{cases} 0 & \text{if } m \text{ is even} \\ (-1)^{(m-1)/2} \frac{2}{m\pi} & \text{if } m \text{ is odd} \end{cases}, \tag{9.2}$$

$$\phi_m = 0, \tag{9.3}$$

so our square wave can be written in terms of sinusoids as

$$F(t) = \frac{2}{\pi} \cos(2\pi t/T) - \frac{2}{3\pi} \cos(6\pi t/T) + \frac{2}{5\pi} \cos(10\pi t/T) - \cdots . \tag{9.4}$$

Because all of the cosines making up $F(t)$ are periodic, this function reproduces the square wave over all time.

Do all the cosine functions in Eq. (9.4) really add up to a square wave? Figure 9.5 shows how this Fourier series works. In Fig. 9.5(a) we see the frequency components for $m = 1$, 3, 5, 7, and 9 plotted individually, and in Fig. 9.5(b) their sum is compared with the square wave. As Fourier would have predicted, the sum approximates a square wave, but, because we've omitted the frequency components for $m = 11$ and above, the agreement is crude. In fact, Fig. 9.5(b) illustrates why electrical engineers use square waves to test amplifiers. If an amplifier fails to transmit high frequencies, then applying a square wave at the input yields the sort of oscillatory output produced by a partial sum of

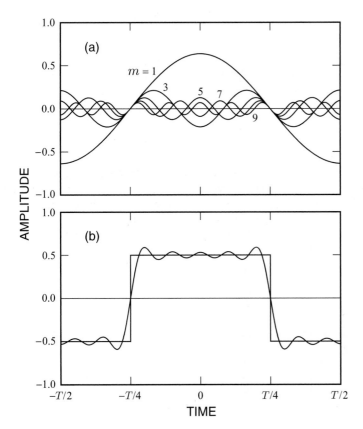

Fig. 9.5 The Fourier decomposition of a square wave. Frame (a) shows the first five non-zero sinusoidal components plotted individually, and frame (b) shows how their sum compares with a square wave.

the Fourier series. A quick glance at the output tells an engineer how well an amplifier is working.

For our present discussion, the information of greatest interest is the set of Fourier amplitudes a_m. These amplitudes tell us how much of each frequency is present in a given waveform. Traditionally, scientists plot the square amplitude a_m^2 as a function of frequency to create a "power" spectrum of a waveform. The power spectrum of the square wave is shown in Fig. 9.6. As we might expect, the power of the $m = 1$ component, the sinusoid with the same frequency as the square wave, is by far the largest. Although the higher-frequency sinusoids appear with much less power, they are essential to produce the square corners of the square wave. Nevertheless, it's remarkable that a square wave can be thought of as a sum of sinusoids at all.

In our discussion of frequency analysis, we haven't explained exactly how the Fourier coefficients a_m and angles ϕ_m are calculated for a given waveform. However, I hope you can see the possibility of representing an arbitrary function in terms of sinusoids, just as Fourier saw it two centuries ago. Unfortunately for Fourier, not all of his contemporaries were willing to trust him on this one. When his treatise on the subject was presented before the Institute of France in December of 1807, there were immediate objections. Laplace and Lagrange both worried that the convergence of Fourier's series was not rigorously proven. In hindsight, Fourier's work was of a genre that had yet to be established, namely mathematical physics. The success of a physical theory is contingent on agreement with experiment, not formal proof. Sadly for Fourier, the convergence of his series would not be fully understood until a century after its invention.

As a consequence of the objections of Laplace and Lagrange, the publication of Fourier's treatise on heat flow was delayed. In 1811, Fourier revised and resubmitted his work in response to a prize competition announced by the Institute of France. Although he was awarded the prize by a committee including Laplace and Lagrange, publication of his work was again delayed. However, in 1822 Fourier was elected Secretary of

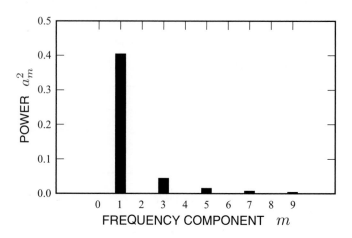

Fig. 9.6 Power spectrum of the square wave shown in Fig. 9.4.

the Academy of Sciences, and shortly thereafter the Academy published his *Analytical Theory of Heat*. Fourier analysis was finally announced to the world.

9.4   Music to the ear

During the years 1804–1807, when Fourier first investigated trigonometric series, Ludwig van Beethoven (1770–1827) was simultaneously composing his Fifth Symphony. Thus, it seems especially fitting to apply Fourier's mathematics to Beethoven's music. We now ask, what frequencies are found in the G and E♭ of the Fifth's famous motif, and why do these notes sound as they do to our ears?

To answer these questions, we apply Fourier analysis to the intervals labeled "a" and "b" in the sound wave plotted in Figs. 9.2 and 9.3. Interval a includes most of the second G of the motif, and interval b is a slice of the same length taken from the E♭. The frequency content of these time slices can be determined by applying Fourier analysis to the respective $F(t)$'s, assuming that each interval defines a range of time from 0 to T. The Fourier series will duplicate the sound wave only in the chosen time slice, and the coefficients a_m will tell us what frequencies are present during that interval.

From the score, we might expect the frequency spectra for time slices a and b to include strong components at 392.00 Hz and 311.13 Hz respectively, since these are the nominal frequencies of the G and E♭ above middle C. The actual spectra, shown in Fig. 9.7, confirm this expectation but reveal other significant components. What's going on here? There are at least two complications. First, the G and E♭ are played by a total of five different instruments—violin, viola, cello, bass

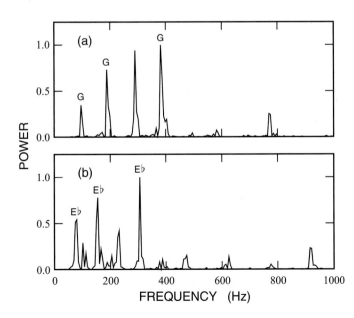

Fig. 9.7 Power spectra of the time slices a and b selected from the sound wave of Beethoven's motif, as indicated in Figs. 9.2 and 9.3.

violin, and clarinet—and some instruments play the notes one or two octaves below middle C. Thus, in Fig. 9.7(a), we find, in addition to the G at 392.00 Hz, frequency components at 196.00 Hz and 98.00 Hz, the G's one and two octaves lower. (Remember, two notes separated by an octave differ by a factor of 2 in frequency.) Second, musical instruments never produce pure sinusoids, but more complex waveforms that, like the square wave, include components at multiples of the fundamental frequency. This "harmonic content" gives each instrument its particular tonal quality. In Fig. 9.7(a), harmonic generation explains components such as that near 294 Hz, a frequency three times the G at 98 Hz. Similar considerations apply to the E♭ in Fig. 9.7(b), which includes instruments playing fundamental E♭'s at 311.13, 155.56, and 77.78 Hz.

The power spectra in Fig. 9.7 provide a very different picture of the sound wave plotted in Fig. 9.3. All the same, when supplemented with the angles ϕ_m, the component amplitudes plotted here allow the original slices of the waveform to be reconstructed exactly as functions of time. Since Fourier, scientists have made good use of this equivalence of information between $F(t)$ in the "time domain" and a_m and ϕ_m in the "frequency domain." Switching between these two domains often yields new insights.

In the case of sound, the frequency domain has a distinct advantage because the hair cells of the cochlea respond to specific frequencies. Thus, while it might be difficult to imagine the exact sound of the orchestral G from the power spectrum of Fig. 9.7(a), the spectrum directly indicates which frequencies to expect. Indeed, given a frequency map of the cochlea, we could determine which hair cells are stimulated by Beethoven's orchestral G. In this sense, our ears naturally speak the language of Fourier analysis.

9.5   White noise

Having learned the basics of frequency analysis, we now return to the velocity waveforms of the pendulum shown in Fig. 9.1. What are the sounds of these oscillatory functions? As usual, we apply Fourier's technique to discover the hidden frequency components and plot the result as a power spectrum.

Recall that Fig. 9.1 traces the behavior of the pendulum as the drive frequency f_1 is lowered from above to below the pendulum's natural oscillation frequency f_0, and the steady-state motion of the pendulum changes from periodic to chaotic. In Fig. 9.1(a) $f_1/f_0 = \omega = 2.5$ and the motion has a period equal to the drive period; in (b) $f_1/f_0 = 1.4$ and the motion's period is twice that of the drive; and in (c) $f_1/f_0 = 0.8$ and the motion is chaotic.

These changes are reflected perfectly by the power spectra plotted in Fig. 9.8. Consider first the spectrum of Fig. 9.8(a) for the case of motion with the same period as the sinusoidal drive. In this plot, we have adopted scales for both power and frequency in which the intervals between tic marks correspond to changes by a factor of 10. This strategy allows us to compress wide variations in power and frequency into a single graph. Note also that the component frequencies f of the Fourier series are normalized to the drive frequency f_1. Thus, $f/f_1 = 10^0 = 1$ corresponds to a component frequency equal to the drive frequency.

Figure 9.8(a) reveals a very strong Fourier component in the velocity waveform at $f = f_1$. The presence of this component isn't at all surprising because the motion has exactly the same period as the drive. Indeed, if the pendulum were linear, it's response to a sinusoidal drive would be an exact sinusoid at the same frequency, and Fourier analysis wouldn't reveal any other component. However, the pendulum isn't linear and the waveform shown in Fig. 9.1(a) departs slightly from a sinusoid. Thus,

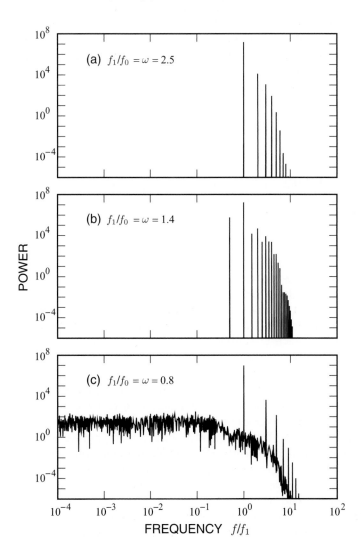

Fig. 9.8 Power spectra of the three velocity waveforms plotted in Fig. 9.1 for the driven pendulum: (a) $\omega = 2.5$ and $\tilde{\tau}_1 = 12$, (b) $\omega = 1.4$ and $\tilde{\tau}_1 = 4.9$, and (c) $\omega = 0.8$ and $\tilde{\tau} = 1.6$. In each case the damping is $\rho = 0.1$ and the spectrum is computed over a time span of 524,288 drive cycles.

as with the square wave, we find frequency components in the power spectrum at multiples of f_1. In particular, Fig. 9.8(a) includes Fourier components at 2, 3, 4, 5, 6, and 7 times f_1. Because of our compressed frequency scale, the spacing between these components doesn't appear to be uniform, but they are analogous to those of the square wave shown in Fig. 9.6. All the same, the waveform of Fig. 9.1(a) is nearly sinusoidal and so even the harmonic at $2f_1$ has a power only about one thousandth that of the fundamental. As far as this spectrum is concerned, the pendulum appears to be acting almost like a linear system.

In contrast, the spectrum shown in Fig. 9.8(b) is far from that expected for a linear system. The nonlinear element present here but not in Fig. 9.8(a) is the strong Fourier component at half the drive frequency, associated with period doubling and symmetry breaking. As discussed in Chapter 7, in this case, the pendulum breaks the time symmetry of the drive by adopting motion that is different on alternate drive cycles even though the drive is the same on each cycle. This leads to motion with a period twice that of the drive and thus to a frequency component $f = f_1/2$ at half the drive. Thus, while Fig. 9.8(b) resembles Fig. 9.8(a), it differs significantly because it reveals period doubling, which requires a strong nonlinearity. In addition to the Fourier component at $f_1/2$, period doubling leads to components at $3f_1/2$, $5f_1/2$, etc., that aren't present in Fig. 9.8(a).

Now we arrive at the spectrum for $f_1/f_0 = 0.8$ in Fig. 9.8(c), corresponding to the chaotic waveform of Fig. 9.1(c). To obtain this spectrum, we applied Fourier analysis to an extended waveform of 524,288 drive cycles rather than the 50 cycles plotted in Fig. 9.1. The extended time slice is of interest because the chaotic waveform never repeats—the broken symmetry now extends over infinitely many drive cycles—so we expect to find Fourier components at very low frequencies. Figure 9.8(c) confirms this expectation. As in previous spectra, the strongest component appears at the drive frequency $f = f_1$ and there are also strong components at odd multiples of f_1. However, chaos introduces a new ingredient to the spectrum. In addition to components at the discrete frequencies f_1, $3f_1$, $5f_1$, etc., there is a broadband component with more-or-less continuous power spanning all frequencies. This qualitative change in the spectrum is a direct consequence of chaotic behavior.

To an engineer, the broadband spectrum of Fig. 9.8(c) is usually understood to reveal the presence of noise, a waveform so mixed up that all frequencies are represented. Indeed, if we consider only frequencies below about $f_1/10$, all the frequencies in Fig. 9.8(c) appear with roughly the same power, and the corresponding noise is commonly called "white noise." Like white light, which usually includes equal contributions from all the colors (frequencies) of the rainbow, white noise results from equal power over a broad band of frequencies.

Does the waveform represented by Fig. 9.8(c) actually sound like noise to the ear? A direct answer to this question can be found in the "sound of chaos" file included on the companion CD. Simply double click on this file to hear the sound of a chaotic pendulum. The sound you'll hear is much the same as that from a radio tuned to a frequency between stations, a persistent hiss that provides little entertainment. So, in the long run, chaos is boring, even though it doesn't repeat. On the other hand, the white noise heard on your radio comes from many distant stations all added together. A small part can even be traced to the cosmic microwave background, an astronomical source of noise left over from the Big Bang. In contrast, our chaotic noise comes from the simple dynamics of a driven pendulum, completely described by Eqs. (7.1)–(7.3). Thus, while the sound of chaos might be boring, we can marvel that so many frequencies, so many possibilities, can come from such a simple system. Like an impossible number of circus clowns emerging from a tiny car, chaos is almost magic.

There is also a chance that the white noise generated by chaos can be put to use. Engineers often use white noise as a test signal, somewhat like a square wave, simply because it includes components spanning a broad range of frequencies. Currently, most noise generators are based on an intrinsically noisy electrical device, such as a diode. In this case, the noise is a product of the random microscopic motion of large numbers of electrons, and amplification is required to transfer the noise to the macroscopic domain. On the other hand, chaotic circuits can directly generate high levels of noise with an exceptionally flat and well-characterized power spectrum, features of interest in microwave metrology.

9.6   Random or correlated?

In this chapter and the previous two, we have looked at the randomness of chaotic motion from a statistical point of view. In particular, we showed that a chaotic pendulum exemplifies two common random phenomena: the random walk and white noise. Although this behavior is certainly surprising for such a simple dynamical system, we must remember that chaotic behavior is never completely random. Thus, while the pendulum's net rotation is largely uncorrelated between well separated drive cycles, the rotations on successive drive cycles are strongly correlated. Similarly, a chaotic pendulum exhibits a random walk over long times but not short ones, and it generates white noise at low but not high frequencies. In general then, the motion of a chaotic pendulum is correlated and predictable over short times, just as we expect from its equation of motion, and apparent randomness and unpredictability enter only in the long term. Of course, the pendulum's motion is always predictable in principle, but in practice, as we begin to explore in the next chapter, the predictability of chaotic motion is strictly limited to short times. There are good reasons why we find statistically random long-term behavior in this deterministic system.

Further reading

Fourier

- Grattan–Guinness, I., and Ravetz, J. R., *Joseph Fourier 1768–1830: A Survey of His Life and Work* (MIT Press, Cambridge, 1972).
- Herivel, J., *Joseph Fourier: The Man and the Physicist* (Clarendon Press, Oxford, 1975).
- Kahane, J.-P., "Joseph Fourier, 1768–1830", *Mathematical Intelligencer* **25**(4), 77–80 (Fall 2003).

Frequency analysis

- Gleason, A., translator, *Who is Fourier?: A Mathematical Adventure* (Language Research Foundation, Boston, 1995).

Noise

- Kautz, R. L., "Using chaos to generate white noise", *Journal of Applied Physics* **86**, 5794–5800 (1999).
- Kosko, B., *Noise* (Viking, New York, 2006).

Sound

- Benade, A. H., *Horns, Strings and Harmony* (Anchor Doubleday, Garden City, New York, 1960).
- Van Bergeijk, W. A., Pierce, J. R., and David, E. E., *Waves and the Ear* (Anchor Doubleday, Garden City, New York, 1960).
- Kock, W. E., *Sound Waves and Light Waves: The Fundamentals of Wave Motion* (Anchor Doubleday, Garden City, New York, 1965).
- Weinberger, N. M., "Music and the brain", *Scientific American* **291**(5), 88–95, (November 2004).

Part IV
Sensitive motion

Edward Lorenz—Butterfly effect

10

In 1959 Edward Lorenz (1917–2008), a mathematician and professor of meteorology at the Massachusetts Institute of Technology (MIT), set himself a challenge that would, by happenstance, shake the foundations of the science of weather prediction. While his ultimate goal was to test statistical prediction techniques, Lorenz's immediate job was to find a simple, nonlinear mathematical model of a weather system that would generate nonperiodic weather patterns. Of course, having explored the chaotic pendulum in previous chapters, we know exactly the kind of thing Lorenz was seeking. In 1959, however, the very existence of nonperiodic solutions to equations lacking any random element was highly suspect. Nonetheless, perhaps bolstered by the thought that turbulence must somehow be hidden in the Navier–Stokes equations, Lorenz boldly promised to deliver a paper on "The statistical prediction of solutions of dynamic equations" at a meteorology conference to be held the following year in Tokyo. Quite a gamble!

Fig. 10.1 Edward Lorenz, 1957. (Courtesy of Edward Lorenz.)

Fortunately, Lorenz had an ace in the hole: a primitive personal computer called a Royal–McBee LGP–30. This desk-sized behemoth was built from vacuum tubes and had a computing power comparable to a modern programmable pocket calculator. All the same, the Royal–McBee was capable of solving a set of weather equations involving a dozen or more variables, time step by time step, printing out a list of updated variables every 10 seconds. In principle, the calculation was just like ours in Chapter 2 for the motion of a falling ball using Eqs. (2.6)–(2.8), with output like that in Table 2.1.

The Royal–McBee was just the thing for exploring the nature of the weather patterns produced by simple models, and with some tweaking of parameters Lorenz soon discovered the nonperiodic patterns he sought. Indeed, the random nature of the output was so convincing that when colleagues gathered around to watch the evolving Royal–McBee printout, they couldn't resist placing small bets on what would happen next. Somehow, a dynamical system without explicit uncertainty was producing apparently unpredictable motion. Lorenz had found the system he needed to write his paper for Tokyo.

At some point in this work, Lorenz fatefully decided to repeat the last part of a calculation to obtain more details. He restarted the computer after typing in a set of intermediate values of the variables as read from the printout. Returning after a coffee break, Lorenz discovered that the weather now predicted by the computer was completely different from that obtained in the earlier run. What was going on? Had one of the vacuum tubes failed? On closer examination, Lorenz saw that the discrepancy between the two runs didn't occur immediately when the simulation was restarted but crept in gradually, affecting the least significant digits first, then higher order digits, and finally giving way to a completely different pattern of weather.

Soon, everything began to make a strange sort of sense. When Lorenz copied the variables from the printout, he had used values rounded to three decimal places. So, when the computer was restarted, it began with slightly different numbers than the original run. The computer hadn't malfunctioned; it had calculated an accurate trajectory for a slightly different set of initial values.

But why was the resulting weather pattern so completely different? In systems lacking friction, an initial offset usually persists forever, perhaps growing slowly with time, but with friction (or viscosity in the case of a fluid) an initial offset usually decays rapidly and is soon lost entirely. Given that viscosity was built into his weather model, Lorenz must have been surprised by this turn of events. Even so, he was confident in the accuracy of his calculation and quickly accepted the fact that the nonperiodic weather patterns he had calculated were highly unstable, being grossly affected by the slightest change in initial values.

More importantly, Lorenz soon realized that this instability might just explain why weather is so very difficult to predict. Meteorologists rely on measurements of temperature, pressure, wind speed, etc. taken at various points around the globe as initial values of the variables they want to predict. Since these measurements are seldom accurate to even three digits, any prediction they might make would be severely restricted if weather dynamics were as unstable as the model Lorenz had used. Judging this a likely possibility, Lorenz began to put the word out to the meteorology community and sounded the alert in his 1960 presentation in Tokyo. Although meteorologists were slow to accept it, Lorenz had made a great discovery that would change their science forever.

10.1   Lorenz equations

Skeptics initially discounted Lorenz's work because his weather model was excessively simple, involving a dozen variables rather than the millions required to insure accuracy. But Lorenz's instinct as a mathematician told him that an even simpler system was needed to understand the nature of the instability he had discovered. In 1961, he found a set of equations with just three variables, usually taken as x, y, and z, that display the same kind of unstable, nonperiodic behavior as his weather model. Now called the Lorenz equations, they can be written in our incremental notation as

$$x_{\mathrm{n}} = x_{\mathrm{o}} + \sigma(y_{\mathrm{o}} - x_{\mathrm{o}})\Delta t, \qquad (10.1)$$

$$y_{\mathrm{n}} = y_{\mathrm{o}} + (x_{\mathrm{o}}(r - z_{\mathrm{o}}) - y_{\mathrm{o}})\Delta t, \qquad (10.2)$$

$$z_{\mathrm{n}} = z_{\mathrm{o}} + (x_{\mathrm{o}}y_{\mathrm{o}} - bz_{\mathrm{o}})\Delta t, \qquad (10.3)$$

which express new values of the variables $(x_{\mathrm{n}}, y_{\mathrm{n}}, z_{\mathrm{n}})$ in terms of the old $(x_{\mathrm{o}}, y_{\mathrm{o}}, z_{\mathrm{o}})$ for a small time step Δt. This system of equations, with the parameters fixed at $\sigma = 10$, $r = 28$, and $b = 8/3$, has become a classic in the study of chaos.

Rather than modeling an extended weather system, the Lorenz equations apply to a single convection cell. Eagles and glider pilots are intimately familiar with the basic phenomenon of convection. When sunlight falls on a plowed field or an asphalt parking lot, the air above is preferentially heated and becomes more buoyant than the surrounding air. This results in a rising column of air, known as a thermal, that birds and gliders rely on for lift. At high altitudes, the warm air cools and falls to Earth again in the surrounding countryside, completing the convection loop.

The Lorenz equations model convection in which fluid is constrained to flow in a toroidal loop that is heated at the bottom and cooled at the top, as in Fig. 10.2. While the definitions of the variables x, y, and z are somewhat abstract (they're actually coefficients of a spatial Fourier expansion), x is roughly the velocity of the fluid as it rotates around the loop. Thus, from the plot of x versus time shown in Fig. 10.3, we see that, for the parameters chosen by Lorenz, the fluid is in constant motion, circulating first one way and then the other, without any obvious long-term pattern.

Why does the fluid rotate at all? Just as with a thermal updraft, the hot fluid is lighter than the cold, so in the absence of initial motion we have an unstable situation much like an inverted pendulum. The hot, buoyant fluid at the bottom of the loop would like to rise and displace the cold, dense fluid at the top, but it can't decide which way to move. Although the system might remain in balance for a while, the slightest force will start the fluid rotating in one direction or the other. Once started, the fluid gains momentum and continues to rotate in the same direction for some time, circulating around and around the loop. Because the fluid always slows as the heavier fluid approaches the top, with time the cold fluid gets colder and heavier while the hot fluid at the bottom simultaneously gets hotter and lighter. Thus, the fluid becomes less uniform and the variations in its velocity gradually increase, as seen over several intervals in Fig. 10.3. Eventually, however, the fluid approaches the balance point with a velocity near zero, and the pendulum is again forced to choose in which direction it will continue to rotate. Like the driven pendulum, the convection loop displays haphazard rotation that continues indefinitely, fed by the heat energy injected at the bottom of the loop.

Do the Lorenz equations exhibit the same instability as his earlier weather model? We can easily answer this question using the same

Fig. 10.2 The Lorenz equations model a convection cell in which fluid is constrained to flow in a toroidal tube that is heated at the bottom and cooled at the top.

test Lorenz first applied accidentally. That is, we can offset one of the variables from its original value and repeat the calculation to see how the result is affected. To get a clear picture of what's happening, we'll look at the net rotation w of the fluid rather than its velocity x. We can calculate w as a freebie just by adding the relation

$$w_{\mathrm{n}} = w_{\mathrm{o}} + x_{\mathrm{o}} \Delta t, \tag{10.4}$$

to the usual Lorenz equations (10.1)–(10.3). Since x_{o} is roughly the velocity, $x_{\mathrm{o}} \Delta t$ approximates the incremental rotation during time Δt.

The wide line in Fig. 10.4 plots $w(t)$ for the $x(t)$ shown in the first frame of Fig. 10.3. Comparing these figures, we see that the rotation increases when the velocity is positive and decreases when it's negative, as expected. Not coincidentally, the rotation of the convection cell seen here follows a rough random walk similar to those of the chaotic pendulum shown in Figs. 7.2 and 8.1. To investigate the stability of this motion, we assume that the $w(t)$ plotted in Fig. 10.4 as a wide line represents the unperturbed motion and now consider what happens if the initial velocity $x(0)$ is offset by a small amount x_{off}. This offset corresponds to the round-off error in Lorenz's accidental experiment. Perturbed trajectories $\tilde{w}(t)$ for various values of x_{off} are plotted by

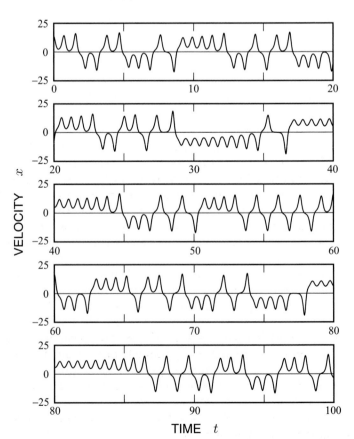

Fig. 10.3 Rotational velocity of the fluid in a convection loop according to the Lorenz equations for $\sigma = 10$, $r = 28$, and $b = 8/3$.

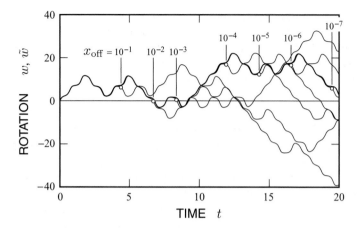

Fig. 10.4 Net rotation of the fluid in a convection loop for $\sigma = 10$, $r = 28$, and $b = 8/3$. The wide line corresponds to the same initial conditions as the trajectory shown in the first frame of Fig. 10.3 and is taken to be the unperturbed trajectory $w(t)$. The narrow lines result when the initial value of x is shifted by various offsets x_{off}. Each offset trajectory $\tilde{w}(t)$ is marked with a circle at the point where it first differs from $w(t)$ by 1 unit.

narrow lines in Fig. 10.4, and all of them eventually diverge radically from the unperturbed motion. This divergence is exactly what Lorenz first observed in his weather model and is a perfect example of extreme sensitivity to initial conditions, or the butterfly effect, as he so famously named it.

Each perturbed trajectory in Fig. 10.4 is marked by a circle at the point where it first differs from the unperturbed motion by 1 unit of rotation. Thus, we see that for $x_{\text{off}} = 10^{-1}$, the divergence it spawns is first perceptible at $t = 4.4$. We'll call 4.4 the growth time t_g of this offset, since it's the time required for the effect of x_{off} to grow from the domain of small quantities to the domain of perceptibly large quantities. An offset of 10^{-1} is not especially small, so we aren't too surprised that its affect becomes perceptible in a short time. Because typical values of the velocity are of order 10, an offset of 10^{-1} might be thought of as rounding off the second digit of x.

The surprise in Fig. 10.4 is the rapidity with which very small offsets become grossly perceptible. As expected, the smaller the offset, the longer it takes before the gross motion of the convection cell is affected. But the growth time t_g increases only slowly as x_{off} is reduced by successive factors of 10. Thus, for $x_{\text{off}} = 10^{-7}$, which corresponds to rounding off the ninth decimal digit of the initial velocity—a seemingly insignificant change—the motion is grossly affected after just $t_g = 19.5$ time units. The tiny offset must have grown very rapidly to change the motion so quickly. To say the least, the outcome is extremely sensitive to the initial velocity.

As Lorenz immediately realized, a trend like that shown in Fig. 10.4 has serious consequences for our ability to predict the future. If we think of the offset as an error in our measurement of the fluid's initial velocity, then the result for $x_{\text{off}} = 10^{-1}$ tells us that a measurement accurate to about 1% will allow us to predict convective motion for about a half dozen cycles of oscillation. At larger times the predicted motion will differ grossly from the actual motion.

To extend the period of prediction, we require a more accurate knowledge of the initial conditions. Suppose that we somehow take the care necessary to measure the initial velocity of the convection cell to 1 part in 100,000. This accuracy corresponds to an x_{off} of 10^{-4}, and in this case we see from Fig. 10.4 that $t_g = 11.9$, allowing the motion to be predicted over more than a dozen cycles of oscillation. However, the difficulty of making highly accurate measurements implies that longer-term predictions for the convection cell are virtually impossible. We can never expect to measure the initial velocity to 1 part in 100,000,000, as required to predict the motion for 19.5 time units. If weather in the real word behaves anything like Lorenz's model, the possibility of long-range weather forecasting will always remain a dream.

While Laplace never said predicting planetary motion was easy, he also would have been chagrined by Lorenz's discovery. To understand Laplace's expectations, let's suppose for a moment that the convection equations are well behaved. In this case, an offset in velocity might have the same effect as it does for a frictionless hockey puck. As noted in Chapter 2, the position D of a hockey puck that begins at $D = 0$ and travels at velocity V is $D = VT$ at time T. If the initial velocity is offset by V_{off}, then the perturbed position is $\tilde{D} = (V + V_{off})T$. In this case, $\tilde{D} - D = V_{off}T$, and the difference between $\tilde{D}$ and D grows in proportion to time as $V_{off}T$.

Translating the behavior of the hockey puck into Lorenz's convection system implies that the perturbed trajectory $\tilde{w}$ will diverge from the original according to $\tilde{w} - w = x_{off}t$. Or, since t_g is defined as the time at which $|\tilde{w} - w| = 1$, can we expect to find

$$t_g \approx 1/x_{off}? \tag{10.5}$$

Figure 10.4 shows that $t_g = 4.4$ for $x_{off} = 10^{-1}$, and Eq. (10.5) predicts $t_g = 10$, in rough agreement. On the other hand, for $x_{off} = 10^{-7}$, the prediction for t_g is 10^7 but we actually observe $t_g = 19.5$. If convection were as well behaved as the motion of a hockey puck or a typical planet, we could expect dynamical predictions to remain accurate far into the future, just as Laplace would have supposed. But Laplace didn't know about chaos, a phenomenon that upsets the apple cart of classical prediction.

Contrary to the regular motion of hockey pucks and planets, a deviation $|\tilde{w} - w|$ in the chaotic Lorenz system grows much faster than $x_{off}t$, and this rapid growth gives rise to extreme sensitivity to initial conditions.

10.2   Exponential growth

To get a mathematical handle on the butterfly effect, we can use the growth times in Fig. 10.4 to plot t_g as a function of the initial offset x_{off}. The result is shown in Fig. 10.5. Here t_g is seen to decrease with increasing x_{off} in an apparently linear way. However, you'll note that

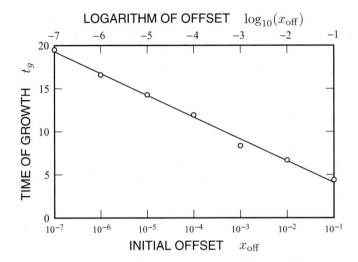

Fig. 10.5 Time t_g required for a small initial offset x_{off} to produce a perceptible divergence from the unperturbed trajectory of the Lorenz equations. The parameters are $\sigma = 10$, $r = 28$, and $b = 8/3$.

x_{off} is plotted on a scale that gives each factor of 10 the same space. This type of scale is said to be logarithmic, and, as we'll see shortly, the linear behavior recorded in Fig. 10.5 implies that the effect of an initial offset grows exponentially with time. This exponential growth is key to the butterfly effect and indeed chaos itself.

Before continuing our analysis of the Lorenz equations, let's take a quick look at the phenomenal phenomenon of exponential growth. A simple example comes from an often told tale about the game of chess. As legend has it, when the game was presented to the king by its inventor, the king was delighted and asked if the inventor would accept a reward. In reply, his subject asked to receive some wheat: enough to put one grain on the first square of the chess board, two grains on the second square, four on the third square, and so on, doubling the number on each successive square through all the squares of the board. The request seemed a modest one, and the king immediately ordered that it be fulfilled.

How much wheat had the king promised? Remembering that 2^n is just 2 multiplied by itself n times, we can calculate the total number N of grains on all 64 squares of the chess board as

$$N = 1 + 2^1 + 2^2 + 2^3 + 2^4 + \cdots + 2^{63}$$
$$= 2^{64} - 1. \tag{10.6}$$

The final result here may seem mysterious. However, if you start adding up the first few terms in the series, you'll discover that the sum at any point is just 1 less that the next term. For example, $1 + 2 + 4$ is 1 less than 8. So by induction the sum through 2^{63} must be 1 less than 2^{64}. Using the fact that 1,000 grains of wheat weigh about 40 grams, we can calculate that 2^{64} grains would weigh 700 billion metric tons, or more than a thousand times the world's annual production today. We can assume that the king never made good on his promise.

As the legendary inventor of chess understood, taking something tiny and doubling it over and over again soon leads to something very large. This is the power of exponential growth. As you know from algebra, x^2 grows faster than x, and x^3 grows faster than x^2, but, if you want very fast growth, it's far better to put x in the exponent. In fact, 2^x eventually outstrips even x^{100} when x is large enough.

Exponential growth may seem crazy, but we occasionally observe it in the world around us, at least for limited periods. For example, if you toss a single bacterium into a jar of nutrient, you'll find that it soon divides into two bacteria, which soon divide again to make four, and so forth. Of course, this process of repeated doubling stops when the nutrient is depleted, but for a while the growth is amazing. For example, the common intestinal bacterium *Escherichia coli* is a microscopic rod about 0.5 μm in diameter and 3 μm long that divides roughly every 30 minutes under optimal conditions. Placed in a liter jar of nutrient, a single *E. coli* would need to spawn roughly $10^{15} = 1,000,000,000,000,000$ bacteria of similar size to fill the jar, but that would happen after just 50 cycles of cell division and require only 25 hours. In the end, the jar would be chockablock with bacteria, all about to die for lack of nutrients.

As famously noted by Thomas Malthus more than 200 years ago, the human population also tends to increase exponentially, limited in the end only by the finite resources of the Earth. In fact, in recent centuries our population has grown even faster than exponentially. For strictly exponential growth, the population would always double in the same number of years. However, in round numbers the human population doubled between 1400 and 1750 (350 years), doubled again between 1750 and 1880 (130 years), doubled again between 1880 and 1960 (80 years), and doubled again between 1960 and 2000 (40 years). To say the least, this ever accelerating rate of growth cannot continue indefinitely. Indeed, many experts think that we are near the maximum population that can be sustained by the Earth. Humans are almost chockablock around the globe.

Not surprisingly, our population growth has driven exponential increases in the production of many commodities. In particular, world oil production doubled nine times during the years between 1880 and 1970, increasing by a factor of 2 every decade: a beautiful example of exponential growth. Since oil isn't a renewable resource, however, it will with certainty become scarce in the not too distant future.

Here's a final curious example of exponential growth. In 1965, Gordon Moore of Fairchild Semiconductor made a prediction that has proven to be prescient. Although integrated circuits were in their infancy, Moore predicted that the number of transistors on a chip could be expected to double at regular intervals. Sure enough, between 1970 and 1998, the capacity of memory chips known as DRAM (dynamic random access memory) increased from 1 kilobyte to 256 megabytes, doubling in capacity 18 times or roughly once every 18 months. Compared with any other technology, this is a phenomenal growth in performance, and it has suddenly landed us in the information age, with computers that are far less expensive and more powerful than the makers of the Royal–McBee could ever have imagined.

10.3   Exponential and logarithmic functions

The mathematics of exponential growth is so important to understanding chaos that we'll pause here to review the basics. We begin with the simplest properties of exponents and gradually expand our knowledge.

As you know from algebra, the quantity a^2 is by definition $a \times a$. Similarly, a^3 is $a \times a \times a$, and in general a^n is a multiplied by itself n times. From this definition, it's easy to see that

$$a^n a^m = a^{n+m}, \qquad (10.7)$$

because n factors of a multiplied by m factors of a gives a total of $n+m$ factors of a. Similarly, we have

$$(a^n)^m = a^{nm} \qquad (10.8)$$

since $(a^n)^m$ is a^n multiplied by itself m times, which gives a total of nm factors of a.

The quantity a^n is easy to understand when n is a positive integer, but what could we possibly mean by something like a^{-3}? Equation (10.7) offers a clue to this puzzle if we assume that it extends to negative exponents. Blindly applying this relation, we obtain $a^5 a^{-3} = a^{5-3} = a^2$. Thus, multiplying a^5 by a^{-3} has the effect of cancelling 3 of the factors of a in a^5. So a^{-3} must equal $1/a^3$. More generally we have

$$a^{-n} = \frac{1}{a^n}, \qquad (10.9)$$

so that a^{-n} is the reciprocal of a^n. Similarly, applying Eq. (10.7) to the case $m = -n$ suggests that $a^0 = a^n a^{-n} = a^n/a^n = 1$ for $a \neq 0$. Thus, the quantity a^n makes sense for any integer n, including 0.

Can we extend exponents to include fractions as well as integers? That is, does a quantity like $a^{1/2}$ make any sense? In this case, a clue is provided by Eq. (10.8). Assuming that this relation holds for fractional exponents, we have $(a^{1/2})^2 = a^{(1/2)2} = a^1 = a$. But this simply tells us that when $a^{1/2}$ is squared we get a itself, so $a^{1/2}$ must be the square root of a. Thus, it seems that even fractional exponents make sense! In fact, a little further thought leads to the general relation

$$a^{n/m} = (\sqrt[m]{a})^n, \qquad (10.10)$$

where $\sqrt[m]{a}$ is the mth root of a. Equation (10.10) allows us to handle any rational exponent: any number that is the ratio of two integers n/m.

Given that irrational numbers can be approximated as closely as desired by a ratio of integers, it's not surprising that a^x is defined for any real number x. However, computing a^x would be very laborious if we were forced to use Eq. (10.10) in every case. Fortunately, there is one particular base, $a = e = 2.71828\ldots$, for which exponentials are easily calculated. The magic formula is the infinite series

$$e^x = 1 + x + \frac{1}{2}x^2 + \frac{1}{6}x^3 + \cdots$$

$$= \sum_{n=0}^{\infty} \frac{x^n}{n!}, \tag{10.11}$$

which converges rapidly for any real number x.[1] What's so special about the number e? In large part, e is important simply because it's the one base for which exponentiation has a simple formula. Indeed, the function e^x is so important that it has acquired the alternate notation $\exp(x)$. As we'll see, any other exponential can be converted to one with base e, so Eq. (10.11) is our key to computing all exponentials.

Even though you may already use the e^x key on your calculator with abandon, it's useful to check out Eq. (10.11) by trying an example or two. You can see immediately that $e^0 = 1$, as expected, and evaluating a few terms of the sum $e^1 = 1 + 1 + 1/2! + 1/3! + \cdots$ reproduces e to high accuracy. Does $e^{-1} = 1 - 1 + 1/2! - 1/3! + 1/4! - \cdots$ really equal $1/e$? Try it and find out! The larger the magnitude of x, the more terms you'll need to calculate e^x accurately, but eventually Eq. (10.11) converges to the right number.

Inevitably, the use of exponential functions such as a^x leads us to the logarithmic function of base a, denoted $\log_a(x)$. The logarithm is the inverse of the exponential, and undoes whatever the exponential does. To be precise, we define $\log_a(x)$ by the relation,

$$x = a^{\log_a(x)}. \tag{10.12}$$

That is, $\log_a(x)$ is the power to which a must be taken to get x. From this definition, you can quickly confirm that $\log_2(8) = 3$ because $8 = 2^3$ and $\log_{10}(100) = 2$ because $100 = 10^2$. Since the exponential 2^x grows rapidly with increasing x, we can expect the logarithm $\log_2(x)$ to grow very slowly with x. Indeed, you can easily confirm that $\log_2(32) = 5$, $\log_2(1024) = 10$, and $\log_2(32,768) = 15$. The argument x has to be very large before $\log_2(x)$ amounts to much of anything.

The relation between the exponential and logarithmic functions is illustrated graphically in Fig. 10.6 for the base e. In (a) we show the modest initial climb of e^x for values of x up to 2, but for lack of space forgo the rapid climb to over 22,000 reached at $x = 10$. In (b) we plot $\log_e(x)$, also known as the natural logarithm or $\ln(x)$. By comparing the two plots, you'll discover that the curves for e^x and $\ln(x)$ are identical in shape but the abscissa and ordinate are interchanged in going between the two. This symmetry explains how rapid exponential growth transforms into slow logarithmic growth.

[1] Although beyond the scope of this book, Eq. (10.11) also defines exponentiation when x is a complex number or even a matrix of complex numbers. Curiously, Leonhard Euler (1707–1783), the Swiss mathematician who gave the numbers $2.71828\ldots$, $3.14159\ldots$, and $\sqrt{-1}$ the names e, π, and i, discovered an identity involving an imaginary exponent that links all three: It's $e^{i\pi} = -1$. The mind reels!

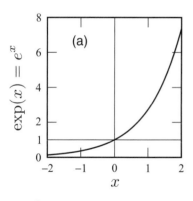

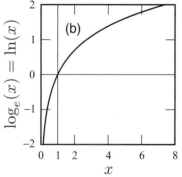

Fig. 10.6 The exponential (a) and logarithmic (b) functions for base e

Like the exponential function, the natural logarithm can be computed from a series expansion. In this case we have,

$$\ln(x) = 2\left(\frac{x-1}{x+1}\right) + \frac{2}{3}\left(\frac{x-1}{x+1}\right)^3 + \frac{2}{5}\left(\frac{x-1}{x+1}\right)^5 + \cdots$$

$$= \sum_{n=0}^{\infty} \frac{2}{2n+1}\left(\frac{x-1}{x+1}\right)^{2n+1}. \qquad (x>0) \qquad (10.13)$$

Again, you can check this formula by noting that it immediately yields $\ln(1) = 0$, and by using it to confirm that $\ln(e) = 1$ and $\ln(1/e) = -1$. Although the series for e^x and $\ln(x)$, Eqs. (10.11) and (10.13), may appear complicated, they explain what happens when you press these keys on your calculator: the process is hidden but it isn't mysterious.

What about the y^x key? How can we take an arbitrary base to an arbitrary power? From the definition of $\ln(y)$, we know that $y = e^{\ln(y)}$, so

$$y^x = \left(e^{\ln(y)}\right)^x = e^{x\ln(y)}. \qquad (10.14)$$

Thus, using both the natural logarithm and the natural exponential, we can compute any power of any number. Similarly, we can use the formula,

$$\log_y(x) = \frac{\ln(x)}{\ln(y)}, \qquad (10.15)$$

to compute the logarithm of any number to any base. (Can you derive Eq. (10.15) using the definition of the logarithm?) The curious number e is the key to all of these calculations.

10.4 Liapunov exponent

Now let's put exponents and logarithms to work understanding chaos. In Figs. 10.4 and 10.5 we discovered that a small offset in velocity eventually causes the perturbed motion of the Lorenz system to diverge radically from its original trajectory. The smaller the offset x_{off}, the longer the time t_g required for its effect to become obvious, and this time increases logarithmically with $1/x_{\mathrm{off}}$. Conversely, we infer that the difference between the perturbed motion $\tilde{w}(t)$ and the original trajectory $w(t)$ grows exponentially with time.

Exactly how quickly does the perturbed trajectory diverge from the original? Assuming for the moment that the difference $|\tilde{w}(t) - w(t)|$ increases exactly exponentially, we can estimate the growth rate from Fig. 10.4. For example, reducing x_{off} from 10^{-1} to 10^{-2} increases t_g from 4.40 to 6.71, so an extra 2.31 time units were required for the deviation to grow by an additional factor of 10. That's very fast growth if we consider that the system experiences only a couple of oscillations during two time units. Of course, this is only a rough estimate of the growth rate, so we must ask if growth continues at the same rate. Decreasing x_{off} from 10^{-2} to 10^{-3} increases t_g by only 1.65 units, so the next factor of 10 in growth occurs over an even shorter time. On the other hand, decreasing x_{off} from 10^{-3} to 10^{-4} increases t_g by 3.57 units, indicating slower growth for the next factor of 10.

The lesson of these calculations is clear. The time required for $|\tilde{w}(t) - w(t)|$ to increase by a factor of 10 is not always exactly the same, but it's typically a few time units. If we define t_{10} as the average time required for $|\tilde{w}(t) - w(t)|$ to grow by a factor of 10, then Fig. 10.4 tells us that t_{10} is roughly two or three time units. Can we be more precise? Yes, from the straight line fit to the data points in Fig. 10.5, we calculate that $t_{10} = 2.5$.[2] This means that, as long as $|\tilde{w}(t) - w(t)|$ is less than 1, we can expect it to increase by a factor of 10 every 2.5 time units on average. With this kind of growth, microscopic offsets can very quickly affect macroscopic motion.

[2]Special techniques can be used to calculate the growth rate even more accurately. Averaged over 10^9 time units, we find that $t_{10} = 2.54151$ for $\sigma = 10$, $r = 28$, and $b = 8/3$.

Now let's write an equation to capture the exponential growth of $|\tilde{w}(t) - w(t)|$. Since the growth isn't exactly exponential, we'll use the sign $\approx$ to indicate approximate equality. As you might guess, the correct exponential form is

$$|\tilde{w}(t) - w(t)| \approx C\, 10^{t/t_{10}}, \tag{10.16}$$

where C is a constant. Why? Because, according to the rules of exponents, the function $10^{t/t_{10}}$ increases by a factor of 10 every time t increases by t_{10}:

$$10^{(t+t_{10})/t_{10}} = 10^{1+t/t_{10}} = (10^1)(10^{t/t_{10}}) = 10 \times 10^{t/t_{10}}. \tag{10.17}$$

Of course, we could equally well have written

$$|\tilde{w}(t) - w(t)| \approx C\, 2^{t/t_2}, \tag{10.18}$$

where t_2 is the average time for the deviation to double, or

$$|\tilde{w}(t) - w(t)| \approx C\, e^{t/t_e}, \tag{10.19}$$

where t_e is the average time for $|\tilde{w}(t) - w(t)|$ to increase by a factor of e. Whether you use Eq. (10.16), (10.18), or (10.19) depends only on how you want to measure the rate of increase: by factors of 10, 2, or e. For the case at hand, we have $t_2 = t_{10}\log_{10}(2) = 0.765$ and $t_e = t_{10}\log_{10}(e) = 1.104$.[3] Thus, on average $|\tilde{w}(t) - w(t)|$ increases by a factor of 2 in 0.765 time units or by a factor of e in 1.104 time units.

[3]You can derive these formulas by equating $10^{t/t_{10}} = 2^{t/t_2} = e^{t/t_e}$, then substituting $10^{\log_{10}(2)}$ for 2 and $10^{\log_{10}(e)}$ for e. Give it a try!

If exponential growth still seems a bit confusing, now is the time to check out Experiment 13 of the Dynamics Lab. This exercise allows you to explore the effect of small offsets on the chaotic dynamics of a driven pendulum and compute its t_{10}.

No matter how you slice it, the exponential departure of a perturbed trajectory from the original indicates extreme instability. Historically, one of the first to explore dynamic stability was the Russian mathematician Aleksandr Liapunov (1857–1918), and in tribute to him the rate of exponential deviation is called the Liapunov exponent. Given the symbol λ, the Liapunov exponent is defined by growth of the form $e^{\lambda t}$.

From Eq. (10.19), we see that $\lambda = 1/t_e = 0.906$ for the system explored by Lorenz. The larger λ, the shorter the time required for growth by a factor of e and the more unstable the solution.

So far, we've only imagined how the microscopic separation between the trajectories in Fig. 10.4 grows with time. The difference $|\tilde{w}(t) - w(t)|$ is simply too small to be seen until it becomes macroscopic. Figure 10.7 fills this gap by plotting $|\tilde{w}(t) - w(t)|$ and $|\tilde{x}(t) - x(t)|$ on a logarithmic scale for the case $x_{\mathrm{off}} = 10^{-7}$. For times less than $t_g = 19.5$, the deviation $|\tilde{w}(t) - w(t)|$ is less than 1, and we observe growth roughly in proportion to $10^{t/t_{10}}$. This growth rate, indicated by a dashed line, is followed on average for $t < t_g$ not only by $|\tilde{w}(t) - w(t)|$ and $|\tilde{x}(t) - x(t)|$, as shown in Fig. 10.7, but also by the deviations in y and z. The perturbed trajectory departs exponentially from the

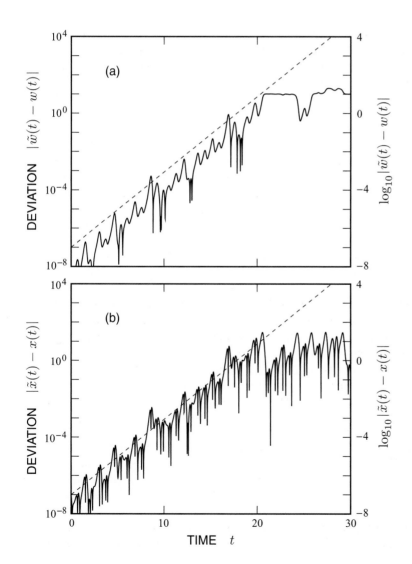

Fig. 10.7 The deviation in (a) rotation $|\tilde{w}(t) - w(t)|$ and (b) velocity $|\tilde{x}(t) - x(t)|$ as a function of time for the Lorenz system with an initial offset of $x_{\mathrm{off}} = 10^{-7}$. Dashed lines show a rate of growth proportional to $10^{t/t_{10}}$, with $t_{10} = 2.54$. The Lorenz parameters are $\sigma = 10$, $r = 28$, and $b = 8/3$.

original in all possible ways. The growth curves are very lumpy, but the exponential trend is clear at times less than t_g.

At longer times, when the deviation becomes macroscopic, the separation between the perturbed and original trajectories no longer grows at an exponential rate. This saturation is apparent in Fig. 10.7 for $t > 20$, and the effect is completely analogous to the limit on exponential growth experienced by bacteria in a jar of nutrient. For the assumed parameters (σ, r, b) the convecting fluid in the Lorenz system doesn't have enough oomph to rotate faster than about $|x| = 17$ (see Fig. 10.3.), so $\tilde{x}(t)$ and $x(t)$ can never differ by more than about 34. While the rotation w has no similar cap, once $|\tilde{w}(t) - w(t)|$ exceeds 1 or so, the two trajectories simply drift apart as separate random walks, and $|\tilde{w}(t) - w(t)|$ increases very slowly, in proportion to $\sqrt{t}$. Thus, although chaotic motion is strange, there are limits on how wild it can be. Just as the rapid growth of the human population on Earth will end someday, the exponential divergence of nearby chaotic trajectories can't continue forever.

In his 1963 paper on "Deterministic nonperiodic flow," Lorenz demonstrated that the random nature of motion in his convection system is inextricably linked to the exponential growth of small perturbations. You can't have one without the other. Chaos implies the butterfly effect and vice versa. In fact, we can conclude that the essential paradox of chaotic motion implies a second paradox. That is, the surprising idea that random motion can be produced by deterministic equations lacking explicit randomness, implies the existence of trajectories that are locally unstable but globally stable. The exponential divergence of two nearby trajectories is, as Liapunov well knew, a sure sign of instability, but once the divergence reaches macroscopic proportions, it suddenly stops, and the two trajectories adopt similar motion while remaining forever distinct. In other words, a chaotic system is globally stable in the sense that the long-term average of any quantity is the same for two trajectories with slightly different initial conditions, while the details of the trajectories may be completely different in the long-term. So we now have two paradoxes to resolve. Maybe understanding one will help us understand the other.

10.5   Exponential decay

To put the butterfly effect in perspective, let's compare the instability of chaotic motion in the Lorenz system with the stability of steady convective motion in the same system. As you might imagine, one way to eliminate chaos in a convection cell is to substitute a fluid with greater viscosity that's not as prone to wild oscillations. Mathematically, this corresponds to lowering Lorenz's parameter r, known to fluid dynamicists as the Reynolds number. Keeping σ and b fixed at 10 and 8/3, respectively, we find that convection changes from chaotic to steady when r is reduced from 28 to 10. In the latter case, the convection is that preferred by glider pilots: a steady circulation, perfect for climbing high in the sky.

For the parameter set $(\sigma, r, b) = (10, 10, 8/3)$, the Lorenz equations predict steady convection with a rotational velocity $x_s = \pm\sqrt{b(r-1)} = \pm 4.899$.[4] Here, the choice of sign reminds us that the fluid in our circular loop can flow in either direction, depending on how the symmetry of the loop is broken. To test the stability of this motion, we offset the velocity x slightly from its steady-state value x_s and solve Eqs. (10.1)–(10.3) by iteration to see how the system responds. As shown in Fig. 10.8(a), the perturbed velocity $\tilde{x}$ quickly approaches the steady-state x_s through a series of damped oscillations similar to those recorded for the damped pendulum in Fig. 5.6.

How fast does $\tilde{x}$ approach x_s? The answer to this question is provided by Fig. 10.8(b), which plots the deviation $|\tilde{x} - x_s|$ as a function of time on a logarithmic sale. As this figure shows, on average the deviation is reduced by a factor of $1/10$ when the time increases by 3.87 time units. That is, perturbations in the case of stable steady-state convection exhibit exponential decay, in contrast to the exponential growth of perturbations in a chaotic situation. In fact, for $(\sigma, r, b) = (10, 10, 8/3)$ any small deviation decays exponentially and steady flow is perfectly stable.

Exponential decay is the opposite of exponential growth and is described by the portion of the curve e^x plotted in Fig. 10.6(a) at

[4]You can discover this steady-state motion for yourself directly from Eqs. (10.1)–(10.3). For the new values of x, y, and z to be the same as the old values, the coefficient of Δt in each equation must be 0. That is, $y - x = 0$, $x(r - z) - y = 0$, and $xy - bz = 0$. Solving these equations for x, y, and z gives the condition for steady motion, although it doesn't tell us if the motion is stable.

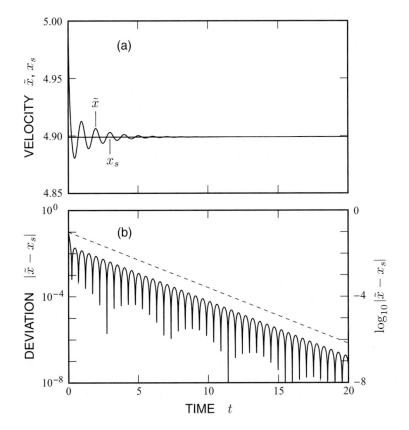

Fig. 10.8 (a) The perturbed rotational velocity $\tilde{x}$ as a function of time compared to the steady-state velocity $x_s = \sqrt{b(r-1)}$, for the Lorenz system with $\sigma = 10$, $r = 10$, and $b = 8/3$. The initial deviation is $|\tilde{x} - x_s| = 0.1$, while $\tilde{y}$ and $\tilde{z}$ are initially set at their steady-state values, $y_s = \sqrt{b(r-1)}$ and $z_s = r - 1$. (b) The deviation $|\tilde{x} - x_s|$ as a function of time. The dashed line shows the expected average rate of exponential decay: $|\tilde{x} - x_s| \propto 10^{-t/t_{0.1}}$, with $t_{0.1} = 3.87$.

negative x: the more negative x, the smaller e^x. Thus, we can represent the decay in Fig. 10.8 by

$$|\tilde{x} - x_s| \approx C\,10^{-t/t_{0.1}},\qquad(10.20)$$

where $t_{0.1} = 3.87$ is the average time required for the deviation to decay by a factor of $1/10$. As with exponential growth, we can also describe this decay by

$$|\tilde{x} - x_s| \approx C\,2^{-t/t_{0.5}},\qquad(10.21)$$

or

$$|\tilde{x} - x_s| \approx C\,e^{-t/t_{1/e}},\qquad(10.22)$$

when it's convenient to measure the rate of decay as the time $t_{0.5}$ required for decay by a factor of $1/2$ or the time $t_{1/e}$ required for decay by a factor of $1/e$. Can you show that $t_{0.5} = t_{0.1}\log_{10}(2) = 1.165$ and $t_{1/e} = t_{0.1}\log_{10}(e) = 1.681$ for the case at hand?

Exponential decay is a common phenomenon that you've probably run into before. Most famously, we observe exponential decay when the nuclei of a radioactive isotope emit their characteristic radiation and transmute to another element. A nuclear physicist measures the decay and tells us that the isotope has a certain half-life, the time $t_{0.5}$ required for the radiation to decay by half.

Examples of exponential decay also abound in systems near a point of stable equilibrium when friction or some other form of dissipation is present. We strike a tuning fork and listen to it ring down; we spin a bicycle wheel and watch it slow to a stop; we hit a speed bump in our car and bounce up and down once before the springs and shock absorbers restore equilibrium. These are examples of a damped response, just like that observed in Fig. 10.8 for the Lorenz system, and they epitomize the exponential decay of perturbations. Before the existence of chaos was recognized, the exponential decay of perturbations was almost universally expected in systems with friction.

If we consider the form $e^{\lambda t}$ proposed by Liapunov, then according to Eq. (10.22) the decay of deviations from steady convection is governed by a negative Liapunov exponent, namely $\lambda = -1/t_{1/e}$. Thus, in the cases of convection explored here, we have $\lambda = -0.5950$ for stable steady flow while $\lambda = +0.9056$ for unstable chaotic flow. In fact, the sign of λ provides a neat and practical test for chaos: a trajectory is chaotic if and only if its Liapunov exponent is positive and perturbed trajectories diverge from it. The butterfly effect is a unique signature of chaotic motion.

10.6   Weather prediction

When Lorenz first issued his alert at Tokyo in 1960, meteorologists were not prepared to consider the possibility that the weather is inherently unpredictable and that long-term forecasting might never be practical.

By 1964, however, numerical experiments with full-scale weather models confirmed the idea that small perturbations grow exponentially in time. Thus, meteorologists began to take Lorenz seriously, realizing that the butterfly effect is a real part of the Earth's weather and a small offset can make a big difference in future events.

Just how much does the butterfly effect limit weather prediction? With his 1960 weather model, which included only a dozen variables, Lorenz observed that on average the difference between the perturbed and original trajectories tended to double over four days of simulated time. Modern weather models, which include millions of variables, also typically yield a doubling time t_2 of a few days. At this rate of growth, errors in our knowledge of the current state of the atmosphere can be expected to prevent accurate forecasts beyond about two weeks. Although only an estimate, this limit of two weeks is a major blow to the optimism with which meteorologists once viewed the possibility of improving long-term forecasts. Lorenz's discovery had permanently altered the agenda of weather research.

On the other hand, when compared to the actual weather, forecasts are often inaccurate beyond two or three days. Why do forecasts go wrong before the two-week limit set by chaos?

One problem is that practical weather models must make many simplifying assumptions in order to run on even the most advanced supercomputers. The models attempt to calculate the weather over the entire Earth or a sizeable portion of it, and they keep track of the speed, pressure, temperature, humidity, and other properties of the air at each point on a grid that covers the area of interest. Because the spacing between grid points might be 50 miles, the models omit phenomena like thunderstorms that are much smaller in scale but are sure to have important effects. In this sense, our best weather models are currently rather crude, and it may be that forecasts will improve as more sophisticated models and faster computers become available. Wouldn't it be great to have an accurate seven-day forecast and know that it won't rain during your backpacking trek?

Until weather models improve, we may be stuck with less than perfect long-term forecasts, but we may at least be able to estimate their reliability. As can be seen from Fig. 10.7, in a chaotic system deviations grow exponentially only on average. Thus, there are periods when an error in initial conditions may not cause our forecast to go awry immediately. In fact, we sometimes observe a daily weather pattern that persists for weeks on end. During such periods, predicting the weather is no problem at all. Indeed, it's well known that the prediction "same as yesterday" has a pretty good track record, even if it isn't much of a prediction.

To determine when the weather is inherently stable, meteorologists generate "ensemble forecasts." That is, they use a model to predict the weather based first on the nominal current conditions, then rerun the calculation a few times with initial conditions offset from nominal. The offsets can be chosen randomly, but they should be within the errors normally found in meteorological data. If all the calculations agree for the first five days, then the weatherman can issue an extended forecast with confidence. If they all disagree on tomorrow's weather, then the forecaster is well advised to pepper his announcement with words like "possibility" and "chance." Thus, ensemble forecasting helps the weatherman to know when chaos is at play and keep the egg off his face.

Now let's turn away from practical weather prediction and consider Lorenz's famous question: "Does the flap of a butterfly's wing in Brazil set off a tornado in Texas?" To the uninitiated this question might seem utterly silly, but, once you know about the butterfly effect, it's quite sensible. In effect, Lorenz proposed the following thought experiment. Suppose we could compare the weather that would result if a certain butterfly in Brazil flapped its wings with the weather that would have resulted if the butterfly had remained still. Assuming that everything else remains unchanged, Lorenz asked if there's a chance that a tornado would be observed in Texas at some later date in the first scenario but not the second.

Of course, we can't answer Lorenz's question by an actual experiment, because we can't return the universe to exactly the same situation, with all of the world and all of the butterflies but one doing exactly the same thing as at an earlier instant. And we can't use a computer to find the answer either. Although resetting the initial condition of a computation is easy, our global weather models are hopelessly inadequate, with both butterflies and tornados omitted entirely.

Nevertheless, we can imagine what difference one butterfly might make. If the exponential growth of atmospheric perturbations extends to the scale of butterflies, then the flap of a wing could cause a divergence in the pattern of flow that would grow rapidly to affect the atmosphere at larger and larger scales. Just as in Fig. 10.4, where a small offset in the initial velocity of a convection cell is at first imperceptible, the flap of the butterfly's wing would at first make no discernible difference in the weather. As the days and weeks pass, however, the difference between the perturbed and unperturbed atmospheres would grow in scale until eventually the global weather on any given day would be distinctly different, depending on whether the butterfly had or hadn't flapped its wings. At that point, a tornado in Texas in one scenario probably wouldn't be matched by a tornado in the other. Amazingly, the answer to Lorenz's silly question is a qualified yes!

We've just argued that a flap of a butterfly's wing can have a big effect. However, you should understand that the butterfly doesn't provide the energy that drives the tornado, and it can't produce a pattern of weather outside the normal range of possibility. Instead, the butterfly provides the slightest offset, which is then amplified by the inherent instability of the weather system to change the specific day-to-day flow of air driven by energy from the Sun. Because the solar heating isn't affected, the weather exhibits the same kinds of events, from tornados to rainstorms, that we're accustomed to. The effect of the butterfly is simply to change the schedule of these events. Given that tornados happen and will inevitably strike at some time or place, we can expect a tornado in Texas whether the butterfly flaps its wing or not. The difference is only in the exact schedule of events. Of course, if you're a tornado forecaster trying to predict that time and place, seemingly insignificant perturbations may be important.

A graphic example of the rescheduling caused by small offsets is provided by Fig. 10.9. Here we show the velocities corresponding to the rotations plotted in Fig. 10.4. The figure compares various trajectories having small initial offsets x_{off} in velocity with the unperturbed trajectory ($x_{off} = 0$). In each case the effect of the offset isn't apparent at first (wide line), but soon the pattern changes completely (narrow line) and never returns to the original. However, we don't find anything in the perturbed trajectories that is at all surprising. Although the exact scheduling is changed by an offset or perturbation, the mix of velocities, large and small, positive and negative, remains the same. An initial offset

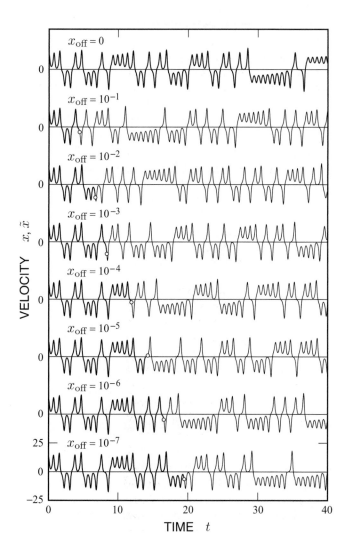

Fig. 10.9 The velocity of the fluid in a convection loop for $\sigma = 10$, $r = 28$, and $b = 8/3$, corresponding to the rotation curves in Fig. 10.4, with the indicated offsets x_{off} in the initial velocity. As in Fig. 10.4, circles mark the time at which an offset trajectory first differs by 1 unit in rotation from the unperturbed trajectory ($x_{off} = 0$). Beyond this time the velocity is plotted with a narrower line.

doesn't change the character of the convective motion, just as a butterfly in Brazil doesn't change the inevitability of a tornado season in Texas. Butterflies don't foul up the weather; they only foul up our ability to predict it.

Further reading

Exponential growth

○ Bartlett, A. A., "Forgotten fundamentals of the energy crisis", *American Journal of Physics* **46**, 876–888 (1978).

○ Deevey, E. S., "The human population", *Scientific American* **203**(3), 194–204 (September 1960).

○ Cohen, J. E., "Human population grows up", *Scientific American* **293**(3), 48–55 (September 2005).

○ Maor, E., *e: The Story of a Number* (Princeton University Press, Princeton, 1998).

○ Schaller, R. R., "Moore's law: Past, present and future", *IEEE Spectrum* **34**(6), 52–59 (June 1997).

Lorenz

○ Gedzelman, S. D., "Chaos rules: Edward Lorenz capped a century of progress in forecasting by explaining unpredictability", *Weatherwise* **47**(4), 21–26 (August/September 1994).

○ Gleick, J., "The butterfly effect", in *Chaos: Making a New Science* (Viking, New York, 1987) pp. 9–31.

○ Taba, H., "The *Bulletin* interviews Professor Edward N. Lorenz", *WMO Bulletin* **45**, 111–121 (1996).

Lorenz equations

• Lorenz, E. N., "Deterministic nonperiodic flow", *Journal of Atmospheric Sciences* **20**, 130–141 (1963).

• Sparrow, C., *The Lorenz Equations: Bifurcations, Chaos, and Strange Attractors* (Springer, New York, 1982).

Weather prediction

• Lorenz, E. N., "The statistical prediction of solutions of dynamic equations", in *Proceedings of the International Symposium on Numerical Weather Prediction* (Meteorological Society of Japan, Tokyo, 1962) pp. 629–635.

○ Lorenz, E. N., "Our chaotic weather", in *The Essence of Chaos* (University of Washington Press, Seattle, 1993) chapter 3.

○ Marchese, J., "Forecast: Hazy", *Discover* **22**(6), 44–51 (June 2001).

○ Matthews, R., "Don't blame the butterfly", *New Scientist* **171**(2302), 24–27 (4 August 2001).

○ Monastersky, R., "Forecasting into chaos", *Science News* **137**, 280–282, (1990).

○ Palmer, T., "A weather eye on unpredictability", *New Scientist* **124**(1690), 56–59 (11 November 1989).

• Read, P. L., "Applications of chaos to meteorology and climate", in *The Nature of Chaos*, Mullin, T., editor (Oxford University Press, Oxford, 1993) pp. 222–260.

○ Reich, E. S., "Making waves", *New Scientist* **180**(2423), 30–33 (29 November 2003).

○ Rosenfeld, J., "The butterfly that roared", *Scientific American Presents* **11**(1), 22–27 (Spring 2000).

Chaos comes of age

11

In his 1963 paper on "Deterministic nonperiodic flow," Lorenz demonstrated that a simple nonlinear dynamical system, lacking any random element, can nevertheless produce persistently irregular motion. Such motion is now aptly described as "chaotic." In addition, he proved that chaos is generally linked to a form of instability, now called "extreme sensitivity to initial conditions," characterized by the exponential growth of small offsets or perturbations. Finally, Lorenz discovered that chaotic trajectories have a complex geometry, now described as a "fractal geometry," which we'll explore in Chapter 17.

Given the depth of his insights, Lorenz is sometimes credited with the discovery of chaotic motion. However, the history of chaos is a tangled tale that spans almost a century and involves the work many people. In fact, it is perhaps fairest to say that chaos was discovered many times, although most discoverers did not understand their discovery as fully as Lorenz. The situation is much like that described by John Saxe in his poem "The Blind Men and the Elephant." In this poem, six blind men touch an elephant and discover six different animals, depending on whether they encounter its side, tusk, trunk, leg, ear, or tail. Similarly, scientists discovered chaos many times from a variety of directions—mathematics, physical observation, computation—each of which revealed a different aspect. Only later did they assemble a complete picture. To explain further, in this chapter we take a break from science and examine the historical development of chaos theory and in the process anticipate many topics from the chapters to come.

Before beginning our historical survey, however, we pause to define some of the varieties of chaos that have been discovered over the years.

11.1 Kinds of chaos

From the examples described so far, we can define chaos as persistent pseudorandom motion in a deterministic dynamical system with exponential sensitivity to initial conditions. By using the words "pseudorandom" and "deterministic" here, we limit ourselves to mathematical models, because any real-world system inevitably includes noise and isn't entirely deterministic. Our models of the Tilt-A-Whirl, driven pendulum, and convection cell (Lorenz system) all display chaotic motion encompassed by this definition, and it's the type of chaos central to this book.

However, the chaotic behavior in these examples has a characteristic that isn't mentioned in our definition. Specifically, the chaotic nature of the motion is unaffected by small perturbations, and the chaos can be described as "dynamically robust." In this case, a perturbation will divert the motion from one chaotic trajectory to another, but the motion remains chaotic. In other cases, primarily of mathematical interest, there are chaotic trajectories of infinite duration that are surrounded by unchaotic trajectories. Here, the slightest perturbation will divert the motion away from chaos, and the chaos is "dynamically fragile." While fragile chaos is never observed in the real world, where noise is always present, mathematical examples have played an important role in the history of chaos theory.

Also, the phrase "transient chaos" is often applied to motion in which sensitivity to initial conditions occurs during an initial interval but doesn't persist. An example is provided by the double pendulum: a system of two pendulums with the pivot point of one pendulum fixed to a support and that of the second fixed to the bob of the first. Mathematically, a frictionless double pendulum is persistently chaotic, but in the real world, where friction rules, chaos is only observed during an initial transient, before the pendulums come to a stop. Nonetheless, a nearly frictionless double pendulum provides an excellent demonstration of chaotic motion. Other examples of transient chaos are found in the sensitive motion of coins, dice, and roulette balls.

With these definitions in mind, let's turn to the history of chaos theory.

11.2   Maxwell

One of the first intimations of chaos is found in an essay presented by James Clerk Maxwell (1831–1879) to the Erănus Club of Cambridge, England in 1873. We briefly introduced Maxwell in Chapter 9 as the author of the equations of electromagnetism, but he also helped found the kinetic theory of gases. In this theory, a gas such as air is envisioned as a collection of molecules moving at high speed and continually colliding with each other. Apparently thinking in this context, Maxwell gives a perfect description of extreme sensitivity to initial conditions in his essay of 1873.

> When the state of things is such that an infinitely small variation of the present state will alter only by an infinitely small quantity the state at some future time, the condition of the system, whether at rest or in motion, is said to be stable; but when an infinitely small variation in the present state may bring about a finite difference in the state of the system in a finite time, the condition of the system is said to be unstable. It is manifest that the existence of unstable conditions renders impossible the prediction of future events, if our knowledge of the present state is only approximate, and not accurate.

Does Maxwell mean to imply that a small variation grows exponentially and that the instability persists indefinitely? Probably so, if he's thinking of the motion of molecules in a gas, and in this case he has described what we now call chaos. In any case, Maxwell clearly identifies unpredictability with sensitivity to initial conditions, an important part of our modern picture of chaos.

11.3 Poincaré

Similar qualitative ideas were expressed later by the French physicist and mathematician Henri Poincaré (1854–1912) in a 1908 essay on "Chance," written for a popular audience.

> If we knew exactly the laws of nature and the situation of the universe at the initial moment, we could predict exactly the situation of that same universe at a succeeding moment. But, even if it were the case that the natural laws had no longer any secret for us, we could still only know the initial situation *approximately.* If that enabled us to predict the succeeding situation *with the same approximation,* this is all we require, and we should say that the phenomenon had been predicted, that it is governed by laws. But it is not always so; it may happen that small differences in the initial conditions produce very great ones in the final phenomena. A small error in the former will produce an enormous error in the latter. Prediction becomes impossible, and we have the fortuitous phenomenon.

Poincaré goes on to give the weather and the game of roulette as examples of situations in which a tiny error in initial conditions can make an enormous difference in the final outcome. Like Maxwell, Poincaré clearly understood sensitivity to initial conditions.

Later in his essay on chance, Poincaré goes beyond Maxwell by recognizing that perturbations can grow at an exponential rate. Considering the molecules in a gas, he points out that a small deviation in the trajectory of one molecule can be greatly amplified by its collisions with other molecules. "And the molecule will not suffer two collisions only, but a great number each second. So that if the first collision multiplied the deviation by a very large number, A, after n collisions it will be multiplied by A^n." Thus, the deviation grows exponentially with the number of collisions, and the gas is chaotic by modern standards.

However, the contribution of Poincaré to chaos theory goes far beyond the simple ideas presented in his essay on chance. Some 30 years earlier, during the 1880s, Poincaré single handedly transformed the science of dynamics by introducing new qualitative methods. To do so, he shifted his focus from the individual trajectories of a dynamic system to the collection of all possible trajectories. As we'll discuss in Chapter 15, Poincaré then applied topology to predict the system's general behavior. Although more abstract in nature, Poincaré's new approach gave a perspective on dynamics that allowed him to see far beyond his predecessors.

Poincaré applied his new methods to Newton's old nemesis, the three body problem. Here, as we'll learn in Chapter 19, Poincaré discovered something mind-boggling, a geometry of dynamic trajectories having infinite complexity. The geometry was so convoluted and strange that Poincaré was taken aback, later remarking that "One is struck by the complexity of this picture, which I do not even attempt to draw." Today, we call Poincaré's complex structure of trajectories a "tangle" and recognize it as the geometry underpinning chaotic motion.

Poincaré's tangle was first described in 1890 in a memoir that was immediately recognized as a giant leap forward in the science of motion. This memoir later became the seed for his great work *Les Méthodes Nouvelles de la Mécanique Céleste*, an instant classic in three volumes. However, *Celestial Mechanics* includes many great discoveries in addition to tangles, and the tangle was largely overlooked by Poincaré's readers. Indeed, the significance of this remarkable geometry would not be fully appreciated for another 70 years.

Poincaré understood that a tangle implied an extreme form of instability. Ironically, however, he apparently never associated it with the persistent chaotic motion that he so accurately described in his later essay on chance. That is, Poincaré understood how chaotic motion could arise among the molecules of a gas and also discovered the underlying geometry of chaos, its heart of darkness, but probably never connected the two. Nonetheless, for introducing the topological approach to dynamics and discovering his troublesome tangle, Poincaré is often called the father of chaos theory.

In Poincaré's wake, most physicists simply forgot about tangles. To physicists, tangles appeared to be monsters of a strictly mathematical kind that could be safely ignored.[1] However, a few mathematicians were fascinated by Poincaré's new ideas and continued to develop them.

11.4   Hadamard

In particular, Jacques Hadamard (1865–1963), another French mathematician, was so impressed by Poincaré's methods that he soon began using them in his own work. Hadamard's encounter with chaos came in a paper published in 1898 that considered motion on surfaces of negative curvature.

The dynamical system studied by Hadamard was a bit like that introduced later by Einstein in his theory of general relativity. According to Einstein, the mass associated with stars and other heavenly bodies warps the very fabric of our space and time, and a small particle moves along a "geodesic" or shortest path in this warped space-time. While we usually think of a shortest path as a straight line, in the warped space-time around the Sun, planetary orbits are geodesics in the form of ellipses. In Einstein's picture, Newton's gravitational force is replaced by the curvature of space-time, but the planetary motions are nearly the same.

[1] This attitude has long pervaded physics. I remember professors who discounted the possibility of strange mathematical behavior by simply assuming that it couldn't happen in a physical system.

Fig. 11.1 Jacques Hadamard, 1890s. (*Acta Mathematica 1882–1912*, Index of Volumes 1–35, (Paris, 1913), p. 144. By permission of the Institut Mittag-Leffler.)

To understand Hadamard's result, it's easiest to think of a particle moving in a two-dimensional space. While a little artificial, Hadamard considered motion in the absence of friction, gravity or other applied forces. The only force experienced by the particle comes from the curvature of the surface that constrains it. Because this force acts only to deflect the particle, it travels at a constant speed and always proceeds along a geodesic.

A flat surface, like a table top, is a two-dimensional space with zero curvature, and the geodesics are simply straight lines. On the other hand, the surface of a spherical ball is obviously curved, and its curvature is taken to be positive. On a sphere, the geodesics are great circles, like the Earth's equator. If you've ever flown across the Atlantic, you almost surely flew along a great circle, because it's the shortest route. If you were surprised to hear the pilot pointing out Newfoundland at some point, try stretching a string between New York and London the next time you see a globe.

But Hadamard calculated motion on a surface of negative curvature. What is negative curvature? Strangely, it's a property of a surface shaped like a saddle or a potato chip. Clearly, a globe and a saddle are curved differently in some way, yet it's hard to put one's finger on the difference. The distinction is apparent in Fig. 11.2, which illustrates surfaces with zero, positive, and negative curvature. To test the curvature at any point on a surface, we simply draw a small circle around the point using a string to fix the radius r. If the string is constrained to lie in the surface but otherwise drawn taut, then the circumference of the resulting circle is $C = 2\pi r$ (as in elementary geometry) if the surface has zero curvature, but $C < 2\pi r$ if the curvature is positive, and $C > 2\pi r$ if it's negative. We can envision making a surface of positive curvature from a flat disk of cloth by removing a pizza slice and sewing the remaining cloth back together. Alternatively, we can make a surface of negative curvature by cutting a slit in the cloth and sewing in an extra pizza slice.

In his analysis, Hadamard applied topology to surfaces having a negative curvature at every point. Unlike positively curved surfaces like the sphere, negatively curved surfaces tend to have strange shapes. In discussing Hadamard's work, his contemporary, Pierre Duhem (1861–1916), described a surface of negative curvature as follows.

> Imagine the forehead of a bull, with protuberances from which the horns and ears start, and with the collars hollowed out between these protuberances; but elongate these horns and ears without limit so that they extend to infinity; then you will have one of the surfaces we wish to study.

Duhem goes on to enumerate the various types of geodesics that Hadamard discovered, including some that are periodic, some that go off toward infinity, and a third class that he describes as follows.

> There are some [geodesics] also which are never infinitely distant from their starting point even though they never exactly pass through it

(a) $C = 2\pi r$

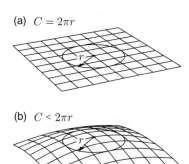

(b) $C < 2\pi r$

(c) $C > 2\pi r$

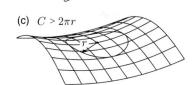

Fig. 11.2 Surfaces with curvatures that are (a) zero (b) positive and (c) negative.

again; some turn continually around the right horn, others around the left horn, or right ear, or left ear; others, more complicated, alternate, in accordance with certain rules, the turns they describe around one horn with the turns they describe around the other horn, or around one of the ears.

The "more complicated" geodesics correspond to motion much like the chaotic trajectories computed by Lorenz for his convection cell, in which the fluid erratically alternates between rotating in the clockwise and counterclockwise directions. In analyzing these trajectories, Hadamard introduced a tool, now called "symbolic dynamics," which we'll learn more about in Chapter 18. Symbolic dynamics is the key to understanding just how random a chaotic trajectory can be.

Hadamard also discovered that all trajectories on a surface of negative curvature exhibit sensitivity to initial conditions, the hallmark of chaotic motion. Thus, it would seem that Hadamard had truly discovered chaos way back in 1898. However, as Hadamard knew, an arbitrarily small perturbation could convert one of his "more complicated" geodesics into one that headed off toward infinity. The pseudorandom motion that he had found was locally unstable but also globally unstable. Unlike Lorenz's convective chaos, in which a small perturbation leads to a new trajectory with the same average properties as the old one, small perturbations completely destroy Hadamard's pseudorandom motion. That is, Hadamard had discovered an instance of dynamically fragile chaos, a phenomenon that we'll meet again in Chapter 18.

As a final observation on geodesic dynamics, it's interesting to note that dynamically robust chaos can result when the surface includes regions of both positive and negative curvature. For example, the geodesics of a double torus, the surface of a donut with two holes, can be persistently chaotic even in the presence of small perturbations. Why does the added positive curvature make a difference? In contrast to regions of negative curvature, where initially parallel geodesics diverge, in a region of positive curvature such geodesics always converge. As Fig. 11.3 suggests, this concept may be well known to space travelers. In any case, the negative curvature of the double torus makes nearby trajectories diverge, while the positive curvature gives the double torus a closed surface, which prevents trajectories from going off to infinity. The positive curvature provides global stability, and the negative curvature creates local instability, two essential elements of robust chaotic motion.

11.5   Borel

Another French mathematician, Émile Borel (1871–1956), supplied the missing example of a simple, stable chaotic system when he further developed the ideas of Poincaré on the sensitive dependence of molecular

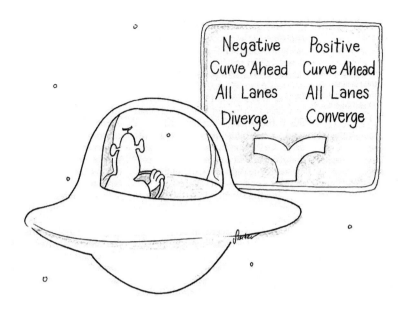

Fig. 11.3 A space traveler headed for a chaotic ride. (© Catherine Fischer, by permission.)

trajectories in a gas. In a brief paper of 1913, entitled "Statistical mechanics and irreversibility," Borel modeled a single gas molecule by a billiard ball bouncing around in the absence of friction on a rectangular table. By adding circular obstacles to the table, he estimated that a small offset in the trajectory of a molecule will grow by a factor of about 10 each time it collides with another molecule. Because the molecules of a gas typically collide 1000 times in a microsecond, Borel argued that the exponential growth postulated by Poincaré can amount to a factor of 10^{1000} in a very short time. Borel thus uncovered a dramatic example of sensitivity to initial conditions.

To further explore this extreme sensitivity, Borel considered the gravitational disturbance produced by displacing 1 gm of mass by 1 cm on a distant star. He concluded that even this minute perturbation would completely alter the microscopic state of a gas on Earth within a microsecond.[2] As Borel realized, this amazing example of the butterfly effect implies that we can't predict the detailed motions of molecules in a gas without "introducing the whole universe into our calculations." Moreover, because Borel's gas molecules could not escape their container, the chaos that he envisioned was robust.

While brief and largely qualitative, Borel's paper goes beyond estimating the butterfly effect. In particular, he pictured the motion of his billiard ball using Poincaré's birds-eye view of all possible trajectories and arrived at a strange, "infinitely laminated structure" of the sort that we now call a fractal. Indeed, Borel's description of the fractal properties of chaotic trajectories is probably the first of its kind. We'll return to this and other aspects of molecular chaos in Chapters 13 and 16.

Fig. 11.4 Émile Borel. (*Acta Mathematica 1882–1912*, Index of Volumes 1–35, (Paris, 1913), p. 132. By permission of the Institut Mittag-Leffler.)

[2]Borel's paper appeared before Einstein's theory of general relativity, so he was unaware that gravitational disturbances travel at the speed of light and require several years to reach the Earth from even a nearby star. However, once the disturbance arrives, the gas is upset almost instantly.

11.6   Birkhoff

While Poincaré's compatriots, Hadamard and Borel, applied his qualitative methods to their own problems, the American mathematician George Birkhoff (1884–1944) followed even more closely in Poincaré's footsteps by working directly with dynamical tangles. Birkhoff completed his doctorate on dynamical systems at the University of Chicago in 1907 and joined the faculty of Harvard University in 1912. During his career at Harvard, Birkhoff quickly rose to preeminence among American mathematicians, working principally in dynamics.

As we'll describe in Chapter 19, Birkhoff was the one who largely decoded the dynamical consequences of a Poincaré tangle. In doing so, he obtained his most important results by proving a number of topological theorems of a type called fixed-point theorems and by applying Hadamard's symbolic dynamics. In papers written between 1912 and 1935, he repeatedly returned to dynamical tangles, gradually establishing their importance in a range of situations and mapping out much of what we now recognize as the mathematical heart of chaos. In 1932, for example, Birkhoff discovered in a simple dynamical system "some remarkable closed curves" that he argued must double back on themselves an infinite number of times. These curves are the fractal traces of a topological tangle. Although the system studied in Birkhoff's 1932 paper is especially simple in that time is assumed to advance in discrete steps, it probably established the first complete mathematical example of robust chaos.

Fig. 11.5 George Birkhoff, 1913. (G. D. Birkhoff, *Collected Mathematical Papers*, vol. I (American Mathematical Society, New York, 1950), frontispiece, by permission of the AMS.)

11.7   Chaos sleeps

Before the turn of the twentieth century, Poincaré had driven the deep mathematical pilings that would become the foundation of nonlinear dynamics in general and chaos theory in particular. He had also identified the complex geometry called a tangle that underlies chaotic motion. Just as the nineteenth century came to a close, Hadamard discovered a dynamic system that exhibits fragile chaos. Borel's paper of 1913 took a step closer toward robust chaos with qualitative arguments suggesting that it happens in an ideal gas. Finally, by 1927 Birkhoff had decoded Poincaré's tangle, showing that it implies the presence of an infinite number of periodic orbits, and in 1932 he identified robust chaos in a discrete-time system. With these pieces of the puzzle in place, a full understanding of chaotic motion would seem to have been just around the corner. As it happened, however, the picture would only be completed in the 1960s, after a world war and a computer revolution. In the interim, mathematicians continued to develop the theory of nonlinear dynamics, building on the foundation laid by Poincaré and occasionally discovering a new aspect of chaotic motion.

While mathematicians followed an abstract path to chaos, others occasionally observed the phenomenon itself in a physical system. As noted in Chapter 1, Herbert Sellner gave an accurate and perceptive description of chaotic behavior in the 1926 patent application for his carnival ride, the Tilt-A-Whirl. In Chapter 12, we'll mathematically model the Tilt-A-Whirl and confirm Sellner's claims. Another example of chaos in the real world was discovered in 1927 by the Dutch electrical engineer Balthazar van der Pol (1889–1959). In trying to understand the behavior of an electronic oscillator in the days before oscilloscopes, van der Pol and his colleague van der Mark simply listened to the oscillations with a telephone receiver. When driven with a sinusoidal voltage at 1000 Hz, they reported hearing this frequency for some parameter values, but at other values they heard subharmonics such as 500 Hz (period doubling), and at still other values they heard "irregular noise." After our exploration of the sound of chaos in Chapter 9, we can guess that van der Pol's irregular noise resulted from chaos. Although neither Sellner nor van der Pol fully understood the significance of their observations, both had in fact discovered robust chaotic motion.

The van der Pol oscillator is particularly significant because it was later studied in depth by the English mathematicians Mary Cartwright (1900–1998) and John Littlewood (1885–1977). Their interest arose in connection with work on radar during World War II but went far beyond any practical need. In 1945, for example, they published a paper in which they noted that the equation of motion for van der Pol's driven oscillator allows the possibility of stable periodic motion, with a period equal to that of the driver or one of its subharmonics, just as van der Pol observed. They also noted that "In the general topological theory, however, other possibilities, indeed very 'bad' ones, have to be contemplated, and it is found very difficult in any given case to rule them out—for the best of reasons . . . ." Cartwright and Littlewood went on to sketch arguments indicating that robust chaotic motion can occur in the driven van der Pol oscillator.

Inspired by the 1945 paper of Cartwright and Littlewood, Norman Levinson (1912–1975), an MIT mathematician, discovered that robust chaos could be more easily proven for a related but simpler driven oscillator, and published a full proof in 1949. And in 1957 the more difficult van der Pol oscillator succumbed to the efforts of Littlewood, who rigorously proved the existence of robust chaos in this system as well. These breakthroughs were significant because they extended robust chaotic motion from the discrete-time system of Birkhoff to continuous-time systems and moreover to systems of potential technological importance.

Chaotic behavior also popped up in the work of two veterans of the atomic bomb project at Los Alamos, New Mexico. Shortly after the war, the Polish–American mathematician Stanislaw Ulam (1909–1984) and the Hungarian–American mathematician John von Neumann (1903–1957) began to apply electronic computers, just then becoming available, to the analysis of nuclear explosions. Ulam and von Neumann are both famous for their work at Los Alamos. Ulam (together with Edward Teller) was the father of the hydrogen bomb and von Neumann was one of the great mathematicians of the twentieth century, making important contributions to set theory, game theory, and computer science. However, incidental to the task of simulating explosions, they needed a source of random numbers, and in 1947 they proposed obtaining them from the equation,

$$X_{\text{new}} = 4X_{\text{old}}(1 - X_{\text{old}}), \qquad (11.1)$$

by the process of iteration. That is, beginning with some X_{old} in the range $0 < X_{\text{old}} < 1$, Ulam and von Neumann could use Eq. (11.1) to generate a new number X_{new} in the same range, and the process could be continued *ad infinitum* by choosing $X_{\text{old}} = X_{\text{new}}$ to calculate another X_{new}. They discovered that the numbers generated by this simple algorithm, usually called an iterated map, are quite random.

In fact, Eq. (11.1) defines a chaotic process that is among the simplest possible. Thus, we shouldn't be surprised that it produces random numbers. Although Eq. (11.1) is similar to the iterative equations we have used to describe the motion of mechanical systems like the pendulum, it does not involve a continuous time variable—no small Δt is required. Instead, Eq. (11.1) comes from another class of systems, for which we can think of time as passing in discrete jumps. Because of their simplicity, however, iterated maps provide insights into some aspects of chaotic behavior that are obscured in continuous-time systems. We'll return to the example uncovered by Ulam and von Neumann in Chapter 14.

11.8   Golden age of chaos

As we've seen, during the first half of the twentieth century, scientists and mathematicians occasionally bumped into various aspects of chaotic behavior. During the 1960s, however, chaos became the subject of serious investigation. The work of Lorenz discussed in Chapter 10 was a full frontal attack, aimed at understanding the origin, nature, and consequences of chaotic motion. But Lorenz was not alone. Simultaneously and independently, the American mathematician Stephen Smale (born 1930) attacked chaos from a completely different point of departure. Whereas Lorenz came from the practical world of meteorology and used a digital computer as his chief investigative tool, Smale was a pure mathematician and applied rarefied abstraction to reveal the heart and soul of chaos. Again simultaneously and independently, the Japanese electrical engineer Yoshisuke Ueda (born 1936) discovered chaotic motion in a driven nonlinear oscillator, a discovery that set him on a quest to explain and understand exactly what he had observed. Thus, in the early 1960s, three scientists independently set themselves the challenge of explaining chaotic motion, each using a different tool: digital computation, mathematical abstraction, and careful experiment. So began what we might call the golden age of chaos theory.

Not surprisingly, the discoveries of Lorenz, Smale, and Ueda all owe a debt to Poincaré. By the 1960s, Poincaré's "méthodes nouvelles," his bird's eye view of trajectories and their topology, had become the standard paradigm of nonlinear dynamics, an edifice elaborated by generations of mathematicians. When Lorenz proved the connection between pseudorandom motion and sensitivity to initial conditions, he relied on a treatment of dynamical systems due to Birkhoff—his own thesis advisor and a disciple of Poincaré. Later, he discovered that elements of his proof had been anticipated by others working in the tradition of Poincaré. Similarly, Ueda worked fully within Poincaré's vision of dynamics and indeed discovered chaos in his driven oscillator by plotting its "Poincaré section," a technique we'll explore in Chapter 16. Smale's work, on the other hand, directly concerned Poincaré's tangle, the geometric structure at the root of chaotic motion. In 1960, Smale discovered a simple way to model the topology of trajectories associated with a tangle and in doing so opened the path to answering questions regarding the occurrence and stability of chaos. His model, which reinvigorated research in the mathematics of tangles, is explored in Chapter 18.

11.9   Ueda

Because we won't return later to the work of Ueda, let's take a brief look now at his contribution. Ueda first encountered chaos in November of 1961 while working as a graduate student in the laboratory of Chihiro Hayashi (1911–1986) in the Department of Electrical Engineering at Kyoto University. His discovery was made using an analog computer, a carefully crafted electronic circuit configured to accurately model a specific dynamic equation. The computer, made with vacuum tubes, operated at slow speed, allowing the changing voltages to be plotted by pen on a chart recorder. For the driven oscillator under study, the voltages varied continuously and the Poincaré section of the trajectory, plotted over an hour or more, would often assume a smooth egg-like shape, indicating quasiperiodic motion. However, for the circuit parameters chosen on one fateful day in 1961, Ueda observed a "shattered egg" that was "totally irregular and seemingly inexplicable."

Like Lorenz, Ueda at first thought that his computer was broken, but several checks soon convinced him that the shattered-egg phenomenon was real. The voltage waveforms were still smooth and continuous, they just didn't settle into any kind of repeating or almost repeating pattern. Moreover, the shattered egg occurred over a range of circuit parameters, so it wasn't a fluke. Of course, unlike the digital simulations of Lorenz, Ueda's analog computation couldn't be repeated with precisely the same initial conditions, so it was impossible to prove that his shattered egg wasn't the product of some unaccounted noise in the circuit. Nonetheless, Ueda was convinced that the shattered egg was an accurate solution of the equation simulated by his circuit, and in the coming years he would encounter the same type of behavior in other circuits.

Fig. 11.6 Students of Chihiro Hayashi at Kyoto University in front of their analog computer, 1960. Yoshisuke Ueda is at the far right. (Courtesy of Yoshisuke Ueda.)

Unfortunately, Professor Hayashi was never convinced that Ueda's shattered egg could be anything but the product of noise or a long-lived transient. Because Ueda continued to work under Hayashi, his discovery of chaos was blocked from publication until 1970, when he became bold enough to mention it in a paper submitted while Hayashi was traveling. Finally, in 1973 as an associate professor, Ueda was able to publish a more complete account of his work on "nonperiodic oscillations." By then, however, chaos was being discovered in other laboratories and was on its way to becoming a sensation in the scientific community. All the same, Ueda's 1961 experiment, later confirmed by digital simulations, can be taken as the first explicit recognition of chaotic motion in a mathematically well-characterized, real-world system. In retrospect, we can say that Ueda was the first to discover chaos by experiment, a stunning accomplishment for a graduate student.

11.10   What took so long?

Roughly a decade after the work of Lorenz and Smale appeared in print, the larger world of science suddenly perceived its import. Word that Newton's equations, the epitome of determinism, might also explain much that is random swept through the scientific community like wildfire, producing an explosion of activity. We'll return to this explosion in Chapter 14, but here we note that the term "chaos" was first applied to predictable random motion by the American mathematician James Yorke (born 1941) in a 1975 paper. Whether or not the term is perfectly appropriate, it was soon adopted so widely that alternatives became untenable. Thus, the paradox of motion that is persistently random yet entirely predictable would capture the imagination of scientists and the general populace alike under the name "chaos." It was a long road from Maxwell and Poincaré, but chaos had at last come of age.

Why were scientists so slow to recognize a phenomenon that is arguably common in the ordinary world? There are many reasons why chaos was difficult to discover, but an unwarranted assumption was part of the problem. From the mathematical point of view, the science of dynamics, as developed by Galileo and Newton, was intended to explain simple, regular phenomena such as the motion of a pendulum or a planet. That the same mathematics might describe irregular motion was entirely unexpected, and scientists generally overlooked the possibility. Why would anyone seek an explanation of complex motion in a simple equation?

The mathematical complexity of chaos is also daunting. When Poincaré uncovered a tangle in a simple dynamic equation, the geometry was so convoluted that it's full meaning wasn't immediately apparent. Without a strong practical need to understand the tangle, most mathematicians simply set the problem aside.

All the same, theorists were challenged to explain obviously random motion like that of molecules in a gas or turbulence in fluid flow. Because these systems involve so many particles, however, mathematicians tended to assume that the randomness arose from the complexity of the system rather than from a simple dynamic instability. Thus, they were apt to begin with a statistical approach instead of considering the details of the motion.

There are, however, mathematically simple systems, like the Tilt-A-Whirl and the driven oscillators of van der Pol and Ueda, in which the paradox of random motion is apparent. For an experimentalist, the problem with such systems is that they are, by their very nature, grossly affected by minute perturbations. No matter how carefully Ueda might have prepared his oscillator circuit, repeating the experiment would never produce the same trajectory, although the general shape of the shattered egg would be unchanged. Thus, Ueda couldn't convince his mentor that his results weren't due to unaccounted noise.

Lorenz, on the other hand, didn't have noise as a problem in his digital simulations. Indeed, it was the assured repeatability of a digital computation that tipped him off to the butterfly effect. From the experimental (or computational) point of view, it thus seems that the discovery of chaos was simply waiting for the advent of digital computers. We shouldn't be surprised that Lorenz made his discovery in the early 1960s, just as computers were becoming widely available.

We can think of many reasons why scientists didn't discover chaos sooner. All the same, Smale's deep insight into the tangle might have come to anyone working in Poincaré's wake. Smale's breakthrough required no more than pencil, paper, and the right perspective on the problem. So it's surely a coincidence that his idea came just as Lorenz and Ueda were discovering chaos in other ways.

Further reading

Birkhoff

- Abraham, R. H., "In pursuit of Birkhoff's chaotic attractor", in *Singularities and Dynamical Systems*, Pnevmatikos, S. N., editor (North-Holland, Amsterdam, 1985) pp. 303–312.

- Barrow-Green, J., "Birkhoff and dynamical systems", in *Poincaré and the Three Body Problem* (American Mathematical Society, Providence, 1997) pp. 209–218.

- Birkhoff, G. D., "Sur quelques courbes fermées remarquables", *Bulletin Société Mathématique de France* **60**, 1–26 (1932).

- Diacu, F. and Holmes, P., "A fixed point begins a career", in *Celestial Encounters: The Origins of Chaos and Stability* (Princeton University Press, Princeton, 1996) pp. 51–55.

Cartwright and Littlewood

- Cartwright, M. L. and Littlewood, J. E., "On non-linear equations of the second order: I", *Journal of the London Mathematical Society* **20**, 180–189 (1945).

- Jackson, E. A., "On the Cartwright-Littlewood and Levinsen studies of the forced relaxation oscillator", in *Perspectives of Nonlinear Dynamics*, Volume 2 (Cambridge University Press, Cambridge, 1991) appendix J.

- Littlewood, J. E., "On nonlinear differential equations of the second order: III", *Acta Mathematica* **97**, 267–308 (1957).

- Littlewood, J. E., "On nonlinear differential equations of the second order: IV", *Acta Mathematica* **98**, 1–110 (1957).

- McMurran, S. L. and Tattersall, J. J., "Cartwright and Littlewood on van der Pol's equation", *Contemporary Mathematics* **208**, 265–276 (1997).

Chaos comes of age

- Abraham, R. and Ueda, Y., editors, *The Chaos Avant-Garde: Memoirs of the Early Days of Chaos Theory* (World Scientific, New York, 2000).

- Borel, É., "La mécanique statistique et l'irréversibilité", *Journal de Physique* **3**, 1697–1704 (1913).

- Dresden, M., "Chaos: A new scientific paradigm—or science by public relations", *The Physics Teacher* **30**, 10–14 & 74–80 (1992).

- Hirsch, M. W., Marsden, J. E., and Shub, M., editors, *From Topology to Computation: Proceedings of the Smalefest* (Springer-Verlag, New York, 1990).

- Lorenz, E. N., "Encounters with chaos", in *The Essence of Chaos* (University of Washington Press, Seattle, 1993) chapter 4.

- Mira, C., "Some historical aspects of nonlinear dynamics: Possible trends for the future", *International Journal of Bifurcation and Chaos* **7**, 2145–2173 (1997).

- Poincaré, H., "Chance", in *Science and Method*, translated by F. Maitland (Dover, New York, 1952) pp. 64–90.

- van der Pol, B. and van der Mark, J., "Frequency demultiplication", *Nature* **120**, 363–364 (1927).

Hadamard

- Barrow-Green, J., "Hadamard and geodesics", in *Poincaré and the Three Body Problem* (American Mathematical Society, Providence, 1997) pp. 201–209.

- Duhem, P., *The Aim and Structure of Physical Theory* (Princeton University Press, Princeton, 1954) Part II, chapter III.

- Hadamard, M., "Les surfaces à courbures opposées et leurs lignes géodésiques", *Journal de Mathématiques* **4**, 27–73 (1898).

- Maz'ya, V. and Shaposhnikova, T., *Jacques Hadamard, A Universal Mathematician* (American Mathematical Society, 1998).

- Ruelle, D., "Hadamard, Duhem, and Poincaré", in *Chance and Chaos* (Princeton University Press, Princeton, 1991) chapter 8.

- Steiner, F., "Quantum chaos", in *Festschrift Universität Hamburg 1994: Schlaglichter der Forschung zum 75 Jahrestag*, edited by R. Ansorge, (Dietrich Reimer Verlag, Hamburg, 1994).

Levinson

- Levinson, N., "A second order differential equation with singular solutions", *Annals of Mathematics* **50**, 127–153 (1949).
- Moser, J., "Commentary", in *Selected Papers of Norman Levinson*, Vol. 1, edited by J. A. Nohel and D. H. Sattinger (Birkäuser, Boston, 1997) pp. 65–68.

Maxwell

- Everitt, C. W. F., *James Clerk Maxwell: Physicist and Natural Philosopher* (Scribners, New York, 1975).
- Hunt, B. R. and Yorke, J. A., "Maxwell on chaos", *Nonlinear Science Today* **3**, 1–4 (1993).
- MacDonald, D. K. C., *Faraday, Maxwell, and Kelvin* (Anchor Doubleday, Garden City, 1964).

- Mahon, B., *The Man Who Changed Everything: The Life of James Clerk Maxwell* (Wiley, Chichester, 2003).
- Maxwell, J. C., "Essay for the Eranus Club on science and free will", in *The Scientific Letters and Papers of James Clerk Maxwell*, Vol. II, edited by P. M. Harman (Cambridge University Press, Cambridge, 1995) pp. 814–823.

Ueda

- Johnstone, B., "No chaos in the classroom", *Far Eastern Economic Review* **144**(25), 55 (22 June 1989).
- Ueda, Y., *The Road to Chaos—II*, 2nd edition (Aerial Press, Santa Cruz, California, 2001).

Ulam and von Neumann

- Cooper, N. G., editor, *From Cardinals to Chaos: Reflections on the Life and Legacy of Stanislaw Ulam* (Cambridge University Press, Cambridge, 1989).
- Ulam, S. M. and von Neumann, J., "On combination of stochastic and deterministic processes", *Bulletin of the American Mathematical Society* **53**, 1120 (1947).

12 Tilt-A-Whirl—Chaos at the amusement park

"Wouldn't chaotic motion make a great carnival ride?" asked Bill Dubé, a colleague of mine at NIST, after seeing my animation of a chaotic pendulum. Thinking for a minute, I realized that Bill was not only right but that some of the rides I had enjoyed as a kid probably were chaotic. Could it be that simple examples of chaos had been in plain view for years without attracting scientific attention? As described in Chapter 1, Bret Huggard and I answered this question in the affirmative by analyzing the Tilt-A-Whirl, an ever-popular ride dating from 1926.

Scientists have discovered chaos in many physical systems since the 1970s, but most are either very complex, like the weather, or laboratory curiosities, like Ueda's driven oscillator. Examples of chaos that are simple yet part of everyday life, like the Tilt-A-Whirl, seem to be exceptional. But it's simple, familiar devices that best illustrate chaotic motion. As Zippy can assure you (Fig. 12.1), there's nothing like a ride or two on the Tilt-A-Whirl to inform one about chaos. The great thing about a carnival ride is that it's real, accessible, and subject to relatively simple analysis. When we calculate the Tilt-A-Whirl's Liapunov exponent, we get a number that relates directly to personal experience. In this chapter, we return to the Tilt-A-Whirl and explore the consequences of chaos in this mundane setting.

As it happens, the Tilt-A-Whirl isn't the only chaotic carnival ride. At the time of my work with Huggard, we could equally well have chosen to investigate the Octopus (dating from 1936), the Turbo (1968), or the Zipper (1969), all of which appear to exhibit chaotic motion. Of course, it's not surprising that some rides are chaotic, because chaos adds an element of surprise, enhancing the excitement. Soon after my paper with Huggard was reviewed in *Science News* (February 26, 1994, p. 143), I received a call from Richard Chance saying that his company, Chance Manufacturing in Wichita, Kansas, was in the midst of designing a ride intended to be chaotic. Dubbed "Chaos," this ride made it's debut in February of 1995. The Chaos ride may be the first large-scale commercial device knowingly designed to be chaotic. More than ever, the amusement park is a good place to find mathematically simple chaotic systems.

Another source of simple, accessible chaotic devices is the toy store. The toys in question, often called kinetic sculptures or executive desk toys, are fascinating simply because their motion is smooth and

Fig. 12.1 Zippy discovers chaos. (© Bill Griffith, by permission.)

continuous yet always changing. A number of these battery powered toys, frequently found in airport gift shops, are made in Taiwan by unidentified makers. One imagines that they're created by some wizened Chinese tinkerer. Two toys of this ilk, marketed as the Space Circle and the Space Trapeze, have been subjected to mathematical analysis and found to be truly chaotic. These toys and all of the carnival rides mentioned are essentially pendulums driven by some form of periodic forcing. Thus they're all cousins of the driven pendulum that we've used as our primary example.

More recently, several toys, usually found in science stores, have been marketed primarily as illustrations of chaotic motion. These include the Magnetron, the MagnaSwing, and the Pendulum Man. All are apparently chaotic, but they're not battery powered, so the chaos doesn't persist. After an initial push and a burst of chaotic motion, friction gradually brings the toy to a stop. Nonetheless, it's encouraging that chaos is now readily accessible in forms that could lead to interesting school science projects.

12.1 Sellner

The Tilt-A-Whirl may well be the granddaddy of all commercial ventures based on a mathematically simple form of chaotic motion. Its inventor, Herbert Sellner (1887–1930), was a woodworker, not a mathematician, and he discovered the action of the Tilt-A-Whirl by trial and error. For years Sellner ran a woodcraft factory in Faribault, Minnesota. Initially he produced furniture and lamps in the mission style and later wooden toys: kiddies cars, tricycles, and a coaster sled. But in the 1920s Sellner began to invent devices for amusement parks— including a "water toboggan

Fig. 12.2 Herbert Sellner. (Courtesy of Sellner Manufacturing.)

slide" that launched a rider on a sled at motor-boat speed some 100 feet across a lake.

The Tilt-A-Whirl was invented when a park owner challenged Sellner to come up with something for those too timid to ride the toboggan slide. With his mind set in motion, Sellner began playing with a swivel chair mounted on a table. Rocking the table back and forth, he found that the chair would swing around this way and that with a gentle whirling action that seemed about right. But most important, when Sellner put his son in the chair, he discovered that the contraption was fun—his son loved it. Further experimentation produced a prototype Tilt-A-Whirl that was immediately purchased by the Wildwood Amusement Park in White Bear Lake, Minnesota. That was in 1926, and the ride was such a success that Sellner sold six more units before year's end. At last count, Sellner Manufacturing has produced well over 1,000 Tilt-A-Whirl rides, and more than 600 are currently operating in locations around the world.

Although the Tilt-A-Whirl has seen many improvements over the years, Sellner's original patent application still captures the machine's basic principles. As seen in Fig. 12.3, the ride consists of several cars, each mounted on a circular platform, so that the car is free to rotate about the platform's center. This arrangement replicates the original swivel chair attached to a table. To tilt the platforms, Sellner had them travel in a circle around an undulating track. A platform tilts as it climbs

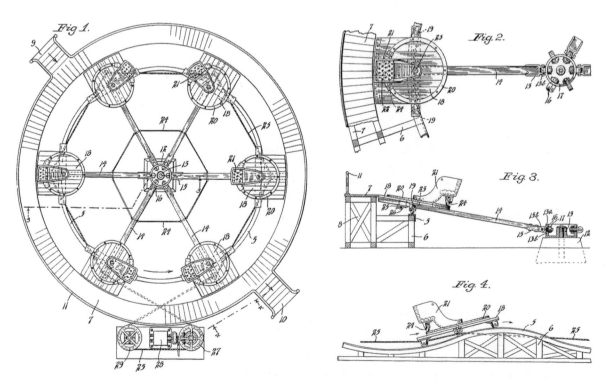

Fig. 12.3 Drawings from the 1926 patent application for the Tilt-A-Whirl.

or descends one of the undulations, and this induces the car to swing one way or another. Moreover, because each platform is tied to the center of the machine by a rigid beam, its direction of tilt may be toward or away from the center as well as in the direction of the track.

Whether or not Sellner's intent was to build randomness into the Tilt-A-Whirl, he certainly recognized the advantage of chaotic motion when he saw it. As noted in Chapter 1, Sellner wrote in his patent application, "A further object is to provide amusement apparatus wherein the riders will unexpectedly swing, snap from side to side or rotate without in any way being able to figure what movement may next take place in the car." Sellner deserves ample credit for this observation, especially since any engineering professor of his day would have guessed that the Tilt-A-Whirl's motion would be entirely regular.

12.2 Mathematical model

Our challenge now is to reduce Sellner's machine to a simple set of equations that we can solve on the computer and see for ourselves whether its motion is regular or random. Unlike Ueda's driven oscillator, which was carefully constructed to match a specific equation, Sellner didn't worry about the mathematical simplicity of the Tilt-A-Whirl. Thus, in deriving a manageable equation, the best we can hope for is an approximation that captures the essence of the ride.

The process of finding such an equation is known as mathematical modeling, and it's as much art as science. Otto von Bismarck (1815–1898), the German Chancellor, once said that "Laws are like sausages—it is better not to see them being made," and the same might be said for mathematical models. All the same, modeling is at the interface between theory and experiment, at the working center of physics, and taking a peek at the process is worth the effort.

We begin by adopting an idealized geometry for the Tilt-A-Whirl. In this simplification, shown in Fig. 12.4 for the classic seven-car ride, each circular platform is connected to the center C of the machine by a rigid beam and each car is represented by a pendulum of radius r_2 that rotates about the center P of the platform. The undulating track is taken to have a radius r_1, equal to the distance between C and P. In this model, the location of the car is completely specified by the angle θ of the beam with respect to a stationary x axis and by the angle ϕ of the pendulum with respect to a moving x' axis, aligned with the beam.

A sense of the Tilt-A-Whirl's full three-dimensional geometry is given by Fig. 12.5(a), where we focus on a single platform and car. In our model, the undulating track, which includes three hills, is taken to be an exact sinusoid and the center of the platform is always assumed to be tangent to the track. (Near a valley, tangency requires that the platform penetrate the track, so our model track is a mathematical fiction.) At any point along the track, the tilt of the platform is specified by the angles

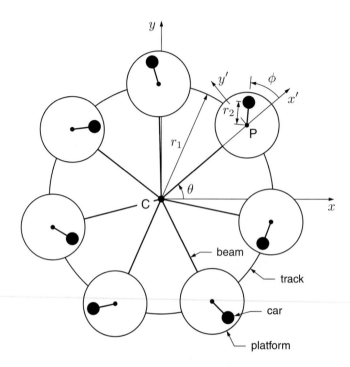

Fig. 12.4 Plan view of the mathematical Tilt-A-Whirl.

α (tilt in the radial direction, Fig. 12.5(b)) and β (tilt in the azimuthal direction, Fig. 12.5(c)). As you might be able to intuit, tangency between the platform and track means that α and β are simple functions of the beam angle θ. Thus, as θ advances and the platform moves along the track, its tilt changes in a regular way.

In adopting a simplified geometry, we make several approximations. For example, the actual beams don't extend all of the way to the center of the machine, the actual track isn't a precise sinusoid, and an actual platform, supported by two wheels, is only approximately tangent to the track. Like a legislature making laws, we have arrived at a compromise—in our case between an accurate description of the machine and mathematical simplicity.

From the geometry, we now make a mathematical leap to the final equations of motion for the Tilt-A-Whirl. Considering just one car, we can describe the motion in terms of the angle θ of the beam and the angle ϕ and angular velocity v of the car. In incremental form, the equations are,

$$\theta_\mathrm{n} = \theta_\mathrm{o} + \omega \Delta t, \tag{12.1}$$

$$\phi_\mathrm{n} = \phi_\mathrm{o} + v_\mathrm{o} \Delta t, \tag{12.2}$$

$$v_\mathrm{n} = v_\mathrm{o} + [\{\epsilon - \cos(3\theta_\mathrm{o})\} \sin \phi_\mathrm{o} - 3 \sin(3\theta_\mathrm{o}) \cos \phi_\mathrm{o} - \rho v_\mathrm{o}] \Delta t. \tag{12.3}$$

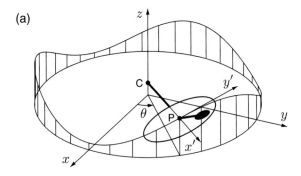

(a)

(b)

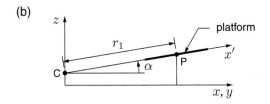

(c)

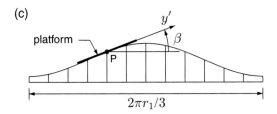

Fig. 12.5 (a) Perspective view of a Tilt-A-Whirl with one platform and car. (b) Cross-section taken through the center of the machine C and the center of the platform P. (c) Side view with the track unrolled in the tangent plane. In all views, vertical dimensions are exaggerated by a factor of 2 for clarity.

Assuming we know the constant parameters ω, ϵ, and ρ, supplying the old angles and the old velocity (θ_o, ϕ_o, v_o) on the left side of these equations yields the new ones (θ_n, ϕ_n, v_n) a short time Δt later.[1]

Although we won't present a derivation of Eqs. (12.1)–(12.3), it's useful to think about them for a moment. For example, Eq. (12.1) states that the beam angle θ advances at a constant rate specified by the parameter ω. For this condition to hold, the motor must be powerful enough to maintain its speed regardless of whether the platforms are being towed uphill or down. Equation (12.2) states that the angle ϕ of the car advances in proportion to the car's angular velocity v, but this is just the definition of v. Finally, according to Eq. (12.3), the car's velocity changes in proportion to the complicated quantity in square brackets, which must be its angular acceleration.

Since the angular acceleration is proportional to the various torques acting on the car, it's natural to ask what torques have been included in Eq. (12.3). First, the terms proportional to $\cos(3\theta)$ and $\sin(3\theta)$ oscillate three times as the platform makes one circuit of the track, so it's natural and correct to assume that these terms account for torques produced by gravity when the platform is tilted. Second, with the platform moving in a circle, there should be a centrifugal force on the car associated

[1] Equations (12.1)–(12.3) are written in terms of dimensionless quantities with angles in radians. In our model, the tilt angles α and β depend on the beam angle θ according to $\alpha = \alpha_0 - \alpha_1 \cos(3\theta)$ and $\beta = 3\alpha_1 \sin(3\theta)$, where α_0 is the average tilt in the radial direction and α_1 is the amplitude of the variation. The time t is in units of $1/\Omega_0 = \sqrt{\alpha_1 g/r_2}$, where g is the acceleration of gravity and Ω_0 is a measure of the car's natural oscillation frequency. The dimensionless parameter ω defines the speed at which the platform circles the track. If T is the time required to complete a circuit, then $\omega = 2\pi/(\Omega_0 T)$. Finally, the parameter ϵ is given by $\epsilon = \alpha_0/\alpha_1 - \omega^2 r_1/r_2$ and the parameter ρ is roughly the inverse of quality factor Q of the car's natural oscillations. For the classic seven-car Tilt-A-Whirl presented here, we have $\alpha_0 = 0.036$, $\alpha_1 = 0.058$, $r_1 = 4.3$ m, $r_2 = 0.8$ m, and $\rho = 0.2$.

with the platform's centripetal acceleration. The torque produced by this centrifugal force is hidden in the term proportional to ϵ. Lastly, we have included a torque proportional to ρv that accounts for friction, just like the one introduced in Chapter 5 for a simple pendulum. And that's all there is to the equations of the Tilt-A-Whirl.

Now, not to pull the wool too tightly over your eyes, it's time to admit that there are a couple more approximations in Eqs. (12.1)–(12.3). To obtain the "simple" form given here, we fudged the geometry somewhat by assuming that the tilt angles α and β are always much less than a radian. But, since neither angle exceeds about 0.17 radian (10°), this condition is pretty well satisfied. More important, it must be admitted that including friction with the viscous-damping term ρv probably isn't justified. Using a viscous-damping term is like throwing an unsavory bit of pig into a sausage.

As Sellner knew well, a major source of friction in the Tilt-A-Whirl is the "rolling friction" of the wheels that support the car on its platform. This friction, sometimes called "rolling resistance," results because the weight of the car deforms the surface on which the wheel rides. As the wheel moves, it must continually climb out of the resulting dimple. Understanding this problem, Sellner took a cue from railroad engineers and made both the wheels of the car and the outer edge of its circular platform from steel. With a hard wheel on a hard track, the dimple is minimized, and both the railroad and the Tilt-A-Whirl experience less friction. Unfortunately, the mathematical form of rolling friction is a little complicated, so (now's the time to avert your eyes if you're squeamish) we have replaced it with simple viscous damping.

All in all, Eqs. (12.1)–(12.3) include at least a half dozen approximations, some more serious than others. In time, we might investigate the adequacy of these equations by experimental tests and additional calculations, but for now let's see if our model predicts motion at all like that of a Tilt-A-Whirl.

12.3   Dynamics

Suppose for a moment that we turn off the motor that drives the Tilt-A-Whirl and consider what happens if we give one of the cars a push while its platform is stopped. In this case, $\omega = 0$, and Eqs. (12.1)–(12.3) reduce to those of a simple pendulum. The equilibrium angle, natural frequency, and quality factor all depend on the tilt angles α and β, but the car swings back and forth with damped oscillations just like any other pendulum. You usually won't observe these oscillations in a real Tilt-A-Whirl because the operator sets a brake as the platforms come to a stop, locking the cars in place to facilitate unloading and loading passengers. Nonetheless, when you ride the Tilt-A-Whirl, you actually ride on a giant pendulum.

You may remember that a pendulum's equation of motion doesn't depend on the mass of the bob. Similarly the Tilt-A-Whirl's equation doesn't depend on the car's mass. This feature has probably contributed to the ride's success because it means that the motion of a car isn't strongly affected by the number of riders. Provided the car's center of mass (which defines r_2) isn't altered significantly by adding passengers, the dynamics of a car will be similar whether it's empty or full. This is a fortuitous feature of the Tilt-A-Whirl.

Now let's turn on the motor and see what happens when the platforms begin to move. First suppose that the platforms move around the circular track very slowly. If a platform moves slowly enough then its car will always be found on the downhill side of the platform, at its equilibrium angle. Figure 12.6 illustrates this case as the platform advances from one valley of the track to the next. For $\theta = 0$ in Fig. 12.6(a), the platform is at the bottom of a valley and tilts slightly toward the outside of the machine, so the car pivots to the outside. As the platform climbs the hill in Fig. 12.6(b), the car swings around and follows the platform up the hill. At the top of the hill in Fig. 12.6(c), the tilt is toward the center of the machine and the car now pivots inward. Coming down the hill in Fig. 12.6(d), the car swings around to lead the platform, and it ends up swinging toward the outside again when it reaches the bottom of the next valley, in Fig. 12.6(e).

Thus, at low speed we expect a Tilt-A-Whirl car to make exactly one revolution with respect to its platform each time the platform goes over a hill. The motion is exactly periodic, repeating itself on every hill as the tilt angles pass through one complete cycle. On the other hand, if the platforms go around the circular track very rapidly, we expect that the centrifugal force will throw all of the cars to the outside of the machine, and the net rotation of the car will be zero on every hill. In either extreme, fast or slow, we get a boring ride with periodic motion synchronized to the tilt cycle of the platform.

Sellner may have anticipated these extreme cases and guessed that something interesting would happen between the fast and slow limits. We can check this possibility by solving Eqs (12.1)–(12.3) on a computer. Figure 12.7(a) displays the nature of the long-term motion of the Tilt-A-Whirl for a range of platform speeds between 1 and 8 rpm. At each speed, the motion was first computed as the platform traversed 100,000 hills to allow the motion of the car to settle into its steady state. The net rotation of the car was then plotted for each of the next 100 hills. At a relatively slow speed, say with the platform making 2.5 circuits of the track in a minute (2.5 rpm), we see from Fig. 12.7(a) that the car rotates backwards by exactly 1 full revolution as it goes over each and every hill. This is the expected result at slow speed, where ϕ decreases by 2π as the platform goes from one valley to the next. Similarly at high speeds like 8 rpm, the net rotation of the car between successive valleys is always exactly 0, due to the large centrifugal force.

In addition to periodic motion synchronized with the tilt cycle of the platform, Fig. 12.7(a) includes behavior demonstrating the Tilt-A-Whirl's essential nonlinearity. For example, at platform speeds near 1.16, 2.00, 4.88, and 6.94 rpm, we find two values of car rotation for each

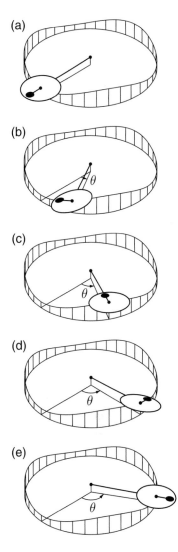

Fig. 12.6 A sequence showing the positions of a Tilt-A-Whirl car as its platform moves slowly from one valley to the next. Equilibrium positions of the car are shown for beam angles $\theta = 0$ (a), $\pi/6$ (b), $\pi/3$ (c), $\pi/2$ (d), and $2\pi/3$ (e).

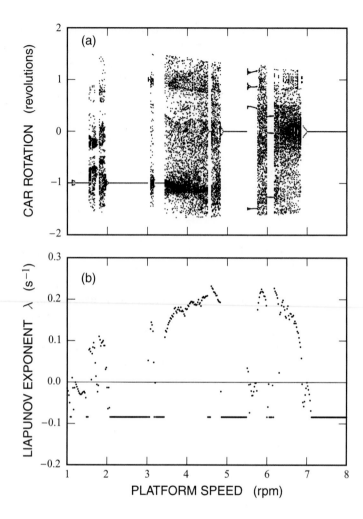

Fig. 12.7 (a) A scatter plot of the net rotation of the car on successive hills as a function of platform speed. For each speed, the motion of the car is calculated for 100,000 hills to eliminate transients, then the rotation on each of the next 100 hills is plotted. (b) The Liapunov exponent as a function of platform speed. Exponents are averaged over 10,000 hills. The parameters for the Tilt-A-Whirl are $\alpha_0 = 0.036$, $\alpha_1 = 0.058$, $r_1 = 4.3$ m, $r_2 = 0.8$ m, and $\rho = 0.2$.

speed. At a speed of 1.16 rpm, the car rotation alternates between values that are just over and just under -1 revolution on successive hills. This corresponds to motion with a periodicity of two tilt cycles, breaking the time symmetry of the Tilt-A-Whirl equations through period doubling. Similarly, at 5.64 rpm, we find four values of car rotation and the car completes exactly one forward revolution each time it passes over four hills, completing four tilt cycles. This extended period would make the ride more fun because it includes both forward and backward rotations.

However, there are also ranges of platform speed for which the car doesn't settle into periodic motion of any kind. At these speeds we find that the sample of 100 rotation values in Fig 12.7(a) is spread between roughly -1.5 and 1.5 revolutions. Of course, this scatter represents the chaotic motion of the Tilt-A-Whirl that we first encountered in Chapter 1. At these particular speeds, the model predicts motion with an apparently random element that continues indefinitely, just as in the actual ride.

While we may have reservations about our mathematical model, the strong resemblance between the observed and predicted motion of the ride suggests that the model is basically sound. Moreover, the band of chaotic behavior in Fig. 12.7(a) between 6.18 and 6.88 rpm matches the typical 6.5 rpm operating speed of the actual Tilt-A-Whirl. Thus, in some respects the model even appears to be quantitatively accurate. To see the Tilt-A-Whirl model in animation, check out Experiment 14 of the Dynamics Lab and try a few different platform speeds for yourself.

12.4 Liapunov exponent

Now that we've had a brief look at the various types of motion exhibited by the Tilt-A-Whirl, it's time to ask what happens when the motion is perturbed by a small disturbance or offset. As we learned in Chapter 10, in the long term the difference between the perturbed and unperturbed trajectories tends to follow an exponential of the form $e^{\lambda t}$, where t is the time and λ is a constant called the Liapunov exponent. For stable trajectories λ is negative and the offset decays in time, while for unstable trajectories λ is positive and the offset grows in time.

Figure 12.7(b) plots λ for each of the trajectories in Fig. 12.7(a). In every case, a negative Liapunov exponent corresponds to exactly periodic motion. The periods of these solutions are related to the number of rotation values recorded in Fig. 12.7(a), and at various platform speeds we find periodicities of 1, 2, 3, 4, 6, 8, and 16 tilt cycles. Likewise, a positive exponent always corresponds to chaotic motion. In most cases, chaos leads to a scatter of car rotations spanning the range from roughly -1.5 to 1.5 revolutions. However, at a few platform speeds, including 1.12, 1.72, 1.98, 3.02, 5.52, 6.3, and 6.86 rpm, the exponent is positive but relatively small and the scatter is restricted to a narrow range of rotations.

The magnitude of the Liapunov exponent tells us how fast an offset grows or decays on average. Thus, at a typical speed for which the Tilt-A-Whirl's motion is periodic, say 2.5 rpm, we find $\lambda = -0.0843$ s^{-1}, and offsets decay by a factor of $1/e$ during a time of $t_{1/e} = 1/|\lambda| = 11.9$ s. At 2.5 rpm, the platform requires 8 s to go over a hill, so only a few tilt cycles are required for the car to return to its periodic steady-state motion.

What about the Liapunov exponent for chaotic motion? To be specific, let's focus on a platform speed of 5.9 rpm. This speed is close to that of the actual machine and the scatter plot in Fig. 12.7(a) suggests that the motion is maximally mixed up. Certainly, the net rotations plotted in Fig. 1.2 for this speed would provide continual surprises for riders of the Tilt-A-Whirl. At 5.9 rpm, we find $\lambda = 0.221$ s^{-1} when the growth of an infinitesimal offset is averaged over 10^6 hills. Thus, the offset between a perturbed trajectory and the original will grow on average by a factor of e in $t_e = 1/\lambda = 4.5$ s or by a factor of 2 in $t_2 = 3.1$ s. Since a platform requires 3.4 s to traverse a hill at 5.9 rpm, an offset trajectory diverges from the original on average by an additional factor of just over 2 on every hill. To learn more about the computation and meaning of Liapunov exponents for the Tilt-A-Whirl, try Experiment 15 of the Dynamics Lab.

The exponential growth of a small offset has interesting consequences for riders of the Tilt-A-Whirl. For example, it assures riders that every time they step aboard, the ride that they receive will be at least slightly different than any previous ride. To back up this statement, let's estimate the number of different experiences that might result. Typically, a ride on the Tilt-A-Whirl lasts between one and two minutes (short enough to avoid too much of good thing) and at 5.9 rpm a car will traverse no more than about 35 hills. But in two minutes we know that any initial offset in the angle of a car will grow by a factor of $\exp(\lambda t) = \exp[(0.221 \text{ s}^{-1})(120 \text{ s})] = 3 \times 10^{11}$. If we suppose that two trajectories (rides) are significantly different if the final angles differ by at least 1 radian, then their initial angles must differ by at least $(1 \text{ radian})/(3 \times 10^{11}) = 3 \times 10^{-12}$ radian, and the total number of initial angles leading to different rides is $(2\pi \text{ radian})/(3 \times 10^{-12} \text{ radian}) = 2 \times 10^{12}$. That's a huge number of different rides, and we haven't accounted for the fact that a car's trajectory is affected by the initial location of the platform as well as that of the car. Thus, it's safe to assume that the exact sequence of forward and reverse whirls that you experience will be different every time you step onto a Tilt-A-Whirl. This conclusion is also supported by the variety of 35-hill rotation sequences seen in Fig. 1.2.

Another consequence of a positive Liapunov exponent has been noticed by aficionados of the Tilt-A-Whirl. In our model it's assumed that the passengers remain fixed within the car and the only time dependence is the regular variation in the tilt angle of the platform as it moves along the track. As long as the passengers sit still, the Tilt-A-Whirl is deterministic and the observed motion can be considered chaotic. However, many riders discover that they can affect the motion of the car by throwing their weight from one side to the other at crucial moments. Such actions can change what might have been a tilt cycle with little or no rotation into one with a good whirl. The ease with which a passenger can modify his trajectory in this way is a consequence of the sensitivity of chaotic motion to small perturbations. As noted above, perturbations grow on average by factor of 2 every 3.1 s. Very probably, perturbations occurring at special times will grow even faster, making the consequences of a well timed lunge almost immediate. Aficionados of the ride can thus take advantage of its extreme sensitivity to control their motion and heighten the excitement.[2] This type of control is even more obvious in rides with lighter cars like the Octopus.

[2]Of course, extreme sensitivity applies only to chaotic motion. On one occasion, my car fell into periodic motion, and I was mystified when my attempts to break the cycle with a well timed lunge were completely ineffective. But with a negative Liapunov exponent, the perturbations that I produced were exponentially attenuated rather than amplified.

12.5   Computational limit

The positive Liapunov exponent of chaotic motion also informs us about the ultimate accuracy of our ability to predict the future. To be sure, our mathematical model of the Tilt-A-Whirl is only approximate, so we can't expect it to accurately predict the motion of the real-world machine. But, even if we assume a perfect correspondence between the model and the machine, a positive exponent places severe practical restrictions on what we can actually compute.

Suppose, for example, that our computer carries 15 digits of accuracy in all of its numbers and calculations. This is more than enough accuracy for simulating most physical processes. However, suppose we'd like to know what a car on the Tilt-A-Whirl is doing—spinning forward, backward, or momentarily stopped—at some point in the future. We'll assume that our model, Eqs. (12.1)–(12.3), is exact and that we know the initial values of θ, ϕ, and v exactly. The only problem is that the computer carries a finite number digits.

Under these ideal conditions, how far into the future can we expect our calculation to remain accurate? The answer is, of course, determined by the Liapunov exponent. If λ is negative, then the fact that the computer rounds all numbers to 15 digits is of little significance. In this case, the error introduced by rounding off tends to be reduced as the computation proceeds and the trajectory hones closer and closer to an accurate periodic trajectory.

On the other hand, if λ is positive, then round-off errors are amplified as the computation proceeds, until even the most significant digit can no longer be trusted. Exactly how long can we compute before our prediction of the car's position and velocity is completely off the mark? The answer is simply the time at which $e^{\lambda t}$ equals 10^{15}, since by then the original round-off error of 10^{-15} will have grown to be of order 1. For a platform speed of 5.9 rpm, this time amounts to $t = (15/\lambda)\ln(10) = 156$ s, or about the time required for the platform to traverse 46 hills. After 46 hills or about 2.6 minutes, our computation can no longer be trusted to tell us whether the ideal Tilt-A-Whirl is rotating forward or backward.

This is an embarrassingly short time for someone like Laplace, who once imagined predicting the entire past and future of the universe! Of course, Laplace only argued that prediction was possible in principle. But still, our inability to know the motion of a simple machine like the Tilt-A-Whirl beyond a couple of minutes into the future is a real setback. Laplace might not have been quite so rash if he had understood the nature of chaotic motion.

However, there's no reason to limit our computer to 15 digits of accuracy. With special programming, we can extend the computational accuracy to as many digits as might be required. Suppose then that we'd like to know a car's direction of rotation an hour after it's set loose with a precise set of initial conditions. The required number of digits N is then given by $10^N = e^{\lambda t}$ or $N = \lambda t/\ln(10) = (0.221 \text{ s}^{-1})(3,600 \text{ s})/\ln(10) = 346$. Could we carry out such a calculation using 346 digits of accuracy? Possibly, but the effort needed would be no less than heroic. Present computer technology simply doesn't allow a practical computation of the Tilt-A-Whirl's state of rotation an hour into the future. In practice, the extent to which we can know the future of a chaotic system is strictly limited.

But wait, didn't we compute the Liapunov exponent by averaging the growth of an infinitesimal offset over 10^6 hills, or the equivalent of 940 hours of Tilt-A-Whirl operation? Wasn't that computation hopelessly inaccurate long before it was complete? The answer to this question is both yes and no. Since we used numbers with just 15 digits of accuracy, the simulation did not accurately predict the correct position of the car beyond the first 46 hills. However, the simulated motion doesn't go crazy after 46 hills—it continues with behavior patterns typical of the machine, exploring more and more of the possible rotation sequences. Thus, by extending the computation to 10^6 hills, we learn more about the range of motion, and quantities like the average rotation of a car or the Liapunov exponent become more accurate. As it happens, at 5.9 rpm the rotation averaged over 10^6 hills is -0.173 revolution per hill.

The difference between calculating the position of a car on the Tilt-A-Whirl and its average rotation is exactly the difference between weather prediction and climate modeling. As we argued in Chapter 10, it's very difficult to produce reliable long-range weather forecasts—the exponential growth of errors quickly clouds the forecaster's crystal ball. However, a climatologist can use the same set of equations to examine the weather year after year and gradually accumulate statistics that will determine the average temperature at Los Angeles in July or the average snowfall at Anchorage in January. A climate modeler can't tell you whether your picnic next Saturday will be rained out, only what will happen on average. On the other hand, by adjusting the parameters of his model, a climatologist may well predict where we'll be growing our food when atmospheric carbon dioxide is twice what is was a century ago. Even in the presence of chaos, average quantities remain useful and are easy to calculate.

12.6   Environmental perturbation

In his 1972 talk before the American Association for the Advancement of Science, Edward Lorenz speculated on the effect a butterfly might have on the weather without reaching a quantitative conclusion. The Tilt-A-Whirl is simple enough, however, that we can estimate how long it would take for a small perturbation to grossly affect its motion. Instead of a butterfly, we'll imagine that a nearby observer takes a step closer to the Tilt-A-Whirl then steps back again, thereby momentarily increasing his gravitational force on the cars. Like Lorenz, we propose a thought experiment—an experiment that could never actually be performed. But, using a little math, we can compare the motion of the Tilt-A-Whirl with and without the fateful steps taken by our observer, assuming that everything else in the universe proceeds identically in the two cases.

To make this comparison, we'll perform a "back-of-the-envelope" calculation, a kind of rough and ready estimate that is a part of every working scientist's tool kit. Our object isn't accuracy but capturing the essential effect with a few simple equations that might be scribbled on a handy scrap of paper.

The first step is to envision the calculation in outline. To begin, we recall that the angular offset ϕ_{off} produced by the observer's gravitational force on the car can be combined with the Liapunov exponent λ to calculate the time t_g required for the offset to grow to an observable size. But how can we estimate ϕ_{off}? We'll use our knowledge of dynamics. In short, the increase in gravitational force F_{ex} effected by our observer will produce an extra torque τ_{ex} on a Tilt-A-Whirl car, the extra torque will lead to an extra angular acceleration α_{ex}, which will yield an extra angular velocity v_{ex}, and finally the extra velocity will yield the small angular offset that we want to know. That's "all" there is to it!

Now that we have a plan, let's put it into effect, beginning with an estimate of the extra force created when the observer takes a step closer to the Tilt-A-Whirl. In Chapter 4, we learned that the gravitational force between two objects is proportional to the product of their masses and inversely proportional to the square of the distance between them: $F = Gm_1m_2/r^2$, where G is the gravitational constant. Suppose that the observer has a mass m and is initially a distance R from a Tilt-A-Whirl car of mass M. When the observer steps a distance δ closer to the car, the increase in gravitational force is the difference between the force at a separation of $R - \delta$ and that at a separation of R,

$$F_{\text{ex}} = \frac{GmM}{(R - \delta)^2} - \frac{GmM}{R^2}. \tag{12.4}$$

This formula would work perfectly well for our purpose, but in the spirit of a back-of-the-envelope calculation, let's simplify it a little by assuming that the observer's step size δ is much smaller than his distance R to the Tilt-A-Whirl. Expanding the square $(R - \delta)^2$ and factoring out R^2, we obtain,

$$(R - \delta)^2 = R^2(1 - 2\delta/R + \delta^2/R^2) \approx R^2(1 - 2\delta/R). \tag{12.5}$$

In the last step, we've dropped the term δ^2/R^2 because we've assumed that $\delta/R \ll 1$ and the square of this quantity is even smaller. Combining Eqs. (12.4) and (12.5) now yields,

$$F_{\text{ex}} = \frac{2GmM\delta}{R^3(1 - 2\delta/R)} \approx \frac{2GmM\delta}{R^3}, \tag{12.6}$$

where in the last step we've again used the fact that $\delta/R \ll 1$ to eliminate this term in the denominator.

Now that we have an expression for the extra force on the car, we can convert it to a torque simply by multiplying by the distance r_2 from the car's center of rotation to its center of mass,

$$\tau_{\text{ex}} = \frac{2GmM\delta r_2}{R^3}. \tag{12.7}$$

This is actually the maximum possible extra torque since we've assumed that the force is applied perpendicular to the vector $\mathbf{r}_2$. On the other hand, our calculation isn't supposed to be exact.

The extra torque can be converted to an extra angular acceleration α_{ex} using the equation of motion for rotating systems developed in Chapter 5. In this case, we have $\alpha_{\mathrm{ex}} = \tau_{\mathrm{ex}}/I$, where I is the car's moment of inertia, $I = r_2^2 M$. Combining these equations with Eq. (12.7), we find

$$\alpha_{\mathrm{ex}} = \frac{2Gm\delta}{r_2 R^3}. \tag{12.8}$$

The extent to which the acceleration affects the car's angular velocity will depend on how long the extra torque is applied. If we suppose that the observer waits for a time t_w before stepping back to his original position, then the extra angular velocity is simply $v_{\mathrm{ex}} = \alpha_{\mathrm{ex}} t_w$ or,

$$v_{\mathrm{ex}} = \frac{2Gm\delta t_w}{r_2 R^3}. \tag{12.9}$$

It's now tempting to say that the angular offset produced by this velocity is $v_{\mathrm{ex}} t_w$, but this assumes that the extra velocity acts over the entire time interval t_w. In fact, the extra velocity builds from 0 to v_{ex} over this interval, so the average extra velocity is half of v_{ex} and the net angular offset is

$$\phi_{\mathrm{off}} = \frac{v_{\mathrm{ex}} t_w}{2} = \frac{Gm\delta t_w^2}{r_2 R^3}. \tag{12.10}$$

With this equation, thanks to our "back-of-the-envelope" philosophy, we've arrived at a simple but reasonable estimate for the angular offset produced by the incidental motion of an observer near the Tilt-A-Whirl.

Now let's plug in some numbers and see how big the offset given by Eq. (12.10) really is. Suppose our observer weighs $m = 50$ kg, is initially stationed $R = 10$ m from the Tilt-A-Whirl, takes a step $\delta = 1$ m closer, and returns to his initial position after $t_w = 2$ s. Given that $G = 6.67 \times 10^{-11}$ m^3/kg/s^2 and $r_2 = 0.8$ m, we find $\phi_{\mathrm{off}} = 1.67 \times 10^{-11}$ radian. Translated into a displacement of the car's center of mass, this angle corresponds to an offset of about 10^{-11} m or $\frac{1}{30}$ the diameter of a typical atom. Given the exceedingly weak gravitational force between objects of human size, we shouldn't be too surprised that the offset produced by the action of our observer is minuscule.

All the same, we know that this tiny offset can be amplified very quickly to macroscopic dimensions by chaotic motion. Because the difference between the original and perturbed trajectories grows as $e^{\lambda t}$, the time t_g required for the initial angular offset to reach 1 radian is given approximately by $\phi_{\mathrm{off}} e^{\lambda t_g} = 1$ or

$$t_g = \frac{1}{\lambda} \ln\left[\frac{1}{\phi_{\mathrm{off}}}\right]. \tag{12.11}$$

For $\phi_{\mathrm{off}} = 1.67 \times 10^{-11}$ and $\lambda = 0.221$ s^{-1}, the growth time is $t_g = 112$ s or just under two minutes. Thus, if an observer standing near a Tilt-A-

Whirl happens to step closer to the machine for a couple of seconds, two minutes later the riders would be whirling completely differently than they would if the observer had remained still.

While not exactly the flap of a butterfly's wing, the incidental motion of our observer has a far larger effect on the Tilt-A-Whirl than we might reasonably have expected. Of course, if the Tilt-A-Whirl were operated at a speed where its motion is periodic and the Liapunov exponent is negative, then the observer's influence would be completely negligible—the initial angular offset would be the same, but it would rapidly decay to nothing. The power of our observer is completely contingent on the chaotic motion of the Tilt-A-Whirl.

But let's take this scenario one step further by moving the observer farther from the Tilt-A-Whirl. Suppose that the Tilt-A-Whirl is located in Denver but the observer is in New York City. Will his fateful step still have a significant effect? If we combine Eqs. (12.10) and (12.11), the growth time can be written as

$$t_g = \frac{1}{\lambda} \ln \left[\frac{r_2 R^3}{Gm\delta t_w^2} \right], \tag{12.12}$$

and increasing R from 10 m to 2.6×10^6 m (the distance for Denver to New York) increases t_g from 1.9 to 4.7 minutes. That is, even if the observer is in New York, his step will grossly affect the Tilt-A-Whirl in Denver 4.7 minutes later. What if we make our observer an astronaut on the Moon? In this case, the observer is 3.8×10^8 m from the Tilt-A-Whirl, and his step affects the ride after 5.8 minutes. Finally, suppose our observer is an alien on a planet orbiting Alpha Centauri, a star some 4.3 light years or 4.1×10^{16} m distant. The alien's step would reach Earth as a gravitational wave 4.3 years after it was taken, but once the wave arrives the Tilt-A-Whirl in Denver will have to operate just 10 minutes before riders are grossly affected.

These are astonishing results that reveal the dramatic power of exponential growth. Because offsets in the chaotic Tilt-A-Whirl grow exponentially, as $e^{\lambda t}$, while the perturbation becomes smaller only as the inverse cube of the observer's distance, or $1/R^3$, we really have an unfair competition. Numerically, for $\lambda = 0.221$ s^{-1}, increasing R by a factor of 10 and reducing ϕ_{off} by a factor of 1,000 will extend the time required for grossly affecting the machine's motion by just 31 s. That is, each time the observer is moved 10 times farther away, we need wait only an extra half minute before the Tilt-A-Whirl is affected just as strongly as before. That's the power of exponential growth.

Is there any limit to this crazy sensitivity of the Tilt-A-Whirl? With Eq. (12.12) you could easily estimate the effect of alien motion in a nearby galaxy or halfway across the universe. But we should question whether the classical theory used here is valid for the exceedingly tiny offsets implied for the Tilt-A-Whirl car. Quantum effects may well limit the offset in the car's center of mass to a distance no smaller than some fundamental length. For example, displacements smaller than the Planck length, $\ell_p = \sqrt{hG/2\pi c^3} = 1.6 \times 10^{-35}$ m (where h is Planck's constant and c is the speed of light), might be not be allowed. In this case, our observer's motion would have no effect beyond about $R = 10^9$ m or roughly twice the distance to the Moon. While this limit is sheer speculation, you should be aware that we may be pushing the validity of our physical model.

If our classical calculation is correct even for modest observer distances, the Tilt-A-Whirl is still a remarkably sensitive machine. Is it possible to use this sensitivity to detect some distant event? In a word, no. The problem is that our thought experiment compares two universes that evolve identically except for our observer's incidental step. This scenario allows us to ignore the motion of all the other objects in the vicinity, but it's not realistic. In reality, the Tilt-A-Whirl is sensitive to the motion of everyone and everything within some large distance, so its long-term motion cannot be predicted in any practical sense. According to our calculation, if we wished to know the motion of a Tilt-A-Whirl in Denver just five minutes into the future, it would be necessary at a minimum to account for the motion all masses of human scale within the entire United States. And this assumes a perfect model of the machine itself. In fact, while the ideal Tilt-A-Whirl is deterministic, there is no way to predict the motion of the real machine even one minute into the future. Herber Sellner exaggerated only slightly when he said that riders would be unable "to figure what movement may next take place in the car."

12.7   Long-lived chaotic transients

Before leaving the Tilt-A-Whirl, I want to introduce a special kind of transient chaos, that's tangential to chaos itself, but nicely exemplified by the ride.

The surprising phenomenon of long-lived chaotic transients was first identified in 1983 by Celso Grebogi, Edward Ott, and James Yorke of the University of Maryland. In the Tilt-A-Whirl, such transients are found for platform speeds between 6.00 and 6.16 rpm, where Fig. 12.7 reveals only stable periodic motion. When this figure was constructed, however, we purposely computed the motion for 100,000 hills to eliminate transient behavior before sampling the trajectory at a given speed. Now let's take a look at the transients that we conveniently avoided plotting.

A typical result is shown in Fig. 12.8 for a platform speed of 6 rpm. Here we plot the net rotation of the car on each hill, beginning with hill number 29,701. The first 29,700 hills aren't shown, but on these first hills the motion appears perfectly chaotic, just as in Fig. 1.2. However, as Fig. 12.8 shows, the chaotic behavior comes to a sudden end with hill 29,767. After this hill, the car quickly settles into periodic motion, with a period of three hills, that continues ever after. The chaos, which we now recognize as no more that a long-lived transient, has given way to stable periodic motion.

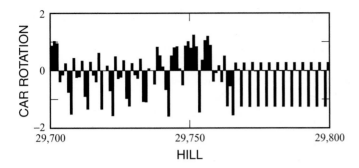

Fig. 12.8 Net car rotation on successive hills for a Tilt-A-Whirl with a platform speed of 6 rpm. Initially, the platform was positioned at the bottom of a valley $(\theta = 0)$ with the car stopped $(v = 0)$ at an angle of $\phi = 1$ radian. The motion of the car is chaotic for more than 29,700 hills but becomes periodic before hill 29,800. The parameters for the Tilt-A-Whirl are $\alpha_0 = 0.036$, $\alpha_1 = 0.058$, $r_1 = 4.3$ m, $r_2 = 0.8$ m, and $\rho = 0.2$.

For anyone experienced with linear systems, this transient behavior is very strange. In a linear system with friction, transients always decay exponentially. While the decay may be fast or slow, a transient disappears bit by bit in a regular way, like the small-amplitude oscillations of the damped pendulum discussed in Chapter 5. Indeed, if we consider the Tilt-A-Whirl for a fixed platform position, the motion of a car reduces to that of a pendulum, and small oscillations decay in proportion to $\exp(-t/t_e)$, with a time constant of $t_e = 11.9$ s. Thus, it's puzzling to observe a chaotic transient that persists for 29,767 hills or about 27 hours without apparent change and then suddenly disappears.

Although we won't resolve the puzzle in this chapter, further numerical experiments shed light on the nature of these long-lived transients. The transient in Fig. 12.8 resulted from an initial state with the platform at the bottom of a valley $(\theta = 0)$ and the car stopped $(v = 0)$ at an angle of $\phi = 1$ radian. Do we obtain the same sort of transient for other initial conditions? Assuming the initial platform angle is always $\theta = 0$, most values of ϕ and v lead to a similar chaotic transient unless (ϕ, v) is very close to one of the three initial conditions, $(1.15, -3.23)$, $(-0.518, -2.32)$, or $(1.29, 2.33)$, that lead directly to the periodic solution. With these exceptions, we always observe a long-lived chaotic transient that suddenly gives way to periodic motion. And simulations reveal that the duration of the transient depends on the initial conditions in a seemingly random fashion.

To investigate further, imagine a collection of 1,000 identical Tilt-A-Whirl machines, each with a single car. Physicists usually call such an imaginary collection of systems an "ensemble." In the ensemble considered here, we assume that initially the platform angles are all $\theta = 0$, the car velocities are all $v = 0$, and the car angles ϕ are distributed uniformly between 0 and 2π. In our numerical experiment, we start all of the machines simultaneously and keep track of when each car switches from chaotic to periodic motion, as if we were watching kernels of corn exploding in a popcorn popper. The results are shown in Fig. 12.9, where we plot the number of cars remaining in the chaotic state as a function of time. This figure uses a logarithmic scale for the number remaining, which makes it easy to see that the number drops by a factor of roughly two after every 34,300 hills. That is, just like a radioactive substance

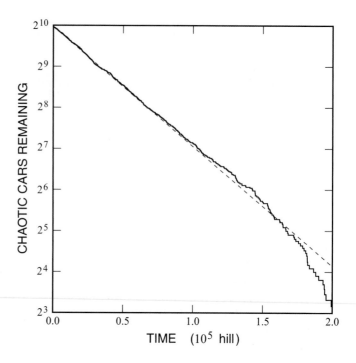

Fig. 12.9 The number of cars remaining chaotic as a function of time for an ensemble of 1,000 identical Tilt-A-Whirl cars operated at 6 rpm. Initially, all platforms are positioned at the bottom of a valley ($\theta = 0$), and all cars are stopped ($v = 0$), but the initial angles of the cars are distributed uniformly between 0 and 2π. For comparison, a dashed line shows exact exponential decay with the same average lifetime as the ensemble. The ride parameters are $\alpha_0 = 0.036$, $\alpha_1 = 0.058$, $r_1 = 4.3$ m, $r_2 = 0.8$ m, and $\rho = 0.2$.

in which the number of atoms that retain their original identity decays exponentially, the number of cars in the chaotic state decays with a half-life of 34,300 hills or 31.8 hours.

Because the half-life is so long, it's unlikely that we'd discover the stable periodic motion at 6 rpm in a real Tilt-A-Whirl. Rides are seldom longer than a couple of minutes, so transient chaos is just as good a stable chaos at the amusement park. On the other hand, this line of reasoning leads us to ask whether the chaos at 5.9 rpm might also be metastable and give way to periodic motion if we waited long enough. Numerical solutions can never prove the long-term stability of a chaotic state, but all indications are that the chaos at 5.9 rpm is truly stable.

Exponential decay in an ensemble of deterministic systems is a new possibility brought to science by chaos theory. In contrast to radioactive decay, which is explained by the intrinsic quantum-mechanical randomness of nuclear tunneling processes, long-lived chaotic transients occur in the absence of explicitly random events. Previously we have shown how other random phenomena like an absence of correlation (Chapter 7), random walks (Chapter 8), and white noise (Chapter 9) are mimicked by chaotic systems. Now we see that chaos is also a possible explanation for the exponential decay of an ensemble of metastable systems. There is no reason to think that the radioactivity of atomic nuclei is due to the chaotic motion of the protons and neutrons inside, but chaotic transients certainly provide a new mechanism for explaining similar decay processes.

Further reading

Chaotic rides

- o Barber, G. L., "Amusement ride", U.S. Patent number 3,495,823, February 17, 1970, filed January 23, 1968 ("Turbo" ride).

- o Brown, J. M., "Plural horizontal axis roundabout having sheave driven carriage", U.S. Patent number 3,596,905, August 3, 1971, filed May 2, 1969 ("Zipper" ride).

- o Emrie, M. W., "Amusement ride", U.S. Patent number 5,688,178, November 18, 1997, filed February 6, 1996 ("Chaos" ride).

- o Eyerly, L. U., "Rotating amusement device", U.S. Patent number 2,113,131, April 5, 1938, filed February 24, 1936 ("Octopus" ride).

- o Page, D., "Formula for fun: The physics of chaos", *Funworld* **12**(3), 42–46 (March 1996).

Chaotic toys

- o Andrews, M. R. and Andrews, R. W., "Magnetic spinner toy", U.S. Patent number 5,135,425, August 4, 1992, filed July 19, 1990 ("Magnetron" toy).

- • Berry, M., "The unpredictable bouncing rotator: a chaology tutorial machine", in *Dynamical Systems: A Renewal of Mechanism*, edited by S. Diner, D. Fargue, and G. Lochak (World Scientific, Singapore, 1986) pp. 3–12 ("Space Trapeze" toy).

- o Samson, I., "Magnetically activated amusement device", U.S. Patent number 5,026,314, June 25, 1991, filed January 4, 1990 ("MagnaSwing" toy).

- • Wolf, A. and Bessoir, T., "Diagnosing chaos in the Space Circle", *Physica D* **50**, 239–258 (1991).

Friction

- o Bowden, F. P. and Tabor, D., *Friction: An Introduction to Tribology* (Anchor Doubleday, Garden City, 1973).

- o Whitt, F. R. and Wilson, D. G., "The wheel and its rolling resistance", in *Bicycling Science* (MIT Press, Cambridge, Massachusetts, 1974) chapter 6.

Tilt-A-Whirl

- • Kautz, R. L. and Huggard, B. M., "Chaos at the amusement park: Dynamics of the Tilt-A-Whirl", *American Journal of Physics* **62**, 59–66 (1994).

- o Sellner, H. W., "Amusement device", U.S. Patent number 1,745,719, February 4, 1930, filed April 24, 1926 ("Tilt-A-Whirl" ride).

Transient chaos

- • Grebogi, C., Ott, E., and Yorke, J. A., "Fractal basin boundaries, long-lived chaotic transients, and unstable-unstable pair bifurcation", *Physical Review Letters* **50**, 935–938 (1983).

13

Billiard-ball chaos—
Atomic disorder

In all of the chaotic systems studied so far, we have included some kind of friction, like fluid viscosity in Lorenz's convection cell or viscous damping in the Tilt-A-Whirl. In fact, friction is virtually always present in the world of macroscopic objects. Even the Solar System, which is usually assumed to be frictionless, is encumbered with forces that dissipate motion. Ocean tides, for example, act as a brake that gradually slows the Earth's rate of rotation. Although you may not have noticed, tidal friction increases the length of a day by 2.3 milliseconds every century. However, we are about to enter the realm of atoms and molecules, where friction is entirely absent and motion can persist indefinitely without a driving force. In fact, in this chapter we will be preoccupied with explaining the macroscopic world in terms of microscopic events, and chaos is at the heart of the story.

Understanding how the macroworld arises from the microworld requires resolving a paradox associated with the arrow of time. When observing the macroworld, we commonly see events, like a cup of coffee being spilled or a baseball crashing through a window, that are perfectly ordinary but look silly when a movie of them is run backwards. Such events define an unmistakable arrow of time, the direction from past to future, that is an integral part of everyday life. There's no use in crying over spilled milk—what's done is done and can't be reversed.

On the other hand, when we inspect the equations of motion for the atoms and molecules of the microworld, there is no corresponding asymmetry between past and future. According to the equations, everything that happens in the microworld is completely reversible. If we could film two oxygen molecules in the air crashing together then heading off in new directions, there would be nothing remarkable to see if the movie were played backwards. According to our intuition and to the equations of motion, the reverse scattering process is just as natural as the forward process. Thus, we arrive at a paradox. If the macroworld is nothing more than a collection of many atoms and molecules whose motions are entirely reversible and lack any hint of a distinction between past and future, why is time's arrow so obvious at macroscopic scales?

The paradox of time's arrow was first recognized in the nineteenth century when the practical, macroscopic science of thermodynamics, literally the dynamics of heat, was recast in microscopic terms as the science of statistical mechanics. Could the reversible dynamics of atoms and molecules really explain the macroscopic world of friction and heat? At the time, the paradox of time's arrow was an open question, but, with a little help from what we now call chaos, theorists ultimately found an answer. To understand the role of chaos in the problem, let's take a brief detour to look at how heat is related to the motion of molecules. We begin with the macroscopic world of thermodynamics.

13.1   Joule and energy

By now the concept of force is familiar—forces are the origin of acceleration and the key to predicting motion. But we have entirely ignored the related concept of energy, and it's time to redress this oversight. Because energy takes many diverse forms, the idea behind it is more elusive than that of force. Indeed, historically a full understanding of energy was slow to develop precisely because of its diversity. We now recognize energies in the macroworld associated with motion, light, heat, electricity, magnetism, elasticity, gravity, and chemistry, to name a few of its venues. During the nineteenth century, scientists began to understand this diversity and recognize an important fact: when all of its forms are duly tabulated, energy is not created or destroyed in any physical process. The chemical energy in a battery can be converted into electrical energy, and the heat energy of steam can be converted into the motion of a train, but the total energy in the universe doesn't change. This remarkable fact, known as conservation of energy or the first law of thermodynamics, is now a basic pillar of physical science.

To introduce energy more formally, let's reconsider Galileo's free-fall experiment. Assuming that the resistance of the air is negligible, this experiment includes two fundamental types of energy, potential and kinetic, which underlie most other forms of energy. The potential energy E_p of a cannonball of mass M positioned at a height H above the ground is given by the simple formula,

$$E_p = MgH, \tag{13.1}$$

where g is, as usual, the acceleration of gravity. Although we won't try to justify this formula, it makes intuitive sense that a cannonball's potential for destruction goes up in proportion to its mass, its height, and how fast it accelerates when allowed to fall. We don't worry about a cannonball if it's very small or if it's lying on the ground, but a massive cannonball falling from a great height is something to be avoided.

Kinetic energy, the second fundamental type, is the energy associated with an object's motion. For a cannonball or any mass moving with a velocity V, the kinetic energy E_k is simply,

$$E_k = \tfrac{1}{2}MV^2. \tag{13.2}$$

Again, we won't justify this formula, but it does make sense that the kinetic energy increases with both the mass and the velocity of an object.

On the other hand, assuming that energy is conserved, we can easily check that Eqs. (13.1) and (13.2) are consistent with Galileo's analysis. In the absence of friction, the initial potential energy of the ball must all be converted to kinetic energy by the time it hits the ground. Thus, if a ball is dropped from a height $H = B$ then its velocity at $H = 0$ is given by $\frac{1}{2}MV^2 = MgB$ or $V^2 = 2gB$. Can you show that the same formula results from Galileo's equations for free-fall, Eqs. (2.9) and (2.10), by setting $H = 0$ and eliminating T?

The conservation of energy in frictionless motion is implicit in Newton's equations of motion. However, energy apparently isn't conserved when friction acts to slow the motion. Thus, a frictionless pendulum can swing back and forth forever, with the potential energy of the bob at the top of its swing converted to kinetic energy at the bottom then back to potential energy as it rises again on the opposite side. But a little friction will gradually bring the pendulum to a stop, dissipating both its kinetic and potential energy. Does friction actually destroy energy? No, as scientists realized in the nineteenth century, friction simply converts kinetic energy into heat, another form of energy.

Earlier, heat was not thought of as energy but rather as a colorless, weightless fluid called caloric. When a pot of soup was heated on a wood stove, scientists envisioned caloric flowing from the stove into the soup, having been released initially by the fire inside. Moreover, caloric was thought to be indestructible, never created or destroyed, with fire or friction only releasing the caloric already stored in a body. However, this picture came under suspicion in the 1790s, when the American-born scientist Benjamin Thompson or Count Rumford (1753–1814) studied the heat produced in the boring of cannons. As Inspector General of Artillery for the Bavarian Army, Thompson had oversight of the production of cannons at the Munich arsenal. Thompson was so impressed by the amount of heat produced in boring—the barrel, tool, and shavings became extremely hot—that he wondered where all of the caloric came from and why it was available in apparently limitless supply. In fact, Thompson had cannons bored while submerged in water and discovered that the water quickly came to a boil and continued to boil as long as boring proceeded. He published these observations in 1798, raising doubts about the conservation of caloric.

However, the demise of caloric and the rise of energy as a conserved quantity would require another half century. The experimental basis for this paradigm shift can be attributed to the patience and skill of an English brewer of ale by the name of James Joule (1818–1889). In 1840 at the age of 22, Joule began a series of experiments akin to those of Thompson but meticulous in their design and execution. Joule built what is now called a calorimeter, an apparatus to monitor the temperature of a volume of water that is thermally insulated from its surroundings but subject to some form of heating. As a measurement unit, he adopted what would later be called the British thermal unit or BTU: the amount of heat required to raise the temperature of a pound of water by one degree fahrenheit. To achieve the desired level of precision,

Joule routinely measured the temperature of his water bath to better than one thousandth of a degree. Using various calorimeters, Joule was able to compare the heat produced by a wide range of effects, including chemical reactions, electricity, and mechanical work. For example, he found that the heat generated by a current flowing through a resistor (now called Joule heating) is proportional to the resistance, the square of the current, and the time it flows. Similarly, when he arranged a falling mass to turn a paddle wheel submerged in the water bath, Joule discovered that the heat produced is proportional to the weight Mg of the mass and the height H through which it falls.

Joule's experiments eventually established equivalences between thermal, electrical, chemical, and mechanical effects that were completely consistent. However, Joule's publications were initially ignored by the scientific community, and it was only in 1847 that his ideas caught the attention of a 22-year-old theoretical physicist named William Thomson (later Lord Kelvin, 1824–1907).[1] Initially skeptical, Thomson eventually recognized the veracity of Joule's experiments and in 1851 published a long paper, "On the dynamical theory of heat," that recast thermodynamics without the concept of caloric. This paper introduced the word "energy" and elevated Joule's equivalences to the principle of energy conservation.

Almost simultaneously, the results of Joule and Thomson were mirrored in Germany by Robert Mayer (1814–1878) and Rudolf Clausius (1822–1888). In 1842, Mayer proposed an equivalence between mechanical work and heat, although without the extensive experimental evidence of Joule. Then in 1850, Clausius incorporated this equivalence into a new theory of thermodynamics that anticipated Thomson by a year in tossing out caloric theory. The first law of thermodynamics, the conservation of energy, had arrived.

[1] Although it includes both Thomson and Thompson, this chapter shouldn't be mistaken for a Tintin adventure.

13.2 Carnot and reversibility

Thomson's initial skepticism toward the work of Joule stemmed in part from the fact that a key tenet of thermal science, now called the second law of thermodynamics, had originally been set in caloric theory. The second law is more subtle than the first and took longer to achieve its final form. However, the essential elements are found in a memoir entitled *Reflections on the Motive Power of Fire* published in 1824 by the French military engineer Sadi Carnot (1796–1832). Although Carnot made his analysis using caloric theory, Thomson ultimately realized that the essential idea doesn't depend on the conservation of caloric, and that the second law could be joined with energy conservation in a consistent theory. This synthesis, independently formulated by Thomson and Clausius, became the foundation of modern thermodynamics.

What is the mysterious second law? One answer is found in Carnot's work, which arose in an attempt to understand the limitations of steam engines. Although practical steam engines had been pumping water out

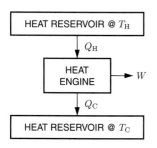

Fig. 13.1 A block diagram of a steam engine, which receives heat Q_H from a hot reservoir at temperature T_H, performs mechanical work W, and gives up heat Q_C to a cold reservoir at temperature T_C.

of mines for more than a century, Carnot was the first to consider their efficiency from a general perspective. In fact, his arguments were so general that he didn't worry about the mechanical nature of a steam engine: its boiler, cylinder, piston, or valves. In Carnot's view, a heat engine is any device that receives heat from a hot reservoir, uses it to perform work, and discards waste heat to a cold reservoir, as shown in the block diagram of Fig. 13.1. He assumed that the engine operates in a cycle in which certain amounts of heat are received Q_H and discarded Q_C and a certain amount of work W is done, leaving the engine in the same state at the end of the cycle as at the beginning.

Viewing a heat engine as analogous to a waterwheel, Carnot saw the temperatures of the hot and cold reservoirs, T_H and T_C, as analogous to the height of the water before and after it passes over the wheel. You can't get work from the water in a mill pond without letting it fall, and according to Carnot you can't get work out of a heat engine without a difference in temperature between the two heat reservoirs. Furthermore, just as you can't get more work from a waterwheel than that required to pump the water back up to the pond, Carnot argued that you can't get more work from a heat engine than that required to pump the heat from the cold reservoir back to the hot reservoir. That is, the highest possible efficiency is obtained from an engine that can be reversed at the end of a cycle to restore the system to the same state as at the beginning.

Thermodynamic reversibility is a key ingredient in Carnot's analysis of heat engines. He observed that, for a system isolated from the rest of the universe, some thermodynamic processes can be reversed to return the system to its original state, while others cannot. For example, an irreversible process results when heat is allowed to flow directly from a hot body to a cold body without extracting work. The irreversible nature of this process is familiar—we always see heat flow directly from hot to cold and never the reverse. Given a source of energy, however, Carnot understood that heat can be moved from a cold body to a hot one, a task now performed routinely by a refrigerator. Thus, Carnot proposed that if heat flows indirectly between hot and cold reservoirs through a heat engine, then the energy derived from an ideal engine would be just sufficient to pump the heat from the cold reservoir back to the hot one. Indeed, the ideal engine could itself be reversed to pump the heat backwards when driven with the stored energy.

How can one build an ideal heat engine? The trick is to design an engine that uses only reversible processes. Carnot proposed a specific engine cycle, now called the Carnot cycle, that achieves this goal, at least in principle. He imagined an air-filled cylinder with a piston that can move to change the volume of the air space. As was well known at the time, pushing the piston inward to compress the air will raise the air's temperature, while extracting the piston to expand the air will lower its temperature. The cycle that Carnot envisioned begins with the air compressed and in thermal contact with the hot reservoir. If the piston is now slowly extracted, the air temperature will start to drop, but heat will flow from the hot reservoir to keep the air at T_H (Fig. 13.2(a)). This process is nearly reversible since heat flows between two bodies at almost the same temperature. Once sufficient heat has been extracted from the hot reservoir and a certain amount of work has been done by the piston, thermal contact with the reservoir is broken. Now the piston is allowed to expand further, and the temperature of the air is allowed to drop while yet more work is done by the piston (Fig. 13.2(b)). According to Carnot, this is also a reversible process, because no heat flows between bodies at different temperatures and the work can be stored as energy to reverse the piston's motion. When the air temperature reaches T_C, the piston is stopped, and the cylinder is put in thermal contact with the cold reservoir. Now some of the stored energy is used to slowly push the piston back in. The temperature of the gas starts to rise, but heat is given up to the cold reservoir, keeping the air near T_C (Fig. 13.2(c)). At a certain point, thermal contact with the cold reservoir is broken, and the piston is pushed even further in, causing the air to heat up again (Fig. 13.2(d)). This final compression step is stopped when the air temperature reaches T_H, completing the Carnot cycle.

Because work must be done to recompress the air in the last half of the Carnot cycle, you might wonder if the engine has a net work output. But at any given position of the piston, the air is always cooler and at lower pressure during compression than during expansion, so it requires less energy to move the piston in than is obtained as it moves out. Moreover, because each step of the Carnot cycle is reversible, the net work performed for a given heat input Q_H and given reservoir temperatures, T_H and T_C, is as large as can be obtained with any heat engine. If some super-engine produced more work, then a Carnot engine could be run in reverse using a fraction of the super-output to restore the heat to the hot reservoir. In this case, the entire system would be returned to its original state but with an energy surplus, and we'd have a perpetual motion machine. The Carnot cycle is as efficient as is possible.

We now know that Carnot was wrong in adopting the caloric theory of heat. However, his arguments about reversibility are nonetheless entirely correct. Carnot's only mistake was assuming that caloric is conserved, leading him to suppose that the heat entering the cold reservoir would be the same as that leaving the hot reservoir, $Q_C = Q_H$. We now know that energy is actually the conserved quantity, which leads to the result

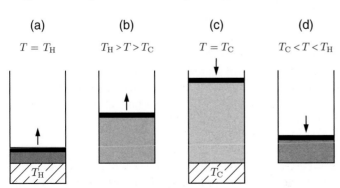

Fig. 13.2 The Carnot cycle. (a) The gas expands while in thermal contact with a hot reservoir at T_H. (b) The gas expands further while thermally insulated, and its temperature drops to T_C. (c) The gas is compressed while in thermal contact with a cold reservoir at T_C. (d) The gas is further compressed while thermally insulated, and its temperature rises to T_H.

$Q_H - Q_C = W$ (assuming units that reflect Joule's equivalence between heat and work). That is, the work output is the difference between the heat extracted from the hot reservoir and the heat discarded to the cold reservoir.

The amazing conclusion to this story, reached when Clausius and Thomson corrected Carnot's theory, is that the heat energy rejected to the cold reservoir can never be zero.[2] Why is this amazing? According to the first law of thermodynamics, the work produced by a waterwheel can't be greater than the difference in the potential energy of the water before and after passing over the wheel. Similarly, for heat engines the first law implies $W = Q_H - Q_C$, but it doesn't require that any heat actually be discarded to a cold reservoir. As far as the first law is concerned, we could just as easily have $Q_C = 0$ and convert all of the input heat into work. Thus, the fact that Q_C is necessarily greater than zero is something new, an unexpected consequence of Carnot's deceptively simple arguments about reversibility. It is in fact a statement of the second law of thermodynamics, and it places an important constraint on the efficiency of heat engines.

Even at the outset, Carnot's theory couldn't be rejected as nonsense: his reasoning led to quantitative predictions that were borne out by the available data. Moreover, his memoir was easy to read, using only the simplest mathematics. It was also highly practical, warning engine designers to avoid irreversible processes by preventing the direct flow of heat from hot to cold and allowing the working fluid to change temperature only by doing work. By rights, Carnot's 1824 memoir should have found a place in the pocket of every steam engine manufacturer of his day. Nonetheless, although it received an initial warm reception, his memoir was never taken very seriously. Eight years later, when Carnot succumbed to cholera at the age of 36, his memoir was all but forgotten. However, a friend and fellow engineer, Émile Clapeyron (1799–1864), rescued Carnot's work from oblivion. Two years after Carnot died, Clapeyron published an account of his work, augmented with the mathematics that Carnot had feared would scare away readers. Translated into English and German, Clapeyron's paper eventually inspired both Thomson and Clausius.

[2]Clausius demonstrated that $Q_C \geq Q_H(T_C/T_H)$, where temperature is measured on an absolute scale and equality holds for the Carnot cycle.

13.3   Clausius and entropy

In his seminal 1850 paper on the theory of heat (*Über die bewegende Kraft der Wärme*), Clausius stated that heat does not flow spontaneously from cold bodies to hot ones, a form of the second law of thermodynamics. But in 1854 Clausius gave the second law its modern mathematical form. This step required recognizing a new property of matter that he later gave the name "entropy."

Energy and entropy are the quantities at the center of thermodynamics. Like energy, a definite quantity of entropy is associated with a given amount of a substance in a given state—say a specified volume of gas at a particular pressure and temperature. Both the energy and entropy of a substance depend only on the state itself and not on how it arrived at that state. Thus, values of energy and entropy are tabulated for various materials, and an engineer can consult steam tables to find the energy and entropy per volume of water vapor at any pressure and temperature.

However, measuring the difference in entropy between two states is more complicated than measuring the difference in energy. The energy difference is simply the total heat and/or work required to affect the change of state, no matter how it occurs. In contrast, to measure the difference in entropy, the state must be changed by a reversible process, and the required heat and/or work must be separately measured for each small change in state from start to finish. Given this complication, we won't try to state Clausius's definition of entropy but simply admire his insight that entropy is a useful quantitative property of matter.

As the complicated nature of entropy might suggest, the second law of thermodynamics isn't quite as simple as the first law. In a system isolated from the rest of the universe, the first law tells us that the total energy is constant. Energy might be converted from one form to another, but the total energy of the system doesn't change. Similarly, the second law tells us that the total entropy of an isolated system is constant as long as it undergoes only reversible changes. However, the second law also states that the total entropy will increase whenever an isolated system undergoes an irreversible change. That is, entropy isn't always conserved and can only increase with time—a most peculiar property of matter!

In the 1865 paper that proposed the term entropy, Clausius famously stated the first and second laws of thermodynamics as they apply to the entire universe. In this view, the two laws are "The energy of the universe is constant" and "The entropy of the universe tends to a maximum." We can't create or destroy energy, but entropy is created throughout the universe every day. In fact, most thermal processes that we observe around us, like the melting of an ice cube or the cooling of a cup of tea, involve the irreversible flow of heat from a hot body to a cold one, and these events produce a net increase in entropy. In other processes, like the conversion of water to ice in a freezer, entropy decreases in one place, but there is an offsetting increase in entropy somewhere else. Thus, the heat absorbed inside a freezer is rejected to the outside air, increasing its entropy. On the whole, however, the second law insures that the entropy of the universe always increases, leading to the simple conclusion that the universe was initially, at the time of the big bang, in a state of relatively low entropy.

Before leaving the thermodynamics of the macroworld, we should note that friction is another hallmark of irreversible processes. The logic is simple. Friction produces heat that immediately flows away to warm the surrounding environment, and the direct flow of heat from hot to cold is always irreversible. Thus, we shouldn't be surprised that the equations of motion for machines with friction aren't reversible: replacing the time t with $-t$ changes everything. When friction brings a pendulum to a stop, we can't expect the thermal energy generated in the process to reverse itself and start the pendulum swinging again. That's the second law of thermodynamics in action. Now let's turn to the science of heat from the perspective of the microworld.

13.4 Kinetic theory of gases

As we've noted, at the turn of the nineteenth century the most widely accepted theory of heat assumed the existence of a colorless, weightless, indestructible fluid called caloric. However, some scientists believed instead that heat might be the microscopic motion of atoms and molecules. Although the existence of atoms was merely an hypothesis at the time, the kinetic theory of heat was advocated by Benjamin Thompson and later by Joule and others. Atoms had also begun to receive support from chemists, who observed that reactants often combine in simple ratios. For example, two volumes of hydrogen combine with one volume of oxygen to make two volumes of water vapor, as if discrete entities were involved. And if atoms were real, it wasn't a great leap to imagine that heat is atomic motion. In this case, friction produces heat because rubbing agitates the atoms, making them move faster. Also, heat might flow from a hot body to a cold one because hot atoms bumping into cold atoms speed up the cold ones and slow down the hot ones. Indeed, many observers suspected that thermal energy is just the kinetic energy of the atoms that comprise a substance.

In the atomic picture, scientists conjectured that the atoms of a solid are bound tightly together but have enough room to vibrate around their average positions. When heated to higher temperatures, however, the vibrations could be strong enough to weaken the bonds between neighboring atoms, allowing them to move past one another and change the solid into a liquid. At yet higher temperatures, the motion could be strong enough for atoms to break free entirely and fly off into the surrounding space, converting the liquid into a gas. Thus, the atomic picture explained many commonly observed phenomena, but whether atoms were real remained an open question.

To test the atomic hypothesis, theorists began to calculate what would be expected if the material world were actually composed of myriads of atoms in various states of motion. In this endeavor, the simplest case to consider is that of a gas, where the atoms are well separated and the interaction between them doesn't play a big role. Called the kinetic theory of gases, the resulting mathematical construct almost immediately began to yield verifiable predictions and new insight into the nature of heat.

The kinetic theory began with Daniel Bernoulli (1700–1782), one of a famous family of Swiss mathematicians. At the time, the Bernoullis were to mathematics what the Bachs were to music, with each family member a master in his own right. However, the Bernoullis seem to have suffered more from jealous rivalries than the Bachs. Daniel's masterpiece, entitled *Hydrodynamica*, was published in 1738, but his father later published

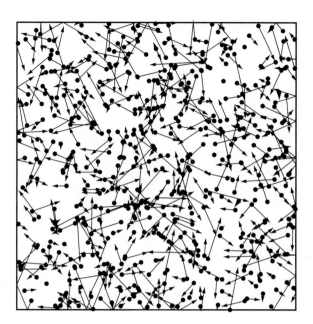

Fig. 13.3 An ideal two-dimensional gas of noninteracting atoms. The atoms are located randomly in space with velocities distributed according to Maxwell's formula.

Hydraulica, which borrowed some of Daniel's ideas and was predated to 1732. Daniel must have been devastated to find his father trying to steal credit for his work.

Bernoulli's theory of gases is developed in one chapter of *Hydrodynamica*. Here he envisioned a gas as a vast collection of minute hard spheres, tiny billiard balls if you like, in constant motion, bouncing around within a container. A two-dimensional version of Bernoulli's box of atoms is shown in Fig. 13.3, where we picture a confused mess of randomly placed balls with random velocity vectors. Bernoulli assumed that energy is conserved in collisions between atoms or between an atom and the container. This assumption makes no sense in the macroworld, where all balls deform inelastically and friction eventually brings all motion to a stop. Even a super-ball will bounce only a few times before coming to rest. But in the microworld, we presume to deal with structureless objects that don't deform and can rebound without losing energy. Due to the quantum nature of atoms, this assumption is actually better than Bernoulli might have guessed.

Using his billiard-ball model of a gas, Bernoulli was able to mathematically derive Boyle's law, a rule already well-known to chemists. Established experimentally by Robert Boyle (1627–1691) in 1662, this law states that at a fixed temperature the volume of a given quantity of gas is inversely proportional to the pressure, $V \propto 1/P$ or $P \propto 1/V$. Pressure in general is force per unit area, and Bernoulli realized that the pressure exerted by a gas on the walls of its container is due to gas atoms bouncing off the walls. Although the force exerted by a single atom is tiny and lasts only an instant, there are so many atoms in a gas that their action is perceived in the macroworld as a steady pressure.

To understand Bernoulli's derivation of Boyle's law, suppose that a gas is confined within a cube and that the cube somehow expands until each side is twice its original length and width. After expanding, the cube's volume will be $2^3 = 8$ times as large. On the other hand, there will be just as many atoms as before, and, if they're at the same temperature, we can assume (along with Bernoulli) that they move just as fast. But because the box is twice as big in all directions, each atom will hit a wall only half as often, so its contribution to the force is reduced by half. Furthermore, the area of each side of the box is $2^2 = 4$ times larger. Thus, the gas pressure (force divided by area) will only be one eighth of the original pressure. In other words, the pressure is reduced by $1/8$ when the volume is increased by 8, in agreement with $P \propto 1/V$. By this argument, the kinetic theory of gases explained Boyle's law and passed its first test.

Today Bernoulli's theory is recognized as fundamentally correct, but it received little attention for almost a century and returned to center stage only with the demise of caloric theory. By the middle of the nineteenth century kinetic theory had begun to flower, becoming ever more sophisticated in its range of predictions, especially in the hands of Clausius and Maxwell. In 1858, for example, Clausius considered the process by which odor molecules diffuse through the air and estimated the average distance a molecule travels before it collides with another. Then in 1860 Maxwell derived an expression for the distribution of molecular velocities and applied this result to calculate the thermal conductivity and viscosity of gases. As these predictions and others were verified in the laboratory, kinetic theory quickly became the preeminent explanation of heat, and the reality of atoms and molecules began to look more and more certain, although doubters remained.

Before proceeding further with kinetic theory, let's step back and consider for a moment a liter of helium gas, as might be found in a child's balloon. Because it's a nobel gas, helium consists of single atoms that attract each other only weakly, and the billiard-ball model fits it well. According to modern knowledge, at room temperature (20 °C) and atmospheric pressure, a liter of helium includes 2.5×10^{22} atoms, each with a diameter of about $d = 2.6 \times 10^{-10}$ m and a mass of 6.65×10^{-27} kilogram. That's a lot of atoms, but each is so tiny that the liter is filled mostly by empty space. The emptiness gives the atoms room to move, and they whiz around with typical speeds of 1,350 m/s or 3,000 miles per hour. At that speed, the atoms are sure to run into each other frequently, and the mean time between collisions for a given atom is only 1.37×10^{-10} s. In that time, an atom travels on average 1.85×10^{-7} m, or about 700 times its diameter. So, if you think traveling on the freeway is hectic, consider an atom of helium gas, which experiences over 7 billion collisions every second. Fortunately, helium atoms are nearly indestructible, so they don't need to visit a body shop between collisions. Nonetheless, the mayhem to be found in a child's balloon or the air around us is astonishing.

13.5   Boltzmann and entropy

When Maxwell calculated the distribution of molecular velocities in a gas, he introduced statistics into kinetic theory for the first time. Previously, theorists had been content to assume that all molecules move with the same speed, if only because it simplified their analyses. Knowing the velocity distribution suddenly changed kinetic theory from a toy into a fully fledged theory with accurate predictive powers. Moreover, by introducing a statistical approach, Maxwell's 1860 paper marked the beginning of what we now call statistical mechanics.

Strangely, Maxwell derived his velocity distribution (related to the bell curve that we met in Chapter 8) not from the detailed physics of molecular collisions but from general arguments based on symmetry properties expected at equilibrium. The derivation was like pulling a rabbit out of a hat, and it left many questions unanswered. Are other equilibrium distributions possible? What kinds of collision processes are required to achieve the Maxwell distribution? And how does an arbitrary distribution evolve in time? Maxwell's derivation proved to be justified, but questions such as these lingered in the wake of his paper, and indeed some still await definitive answers.

However, the biggest unresolved question on the agenda of gas theorists concerned entropy. By 1860, the nature of pressure, temperature, and thermal energy were all understood from a microscopic perspective. Pressure had been conquered long ago by Bernoulli, and for a monatomic gas like helium it was known that the (absolute) temperature is proportional to the average kinetic energy of the atoms and that the thermal energy is the sum of all the kinetic energies. But no one had discovered a microscopic explanation for entropy. Although well defined in the macroworld, entropy remained a mystery in the microworld. What atomic quantity could possibly be found that would only increase with time? And, if all microscopic processes were reversible, wasn't it hopeless to look for something that was irreversible?

The man who ultimately answered these questions was Ludwig Boltzmann (1844–1906), an Austrian theorist who devoted most of his life to understanding the microscopic origin of entropy. Boltzmann cracked the problem open in 1872 at the age of 28. Extending Maxwell's work, Boltzmann developed an equation that specified how any distribution of velocities within a gas evolves in time as the atoms collide. Applying this equation, he discovered that a gas could be characterized by a certain quantity with special properties. This quantity could only decrease with time, and it reached a minimum when the velocity distribution equaled the Maxwellian distribution. This was a remarkable result: it simultaneously put Maxwell's work on a firm foundation, and it amounted to the discovery of a microscopic explanation of entropy.

Boltzmann immediately understood that the quantity he had discovered must be related to entropy, and in an 1877 paper he gave entropy its modern microscopic definition,

$$S = k \log \Omega. \qquad (13.3)$$

Now carved on Boltzmann's tomb in a Vienna cemetery, this equation is the centerpiece of classical statistical mechanics. Here S is the entropy of a given macroscopic state, k is Boltzmann's constant, log is the natural logarithm, and Ω is the number of microscopic states accessible to the system. With Eq. (13.3), the mystery of entropy was at last solved. Boltzmann had discovered how irreversible macroscopic processes result from reversible microscopic processes.

What is the nature of entropy according to Boltzmann? Evaluating Ω and hence S is a matter of counting the number of microscopic states accessible to a system in a given macrostate. Consider, for example, a macrostate consisting of a gas of N atoms in a container of volume V at temperature T. The number of accessible microstates is the number of ways of assigning position and velocity vectors to each atom such that all atoms fall within V and their total kinetic energy is N times the average energy specified by T. Because atoms are always colliding, a gas doesn't remain in a given microstate for more than an instant but frenetically moves from one state to another. If there are a large number of accessible microstates, then the atomic motions are highly disordered. Thus, entropy is often equated with microscopic disorder. If only one microstate is accessible, then $\Omega = 1$, $S = 0$, and the system is completely ordered. But the more accessible microstates, the higher the entropy, and the greater the disorder.

Just how many microstates are accessible to our liter of helium at room temperature and atmospheric pressure? Using the modern quantum mechanical method of counting,[3] we find $\Omega = 10^{1.64 \times 10^{23}}$. This is the number of ways that 2.5×10^{22} atoms can be arranged in a liter container and assigned velocities such that their kinetic energies sum to 152 joule.

In this example, the number of atoms $N = 2.5 \times 10^{22}$ is very large, but the number of microstates $\Omega = 10^{1.64 \times 10^{23}}$ is inconceivably large. To understand the distinction, consider that adding a billion to N is of little consequence,

$$10^9 + 2.5 \times 10^{22} \approx 2.5 \times 10^{22},$$

while multiplying Ω by a billion is similarly inconsequential,

$$10^9 \times 10^{1.64 \times 10^{23}} = 10^{9 + 1.64 \times 10^{23}} \approx 10^{1.64 \times 10^{23}}.$$

Hard as it is to believe, numbers can be so large that multiplication by a billion produces an insignificant change. Boltzmann's understanding of macroscopic irreversibility relies on just such inconceivably large numbers.

To get a better look at irreversibility, let's compare how the microstates of a gas change for two processes, one reversible and one irreversible. In particular, let's see what happens when our liter of helium expands to twice the volume, either by reversibly pushing against a piston or by irreversibly expanding into a vacuum space. These two

[3]For a monatomic noninteracting gas of N identical atoms of mass m confined to a volume V at absolute temperature T, the number of accessible quantum states is $\Omega = \left[e^{5/2} (2\pi m k T)^{3/2} V / (N h^3) \right]^N$, where k is Boltzmann's constant and h is Planck's constant.

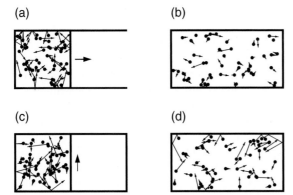

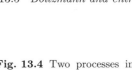

Fig. 13.4 Two processes in which a gas expands to fill twice its original volume. (a) The gas in a thermally insulated cylinder pushes against a piston as it is slowly removed, and (b) the atoms give up some of their kinetic energy in the process. (c) The gas expands into a vacuum space as a partition slides to one side, and (d) the atoms retain their original kinetic energies. The decrease in atomic velocities between (a) and (b) has been exaggerated.

processes, illustrated in Fig. 13.4, might seem almost the same, but they are poles apart from a thermodynamic point of view.

First, consider what happens when gas in a thermally insulated cylinder expands while pushing against a piston, as in Figs. 13.4(a) and (b). How is a helium atom affected when it strikes the receding piston? Imagine standing on the ground in back of a truck and throwing a tennis ball against the rear door. If the truck is stationary, the ball will rebound with essentially the same speed as it was thrown. But if the truck is moving toward you, the ball rebounds with increased speed, and if the truck is moving away it rebounds with reduced speed. Thus, as the piston is withdrawn from the cylinder, each of the atoms striking it gives up energy to the piston and rebounds with reduced speed. If the gas volume increases from 1 to 2 liters, then the average speed of the helium atoms is reduced from 1,350 to 1,070 m/s, the total kinetic energy falls from 152 to 96 joule, and the temperature drops from 20 to $-88\,°C$.

With these changes in volume and energy, we might expect a change in entropy as well. However, the number of possible position vectors goes up with the added volume, and the number of possible velocity vectors goes down with the decrease in energy. Because these effects "happen" to cancel exactly, the number of accessible microstates doesn't change as the piston is removed, and the entropy is constant. Of course, we chose this process, which is part of the Carnot cycle, exactly because it's reversible, and we might have anticipated that the entropy wouldn't change. If the energy given up by the gas were stored, it would be just sufficient to push the piston back and restore the system to its original macrostate.

Now let's consider what happens when removal of a partition allows the liter of helium to expand into a vacuum space, as in Figs. 13.4(c) and (d). In this process, known as free expansion, there is no piston for the atoms to push against, and their average kinetic energy doesn't change when the partition is removed. Thus, the temperature and energy of the gas are unchanged, although the pressure drops by a factor of 1/2 as predicted by Boyle's law. The added volume opens up many new possible position vectors, so we can expect an increase in the number of accessible microstates. In fact, we have $10^{7.5 \times 10^{21}}$ times as many microstates after the expansion. Thus, free expansion leads to a significant increase in entropy, as expected for an irreversible process.

From our microscopic picture, we ought to be able to understand exactly why free expansion can't be reversed. According to Boltzmann, irreversibility is a matter of probabilities. Admittedly, it's possible for all 2.5×10^{22} helium atoms to bounce around within the expanded box and suddenly find themselves, by some stroke of luck, back in the original liter on the left. This wouldn't violate conservation of energy or any other law of physics. However, as Boltzmann argued, the number of microstates corresponding to the original macrostate is only a tiny fraction of the mircostates of the expanded box. In fact, we have already computed the fraction, it's $10^{-7.5 \times 10^{21}}$ since there are $10^{7.5 \times 10^{21}}$ times as many microstates available in the doubled volume. Thus, once the partition is removed, finding the system in a microstate of the original macrostate is possible but inconceivably improbable.

Let's calculate how long we would need to wait to find that all of the helium atoms have by chance returned to the left side of the box. Suppose we inspect the box once every 10^{-13} s, about the time required for an atom to move a distance equal to its own radius. In this case, inspections are frequent enough that we won't miss the awaited return of all atoms to the left side. Because in the doubled volume only one microstate in $10^{7.5 \times 10^{21}}$ corresponds to the outcome that we seek, we'll need roughly this many inspections to find a single instance in which all of the atoms are on the left. Thus, the required wait is

$$10^{7.5 \times 10^{21}} \times 10^{-13} \text{ s} = 10^{7.5 \times 10^{21} - 13} \text{ s} \approx 10^{7.5 \times 10^{21}} \text{ s}.$$

Clearly, the waiting time in seconds is incredibly long, but what would it be in years or billions of years? As you can show for yourself, the answer for all practical purposes is $10^{7.5 \times 10^{21}}$, no matter what reasonable time unit you might choose. Given this, Boltzmann concluded that a macroscopic process accompanied by an increase in entropy will never be seen to spontaneously reverse itself, even when all of the underlying atomic processes are reversible. So, in spite of the apparent paradox, the second law of thermodynamics has a perfectly logical microscopic basis.

13.6   Chaos and ergodicity

Today we recognize Boltzmann's view of entropy as essentially correct, but not everyone was so sure when it first appeared. Many objected that Boltzmann had reduced the second law of thermodynamics from a presumed exact relation, like Newton's laws of motion or the conservation of energy, to a matter of probability and statistics. A law of physics that's true only most of the time seemed like a swindle, even though the probability of its being broken is inconceivably small. In the vernacular of today, Boltzmann might well have replied, "Get over it" to such critics. Two centuries earlier, Newton had apparently banished uncertainty from physics, but a new, less certain era had arrived, and there was no turning back.

Others wondered about an assumption implicit in Boltzmann's theory. In particular, is it true that all of the possible microstates of a macrostate occur with equal probability? This question, with which both Boltzmann and Maxwell wrestled, was eventually settled for practical purposes by the success of statistical mechanics. But as a strictly mathematical question in dynamics, it has yet to receive a complete and rigorous answer.

Boltzmann's assumption, now called the ergodic hypothesis, is often stated as the condition that the average of any quantity over all the microstates of a system equals the time average of that quantity in any one example of the system. In treating free expansion, for example, we assumed that all of the microstates in the doubled volume would be accessed as the atoms zip around and collide with each other. If not, then we can't legitimately compute probabilities based on the numbers of microstates. If so, on the contrary, then the gas is ergodic and our calculation is justified.

For a system to be ergodic, the motions of the atoms must, in some sense, be sufficiently mixed up. Could it be that the chaotic nature of atomic collisions leads to ergodicity? Although the strict mathematical answer to this question is probably no, it's likely that chaos explains the practical success of statistical mechanics. Maxwell and Poincaré may have been thinking as much when they described how collisions between molecules lead to essentially unpredictable trajectories (see Chapter 11). Chaotic motion is good at mixing things up, and that's just what Maxwell and Boltzmann's statistical approach requires.

The difficulty of proving that a thermodynamic system is ergodic stems from the complexity of the underlying molecular motion. With 10^{22} atoms, each colliding billions of times a second, mathematical progress requires making a gross simplification and hoping that it captures the essence of the actual atomic mayhem. In fact, Maxwell began his 1860 paper on the distribution of molecular velocities, the paper that first introduced statistical ideas to kinetic theory, by considering collisions between just two hard spheres. For such collisions, Maxwell was able to show that "all directions of rebound are equally likely," suggesting that collisions may randomize molecular motions enough to justify statistical analysis.

In his 1913 paper Émil Borel investigated the randomizing effect of molecular collisions more thoroughly using another simple model of a gas. Borel first restricted himself to two dimensions, then pictured collisions of a given atom by imagining that the other atoms are fixed circular obstacles arranged randomly on a vast billiard table. In collisions with these obstacles, Borel's atom always rebounds with its speed unchanged, although its direction of motion is altered. In reality, of course, a gas is three dimensional, all of the atoms are in motion, and energy is usually transferred between colliding atoms. Nonetheless, Borel's model gives some clues about the behavior of a real gas.

(a)

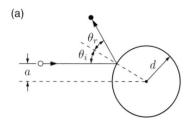

(b)

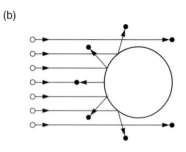

Fig. 13.5 Elastic scattering of a point particle from a fixed obstacle of radius d. (a) The angle of reflection θ_r equals the angle of incidence θ_i. (b) Scattering for particles with various impact parameters a.

To see what Borel discovered, let's first consider a single collision. Instead of picturing a collision between two atoms of equal size, we can equivalently think of a point particle scattering from a circle with a radius equal to the diameter of an atom, as in Fig. 13.5(a). Can you see why? When energy is conserved, as in Borel's scenario, the scattering process is said to be elastic, and it's especially simple. In this case, the particle retains its original speed, and the angle of reflection θ_r equals the angle of incidence θ_i, like light reflected from a mirror. Indeed, for elastic scattering we don't need to keep track of the particle's speed, which is constant, or its acceleration, which is zero except at the instant of collision. Thus, the laws of motion that we labored to understand for the pendulum are replaced in the billiard-ball model by simple geometric reflection. The motion of a billiard ball is very easy to visualize.

When a point particle is scattered from a circular obstacle, the outcome is determined solely by the distance a between the original trajectory and the center of the circle, as in Fig. 13.5(a). This distance is called the impact parameter, and we can see from Fig. 13.5(b) that if $a = 0$ then the particle bounces straight back from the obstacle with its velocity reversed. If $a > d$ then the particle misses the obstacle, and for intermediate values, $0 < a < d$, the net deflection of the particle's trajectory can be anything between 0 and 2π. That is, as the trajectory of the incoming particle is displaced over a range of two atomic diameters, the particle is scattered in all possible directions of the compass. So scattering depends critically on the impact parameter, and we shouldn't be surprised that trajectories involving repeated scattering from circular obstacles are often chaotic.

The fixed obstacles of Borel's gas model may remind you of the quincunx that we met in Chapter 8. In the quincunx, the obstacles are points and the moving particle is a circle, but aside from this reversal, the geometry is basically the same. Ironically, we used the quincunx as an example of a natural random process devoid of chaos, but now it appears as if the quincunx may be a chaotic machine. In fact, Experiment 16 of the Dynamics Lab allows you to verify this proposition and estimate the Liapunov exponent of the quincunx. As chaos theory has developed, more and more processes that were once thought to be purely random have proven to be described by chaotic mechanisms.

Borel argued that the trajectory of his point particle, bouncing off one circular obstacle after another, would be extremely sensitive to initial conditions. To understand why, we consider a particle aimed straight at an obstacle with an impact parameter of $a = 0$, and compare this to the trajectory of a particle with the same speed and direction of motion but offset slightly by $a = \delta$. Before reaching the obstacle, the two particles will travel exactly side by side. However, the first particle will reflect straight back from the obstacle along its original path, while the offset particle will reflect back at a slight angle, and its trajectory will begin to diverge from the first. Recalling that $\theta_r = \theta_i$ in Fig. 13.5(a), the angle of divergence in radians is approximately $\theta_i + \theta_r \approx 2\delta/d$ for $a = \delta$.

Now suppose that the two particles travel a distance D before they encounter another obstacle. Given that the angle between their trajectories is $2\delta/d$, after traveling a distance D the separation between them will be $\delta' = 2\delta D/d$. Thus, one collision has the effect of increasing the separation between the particles by a factor of $\delta'/\delta = 2D/d$, and we can expect further collisions to increase the separation by a similar factor. If D represents a typical distance between collisions, then the separation will increase by a factor of $(2D/d)^n$ after n collisions, as long as the particles remain close enough to collide with the same obstacles. Thus, the offset between the two trajectories grows exponentially with n or with time, and our particle is subject to the butterfly effect. (You may object to assuming that the collisions occur at impact parameters near 0, but trajectories diverge even more rapidly for larger a.)

Anyone who has tried to make a combination shot on a pool table knows exactly what Borel is talking about. Getting the cue ball to strike the first ball is easy enough, but it takes some skill to make the first ball strike the second, and pocketing the second ball is almost impossible. That's sensitivity to initial conditions for you.

To understand the significance of Borel's observation, let's return to our example of helium at room temperature and atmospheric pressure. We noted previously that the diameter of a helium atom is $d = 2.6 \times 10^{-10}$ m, that it travels on average $D = 1.85 \times 10^{-7}$ m between collisions, and that it experiences about 7×10^9 collisions per second. Thus, an offset will grow by a factor of about $2D/d = 1.4 \times 10^3$ with each collision and by a factor of $(2D/d)^{7 \times 10^9} = 10^{2.2 \times 10^{10}}$ after one second. Suddenly we are back in the realm of inconceivably large numbers, and we can appreciate why Borel concluded that even the gravitational effect of 1 gram of matter at an astronomical distance will totally upset the atomic motions of a gas in a fraction of a second. By comparison, the effect of a butterfly flapping its wings on the opposite side of the Earth would be gargantuan.

13.7 Stadium billiards

Borel's analysis of collisions may convince us that atomic motions are probably chaotic, but it doesn't prove that a gas is ergodic with the rigor that would satisfy a mathematician. To obtain such proof, we must make further simplifications in our microscopic model. In fact, although collisions between atoms are a potent source of chaos, they are highly complex and may not be essential to ergodicity. As Maxwell wrote in an 1879 paper, "[I]f we suppose that the material particles, or some of them occasionally encounter a fixed obstacle such as the sides of the vessel containing the particles, then, except for special forms of the surface of this obstacle, each encounter will introduce a disturbance into the motion of the system, so that it will pass from one undisturbed path to another." Thus, perhaps we need only consider a single atom in a suitably shaped container to discover mathematically rigorous ergodicity.

A single atom in a box is only a toy model of a gas with 10^{22} atoms, but even this model is difficult to treat with complete rigor. Thus, it was a breakthrough when in 1924 the Austrian mathematician Emil Artin (1898–1962) gave a rigorous proof of ergodicity for a particle confined to a curved surface, like those studied by Hadamard. This system, now called Artin billiards, is strongly chaotic as well as ergodic, but it only vaguely resembles a gas. In 1963, however, the Russian mathematician Yakov Sinai (1935–) proved ergodicity for a much closer gas analog. Now called Sinai billiards, the system he studied consists of a point particle that moves without friction on a square table with a circular obstacle in the center. Bouncing back and forth between the circle and the square boundary, the particle's trajectory is subject to the same exponential sensitivity as Borel discovered for a particle scattered by an array of circular obstacles. Thus, in accord with the vision of Maxwell, Poincaré, Borel, and others, Sinai established that there is at least one system in which the butterfly effect seems to mix atomic trajectories enough to assure ergodicity. Sinai's proof leaves us a long way from a general proof of the ergodic hypothesis, but it suggests that chaotic motion is an important element underpinning the second law of thermodynamics.

For a closer look at chaos on a billiard table, we'll examine another system of proven ergodicity. In 1979 Leonid Bunimovich (1947–), a former student of Sinai, proved that a type of billiards called stadium billiards is also ergodic. In this system, which has become a popular example of how simplicity leads to chaos, the square ends of an ordinary rectangular table are replaced by semicircles, and there is no central obstacle. Bunimovich chose the stadium shape to see what happens when particles reflect from a concave arc rather than a convex arc as in Sinai billiards. Here we'll compare a stadium, where motion is typically chaotic, with a rectangle, where it's always regular.

Typical trajectories for both types of table are shown in Fig. 13.6, and the difference between the two is immediately apparent. On the rectangular table the particle shuttles back and forth with perfect regularity, while on the stadium table its path is surprisingly mixed up.

The regularity of motion within the rectangle is actually easy to prove. From an alternative perspective, the trajectory of Fig. 13.6(a) is just a straight line extending to infinity. This is illustrated by Fig. 13.7, where our rectangular table is shown, at lower left, as part of an infinite grid of rectangles. Here we see that each segment of a trajectory confined to the table by reflection is equivalent to a segment of a rectilinear trajectory extending across the grid. This equivalence assures us that an initial offset will create a second line across the grid that can diverge from the original only linearly with distance. In other words, motion on a rectangular table is never exponentially sensitive to initial conditions.

On the other hand, no such argument can be made for typical motion on the stadium table, and we might suspect that exponential sensitivity gives rise to the mixed up motion of Fig. 13.6(b). This suspicion is confirmed by Experiment 17 of the Dynamics Lab, which shows that

(a)

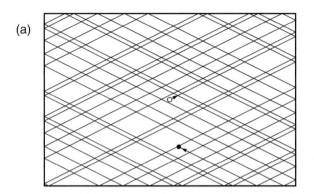

(b)

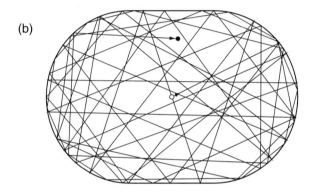

Fig. 13.6 Trajectories of point particles on (a) a rectangular table having a width 2/3 of its length and (b) a stadium table of the same dimensions with semicircular ends. On both tables, particles rebound from the sides with the angle of reflection equal to the angle of incidence. The two trajectories begin at the centers of the tables with an initial direction of motion at 26° from horizontal.

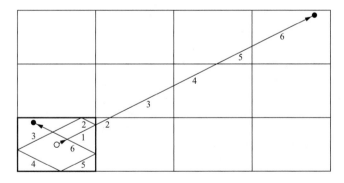

Fig. 13.7 The motion of a particle confined by reflection inside a rectangle (lower left) is equivalent to rectilinear motion on an infinite grid of rectangles. Corresponding segments are numbered 1 to 6.

small offsets grow by a factor of about 3.8 for each stadium length traversed by a particle. Thus, the stadium is another simple model of a gas for which ergodicity and chaos go hand in hand.

It's remarkable that a dynamical system as simple as a single particle on a billiard table supports Boltzmann's ergodic hypothesis. By discarding collisions between atoms, the complexity associated with large numbers of atoms, and the additional freedom afforded by a third dimension, we have thrown out several mechanisms that would help mix up the atomic motion of a gas and promote ergodicity. Thus, the toy models of Sinai and Bunimovich, which by comparison have almost no reason to be ergodic, make the truth of the ergodic hypothesis seem certain for a real gas. Could there be any doubt that chaos plays a role in this conclusion?

13.8   Time's arrow

Today scientists generally accept Boltzmann's explanation of the arrow of time. A process is irreversible when the final macroscopic state of the system permits an inconceivably greater number of microstates than the original macroscopic state. When a gas freely expands into a vacuum space, the probability of it returning to its original volume is inconceivably small because the original microstates, while still accessible, are a tiny fraction of those available in the larger volume. That's all there is to it: events that look silly in a movie played backwards aren't impossible, just inconceivably improbable.

That said, let's look at an objection to Boltzmann's ideas put forward independently by William Thomson in 1874 and by Boltzmann's friend and fellow Austrian Josef Loschmidt (1821–1895) in 1876. Writing before the statistical nature of Boltzmann's argument was fully appreciated, Thomson and Loschmidt pointed out that a supposedly irreversible process could in principle be reversed. In particular, if at some point all of the particles' velocities are instantaneously reversed, then the equations of motion imply that the system will return exactly to its original state. Usually called Loschmidt's paradox, this thought experiment raises the difficulty mentioned at the beginning of the chapter: the seeming impossibility of obtaining macroscopic irreversibility from reversible microscopic equations.

The objection of Thomson and Loschmidt prompted Boltzmann to develop the statistical aspects of his theory more explicitly in an 1877 paper. Boltzmann couldn't rule out the Loschmidt scenario, but, given the ergodic hypothesis, he could show that it was inconceivably improbable.

In his 1913 paper, Borel took another approach to resolving Loschmidt's paradox. He focused on the practical impossibility of the system evolving backwards with the accuracy needed to realize the original low-entropy state. That is, given the presence of any outside influence, even an extremely weak one, the exponential sensitivity of a gas precludes it from returning to its original microstate, even if the atomic velocities are reversed almost immediately.

Borel's idea is illustrated in Fig. 13.8 by a microscopic simulation of the macroscopically irreversible process of free expansion in a gas. To keep our calculation simple, we omit collisions between atoms, the powerful source of chaos considered by Borel, and rely instead on the weaker chaos of Bunimovich's stadium billiards. To emphasize the role of chaos, we compare stadium billiards with motion on a rectangular table. Rather than including an explicit outside influence, we'll let round-off errors provide the small perturbations in our simulation.

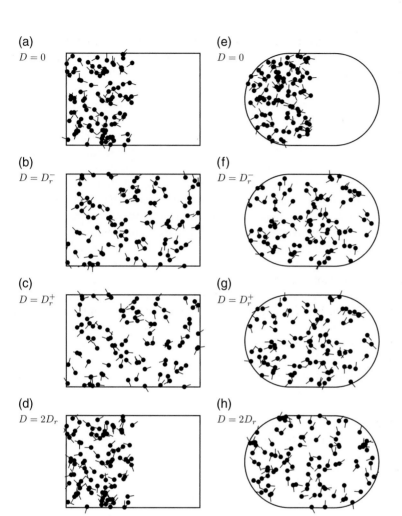

(a) $D = 0$

(b) $D = D_r^-$

(c) $D = D_r^+$

(d) $D = 2D_r$

(e) $D = 0$

(f) $D = D_r^-$

(g) $D = D_r^+$

(h) $D = 2D_r$

Fig. 13.8 Free expansion and microscopic reversal of a two-dimensional gas of 100 point particles on a rectangular table (a)–(d) and on a stadium table (e)–(h). Particles reflect from table boundaries, but collisions between particles are not included. The initial positions of the particles are randomly chosen within the left half of each table. All of the particle's speeds are assumed to be identical, but they are assigned random initial directions of motion, indicated by short lines. From the initial state shown in (a) and (e), the particles first propagate a distance of $D_r = 40$ table lengths ((b) and (f)), then their directions of motion are reversed ((c) and (g)), and finally they propagate an additional distance D_r, ((d) and (h)).

To begin, we assume that 100 point particles are randomly positioned on the left half of each table, as shown in Figs. 13.8(a) and (e). Each particle is given the same speed, but its direction of motion (indicated by a short line) is chosen randomly. After the particles travel a distance of $D_r = 40$ table lengths, we find that they are now scattered over the whole of each table, Figs. 13.8 (b) and (f). This is exactly the expected outcome of free expansion.

Now we implement the suggestion of Thomson and Loschmidt by reversing the direction of motion of each particle, as shown in Figs. 13.8(c) and (g). If the particles now travel an additional distance D_r, they should in principle return to their original positions, thereby reversing a macroscopically irreversible process. The results of the simulation prove that Thomson and Loschmidt are correct for the rectangle, Fig. 13.8(d), but not for the stadium, Fig. 13.8(h). Experiment 18 of the Dynamics Lab allows you to explore this effect as a function of the reversal distance D_r.

The different outcomes result, of course, from the difference between regular and chaotic motion. For the rectangle, atomic motions are regular, and errors grow very slowly with time, so the computation can return the atoms to their original positions after they have traveled a great distance. For the stadium, motion is extremely sensitive to small perturbations, and round-off errors grow so quickly that the simulation cannot accurately retrace its steps. Thus, the atoms on the stadium table quickly forget where they came from and end up in one of the inconceivably more numerous microstates of the full table, rather than one in which the particles are restricted to the left half.

As in our simulation, Borel argued that exponential sensitivity makes it impossible to eliminate outside influences in a real gas and that we should expect irreversibility if Thomson and Loschmidt's thought experiment could actually be performed. From Boltzmann's point of view, on the other hand, Loschmidt's paradox is a clever way of pointing out that the microstates of the expanded gas include the microstates of the original system, a fact accounted for in his statistical explanation of irreversibility. In either case, we have a resolution of Loschmidt's paradox in which chaos plays a central role. Borel's argument relies on chaos explicitly through exponential sensitivity, while Boltzmann's argument relies on the ergodic hypothesis, which is closely associated with chaotic behavior. Either way, chaos appears to be central to time's arrow and the foundations of statistical mechanics.

13.9   Atomic hypothesis

Before leaving kinetic theory, let's complete the history of the atomic hypothesis, an idea that remained in doubt for centuries.

Throughout his life, Boltzmann had argued strongly for the existence of atoms. Another Austrian physicist, Ernst Mach (1838–1916), was highly skeptical of atoms and an active critic of Boltzmann's work. Sadly, the resolution of this debate came shortly after Boltzmann's death. Suffering all his life from bouts of depression, Boltzmann took his own life in 1906 at the age of 62 while on vacation in Italy. As noted in Chapter 8, however, in the previous year Einstein had published an analysis of Brownian motion based on kinetic theory. This analysis opened the door to what scientists now regard as the final proof of the reality of atoms.

Einstein's theory of Brownian motion was confirmed by the French physicist Jean Perrin (1870–1942) in a series of meticulous experiments begun in 1908. Through careful observations of colloidal particles in suspension, Perrin not only obtained excellent agreement with Einstein's theory, but he was able to derive an accurate value for Avogadro's number. Roughly the number of atoms in a gram of hydrogen, Avogadro's number is a crucial link between the atomic realm and the macroscopic world. All in all, Perrin uncovered irrefutable proof of the existence of atoms, justifying the atomic foundation of statistical mechanics and the work of all the kinetic theorists. Were he still alive, Boltzmann would have been especially pleased, as he had wagered his entire career on the reality of atoms.

In 1913 Jean Perrin published a book, *Les Atomes*, summarizing his confirmation of the atomic hypothesis. In the same year, Ernst Mach noted in the preface to his final book that he remained skeptical of atoms. Perhaps it was with the atomic hypothesis in mind that the physicist Max Planck (1858–1947) later wrote as follows in *The Philosophy of Physics*.

> An important scientific innovation rarely makes its way by gradually winning over and converting its opponents: it rarely happens that Saul becomes Paul. What does happen is that its opponents gradually die out and that the growing generation is familiarized with the idea from the beginning.

Mach died in 1916, the last serious opponent of atoms to leave the stage. In 1926 Jean Perrin was awarded the Nobel Prize in physics.

Further reading

Boltzmann

- Cercignani, C., *Ludwig Boltzmann: The Man Who Trusted Atoms* (Oxford University Press, Oxford, 1998).
- Cercignani, C., "Ludwig Boltzmann: Atomic genius", *Physics World* **19**(9), 34–37 (September 2006).
- Lindley, D., *Boltzmann's Atom: The Great Debate that Launched a Revolution in Physics* (Free Press, New York, 2001).

Chaotic scattering

- Ekeland, I., "From computations to geometry", "Poincaré and beyond", and "Pandora's box", in *The Best of All Possible Worlds: Mathematics and Destiny* (University of Chicago Press, Chicago, 2006) chapters 4–6.
- Rockmore, D., "God created the natural numbers... but, in a billiard hall?", in *Stalking the Riemann Hypothesis: The Quest to Find the Hidden Law of Prime Numbers* (Pantheon, New York, 2005) chapter 12.
- Sweet, D., Ott, E., and York, J. A., "Topology in chaotic scattering", *Nature* **399**, 315–316 (1999).

- Zaslavsky, G. M., "Chaotic dynamics and the origin of statistical laws", *Physics Today* **52**(8), 39–45 (August 1999).

Clausius

- Guillen, M., "An unprofitable experience: Rudolf Clausius and the second law of thermodynamics," in *Five Equations that Changed the World* (MJF Books, New York, 1995) pp. 165–214.

Kinetic theory of gases

- Brush, S. G., *The Kind of Motion We Call Heat: A History of the Kinetic Theory of Gases in the 19th Century* (North-Holland, Amsterdam, 1976).

Statistical mechanics

- Goldstein, M. and Goldstein, I. F., *The Refrigerator and the Universe: Understanding the Laws of Energy* (Harvard University Press, Cambridge, Massachusetts, 1993).
- Lebowitz, J. L. and Penrose, O., "Modern ergodic theory", *Physics Today* **26**(2), 23–29 (February 1973).

○ Lloyd, S., "Information and physical systems", in *Programming the Universe: A Quantum Computer Scientist Takes on the Cosmos* (Knopf, New York, 2006) chapter 4.

○ Ruelle, D., *Chance and Chaos* (Princeton University Press, Princeton, 1991).

○ von Baeyer, H. C., *Maxwell's Demon: Why Warmth Disperses and Time Passes* (Random House, New York, 1998).

Thermodynamics

○ Atkins, P., *Four Laws That Drive the Universe* (Oxford University Press, Oxford, 2007).

• Cropper, W. H., "Thermodynamics", in *Great Physicists: The Life and Times of Leading Physicists from Galileo to Hawking* (Oxford University Press, Oxford, 2001) Part II.

○ Rubí, J. M., "The long arm of the second law", *Scientific American* **299**(5), 62–67 (November 2008).

○ Sandfort, J. F., *Heat Engines: Thermodynamics in Theory and Practice* (Anchor Doubleday, Garden City, 1962).

Thompson (Rumford)

○ Brown, S. C., *Count Rumford: Physicist Extraordinary* (Anchor Doubleday, Garden City, 1962).

Thomson (Kelvin)

○ Flood, R., McCartney, M., and Whitaker, A., editors, *Kelvin: Life, Labours and Legacy* (Oxford University Press, Oxford, 2008).

○ Lindley, D., *Degrees Kelvin: A Tale of Genius, Invention, and Tragedy* (Joseph Henry, Washington, 2004).

○ MacDonald, D. K. C., *Faraday, Maxwell, and Kelvin* (Anchor Doubleday, Garden City, 1964).

Time's arrow

• Borel, É., "La méchanique statistique et l'irréversibilité", *Journal de Physique* **III**(5), 1697-1704 (March, 1913).

• Bricmont, J., "Science of chaos or chaos in science?", in *The Flight from Science and Reason*, edited by P. R. Gross, N. Levitt, and M. W. Lewis, *Annals of the New York Academy of Sciences* **775**, 131–175 (1996).

○ Carroll, S. M., "The cosmic origins of time's arrow", *Scientific American* **298**(6), 48–57 (June 2008).

○ Layzer, D., "The arrow of time", *Scientific American* **233**(6), 56–69 (December 1975).

• Lebowitz, J. L., "Boltzmann's entropy and time's arrow", *Physics Today* **46**(9), 32–38 (September 1993).

Iterated maps—Chaos made simple

14

In earlier chapters, we entered the world of chaos through the portal of Newtonian dynamics. In systems like the driven pendulum and the Tilt-A-Whirl, chaos is particularly paradoxical because the motion is smooth and continuous, even when the trajectory is random in the long run. From a computational point of view, however, smooth motion is difficult to simulate because it requires using many tiny time steps. On the other hand, as we learned in the last chapter, the motion of a billiard ball, also governed by Newton's equations, is far from smooth but relatively easy to calculate. When a ball collides with a fixed obstacle, we model the motion by equating the angle of reflection to the angle of incidence, while between collisions the ball travels in a straight line at constant velocity. In this case, chaos arises simply from the geometry of reflection.

But there are even simpler systems, known as iterated maps, that also exhibit chaotic behavior. These systems don't have a continuous time variable and are not primarily intended as models of Newtonian mechanics. Instead, iterated maps use a discrete time variable, usually taken to be an integer, that advances as $i = 0, 1, 2, \ldots$, with the understanding that we never ask what happens between time $i = 0$ and time $i = 1$. Birkhoff was the first to explore robust chaos in an iterated map in his 1932 paper, "Sur quelques courbes fermées remarquables." However, as our primary example, we'll consider a generalization of the map mentioned earlier in connection with Ulam and von Neumann, namely $X_{\text{new}} = RX_{\text{old}}(1 - X_{\text{old}})$. If we introduce an integer time variable, this map, called the logistic map, becomes,

$$X_{i+1} = RX_i(1 - X_i) \qquad (0 \leq X_i \leq 1 \text{ and } 0 < R \leq 4), \qquad (14.1)$$

where R is a fixed parameter. For a given R and initial X_0, Eq. (14.1) allows us to jump from X_0 to X_1 and then from X_1 to X_2, etc. by iteration and elementary arithmetic. What could be simpler?

In the 1950s, ecologists introduced the logistic map as a model for the dynamics of the population of an animal species, say, for example, the deer in Colorado. In this case, X represents the ratio of the number of deer to some maximum population, and the time integer i numbers successive years. Why does Eq. (14.1) make sense? Suppose that the initial population is much less than the maximum, so that $X_0 \ll 1$. Then the factor $(1 - X_0)$ is nearly 1 and $X_1 \approx RX_0$. Thus, if $R < 1$, an initial small population will be even smaller one year later, and it won't take many years before there aren't any deer in Colorado. On the other hand, if $R > 1$ then the population will be larger by a factor of R one year later $(X_1 \approx RX_0)$, and if $X_1 \ll 1$ then the population will grow to $X_2 \approx R^2 X_0$ after two years. More generally, as long is the deer population is much less than its maximum, then $X_i \approx R^i X_0$, and the deer population either grows or decays exponentially, depending on whether R is greater or less than 1. Given its role in this limit, we'll refer to R as the growth parameter.

Of course, exponential growth is a trend that Malthus pointed out two centuries ago, so Eq. (14.1) captures an essential feature of population dynamics. But Malthus was also aware that exponential growth can't persist indefinitely because finite resources are bound to limit the population. This reality is expressed in Eq. (14.1) by the factor $(1 - X_i)$, which curtails growth as the population approaches its maximum size and $X_i \rightarrow 1$. Thus, the logistic map is a highly simplified model for the dynamics of a population limited by finite resources. Now it's time to find out in detail what the logistic map predicts.

The Oxford University mathematicians Theodore Chaundy (1889–1971) and Eric Phillips were the first to explore the logistic map as a function of the growth parameter R. In 1936 they reported that, in the limit of large time, X_i approaches 0 for $0 < R < 1$, approaches $1 - 1/R$ for $1 < R < 3$, and "oscillates finitely" for $3 < R < 4$. The first of these results is reasonable: the population should vanish if the growth rate is less than 1. The second result also makes sense: due to the limit imposed by finite resources, the population might be expected to approach a constant for R greater than 1. Indeed, if we assume a constant population such that

$$X_{i+1} = X_i \tag{14.2}$$

and combine this condition with Eq. (14.1), then we obtain Chaundy and Phillip's result $X_i = 1 - 1/R$. But the oscillations for $R > 3$ are a curiosity that demands further explanation. In the early 1970s two groups studied periodic oscillations in this regime: Nicholas Metropolis (born 1915), Myron Stein, and Paul Stein at Los Alamos National Laboratory and Robert May (later Lord May, born 1936) at Princeton University. However, both groups knew from the work of Ulam and von Neumann that the oscillations were not always periodic.

14.1 Simple chaos

May was among the first to significantly explore the chaotic dynamics of the logistic map, in part because most scientists assumed that nothing interesting could come from such a simple system. But the apparent simplicity of Eq. (14.1) evaporates when it's written as $X_{i+1} = R(X_i - X_i^2)$. The X_i^2 term in this equation implies that the logistic map is

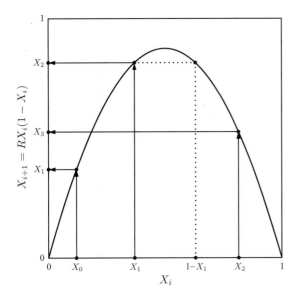

Fig. 14.1 A graphical representation of the logistic map for $R = 3.5$. As the dotted lines indicate, the logistic map isn't invertible because we can't know for sure whether X_2 came from the indicated X_1 or from $1 - X_1$.

nonlinear, so we have ample reason to expect unusual behavior when it's iterated.

Nonetheless, the logistic map does have a simple geometric interpretation. As shown in Fig. 14.1, which plots Eq. (14.1), the map is a parabola that defines the value of X_{i+1} that follows X_i. Thus, if we begin with an abscissa X_0, the map gives the subsequent X_1 as the ordinate. Iteration proceeds if we then plot X_1 as the abscissa and read X_2 as the ordinate, and so forth. In this way, the entire series $X_0, X_1, X_2, \ldots$ is defined by the parabola $X_{i+1} = RX_i(1 - X_i)$, which acts as a graphical lookup table for the next X.

A curious point, illustrated by Fig. 14.1, is that the logistic map can't be inverted to determine previous values of X_i. Suppose, for example, that we know X_2 but can't remember the corresponding X_1. It might seem a simple matter to read X_1 by following the arrows in Fig. 14.1 backwards. But we can't be sure if the answer is the abscissa X_1 indicated in the figure or $1 - X_1$, because both lead to the same X_2. Thus, the logistic map is said to be noninvertible, and, in contrast to typical Newtonian systems, we can't calculate backward in time from the present state of the system. If you're interested in Newtonian dynamics, the logistic map is only a toy, but an instructive toy nonetheless.

An improved graphical representation of the logistic map is shown in Fig. 14.2. Sometimes called a cobweb plot, this version incorporates the process of iteration as a reflection off of the 45° line $X_{i+1} = X_i$, allowing the motion to be followed continuously. Thus, beginning with the abscissa X_0, we move vertically to the parabola to find the ordinate X_1, then move horizontally to the 45° line to convert the ordinate X_1 into an abscissa. From this point the process repeats, alternately moving vertically to the parabola then horizontally to the 45° line, with every other step locating a new value of X_i. Of course, the construction

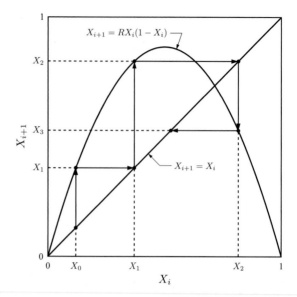

Fig. 14.2 The dynamics of the logistic map as a cobweb plot for $R = 3.5$.

lines between points have no particular significance—only the successive abscissas are meaningful.

When confronted with puzzling behavior in the logistic map, May chose to analyze the motion by computer simulation, a choice not available to Chaundy and Phillips in 1936. A sampling of such simulations is shown in Fig. 14.3 for various growth parameters. In Fig. 14.3(a), we find a trajectory for $R = 2.8$ in which the population quickly settles to a stable equilibrium value of $X = 1 - 1/R = 0.643$, as predicted by Chaundy and Phillips. Increasing R above 3, where Chaundy and Phillips predict "finite oscillations," we discover at $R = 3.1$ motion with a period of two iteration cycles (Fig. 14.3(b)). Thus, the logistic map replicates the breaking of time symmetry previously seen in continuous-time systems. At $R = 3.5$ (Fig. 14.3(c)), the time symmetry is further broken to produce an oscillation with a period of four iteration cycles, and at $R = 4$ (Fig.14.3(d)) we find motion that apparently doesn't repeat at all. Thus, the logistic map looks innocent, but, as May discovered, it exhibits a range of complex behaviors, including chaos. To check out the possibilities for yourself, try Experiment 19 of the Dynamics Lab.

Figure 14.4 provides an overview of steady-state behavior in the logistic map as a function of the growth parameter R. For each value of R, the map is first iterated 1000 times to eliminate transients, and the next 100 values of X are plotted. Over the range $1 < R < 3$, the steady-state population is the same every year, and only one population is plotted. However, as R increases above 3, the population begins to oscillate between two values, and the curve splits into two branches. Mathematicians call a splitting of this type a bifurcation. Similarly, above $R = 3.4495$ the period increases to four years, and the curve splits again for a total of four branches. As May realized, this

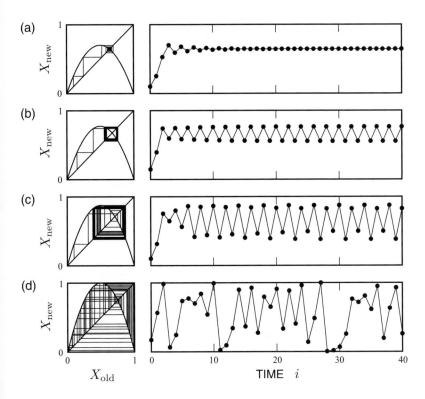

Fig. 14.3 The transient behavior of the logistic map for (a) $R = 2.8$, (b) $R = 3.1$, (c) $R = 3.5$, and (d) $R = 4$. For each growth rate, we show a cobweb plot and the corresponding time series for 40 iterations of the map.

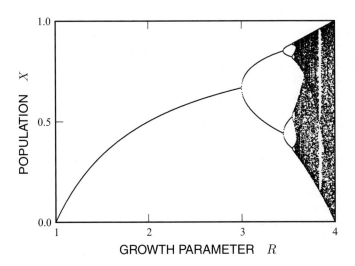

Fig. 14.4 A scatter plot of the steady-state population as a function of the growth parameter R for the logistic map. For each value of R, the map is first iterated 1000 times to eliminate transients, then the next 100 iterates of X are plotted.

"period-doubling" process proceeds more and more rapidly as R increases, such that solutions having periods of 8, 16, 32, ... iterations occur over smaller and smaller intervals of R. Finally, beyond $R = 3.5700$ the period becomes infinite, and the logistic map enters a chaotic regime. Indeed, the steady state behavior proves to be chaotic over most of the range $3.5700 < R \leq 4$, with the exception of several narrow intervals (such as $3.8284 < R < 3.8495$) where periodicity returns.

By exploring in detail the finite oscillations predicted for $3 < R \leq 4$, May uncovered something new and wonderful: a sequence of solutions having periodicities of 2^n iterations, with n increasing *ad infinitum*, followed finally by motion with no apparent periodicity. While the appearance of chaos may seem unprecedented, it is at least a logical extension of the period-doubling sequence. However, we must still ask why period doubling happens in the first place.

14.2   Liapunov exponent

What happens to the constant-population solution above $R = 3$? As Chaundy and Phillips realized, it's a question of stability. Plugging $X_i = 1 - 1/R$ into Eq. (14.1) yields $X_{i+1} = X_i$ for any R in the range $1 \leq R \leq 4$. Thus, the constant solution persists for $R > 3$, but we need to check its stability.

As usual, we test for stability by determining whether a small offset tends to grow or decay. Let's add a small δ_i to X_i to obtain $X_i' = X_i + \delta_i$ and see whether the offset between the X and X' trajectories grows or decays after one iteration. Choosing $X_i = 0.2$ and $R = 3.5$, we find in Fig. 14.5 that $\delta_{i+1} = X_{i+1}' - X_{i+1}$ is slightly greater than the original offset δ_i. On the other hand, if our initial population is $X_j = 0.4$, then we find that the offset δ_j is reduced after one iteration. According to the geometry of Fig. 14.5, an offset grows at points where the map is steeper and decays where the map is flatter. Clearly, stability is related to the slope of the map.

To be quantitative, let $S(X_i)$ be the slope of the logistic map at X_i. In the limit of small δ_i, we have from Fig 14.5,

$$\delta_{i+1} = S(X_i)\delta_i. \tag{14.3}$$

Thus, the offset at X_i grows or decays on the next iteration depending on whether the magnitude of the slope $S(X_i)$ is greater or less than 1. After two iterations we have

$$\delta_{i+2} = S(X_{i+1})\delta_{i+1} = S(X_{i+1})S(X_i)\delta_i, \tag{14.4}$$

so the offset grows as the product of the successive slopes. If we start with an initial X_0, then after N iterations the accumulated growth factor of the offset is,

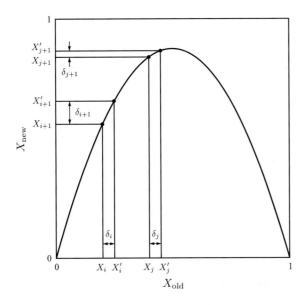

Fig. 14.5 The effect of a small offset δ on the next iterate of the logistic map at $X_i = 0.2$ and $X_j = 0.4$ for $R = 3.5$.

$$\delta_N / \delta_0 = S(X_{N-1})S(X_{N-2}) \cdots S(X_0)$$

$$= \prod_{i=1}^{N} S(X_{i-1}), \tag{14.5}$$

where Π is used to indicate a product of factors. Thus, the stability of a solution depends on the product of the slopes $S(X_i)$ at each of the successive iterates X_i.

In the long term, an offset will generally grow or decay exponentially, and we use this fact to define the Liapunov exponent λ,

$$|\delta_N / \delta_0| \approx e^{\lambda N}, \tag{14.6}$$

or, in the limit of large N,

$$\lambda = \lim_{N \to \infty} \frac{1}{N} \ln |\delta_N / \delta_0|$$

$$= \lim_{N \to \infty} \frac{1}{N} \ln \left\{ \prod_{i=1}^{N} |S(X_{i-1})| \right\}$$

$$= \lim_{N \to \infty} \frac{1}{N} \sum_{i=1}^{N} \ln |S(X_{i-1})|. \tag{14.7}$$

The last step here may seem a little mysterious, but it follows from the fact that the logarithm of a product is the sum of the logarithms: $\log(ab) = \log(a) + \log(b)$. As Eq. (14.7) proves, the logistic map is simple enough to yield an explicit formula for the Liapunov exponent—it's just the average of the logarithms of the magnitudes of the successive slopes.

Now let's see what Eq. (14.7) has to say about the stability of the constant-population solutions, $X_i = 1 - 1/R$. This case is especially simple because, with X_i constant, Eq. (14.7) reduces to

$$\lambda = \ln |S(1 - 1/R)|. \tag{14.8}$$

That is, the Liapunov exponent depends only on the magnitude of the slope at $X_i = 1 - 1/R$. If $|S(1 - 1/R)| < 1$ then $\lambda < 0$ and the constant-population solution is stable, while if $|S(1 - 1/R)| > 1$ then $\lambda > 0$ and it's unstable.

To proceed, we need one more formula. While we won't provide a proof, the slope of the logistic parabola is

$$S(X_i) = R(1 - 2X_i). \tag{14.9}$$

This formula makes sense in that it gives slopes of $S = R$ at $X_i = 0$, $S = 0$ at $X_i = 1/2$, and $S = -R$ at $X_i = 1$, in qualitative agreement with Fig 14.5. For a constant population, we have $X_i = 1 - 1/R$ and the slope is

$$S(1 - 1/R) = 2 - R. \tag{14.10}$$

This equation confirms the 1936 results of Chaundy and Phillips. For $1 < R < 3$, the slope of the constant solution falls in the range $|S| < 1$, the Liapunov exponent is negative, and the solution is stable. On the other hand, for $3 < R \leq 4$, we have $|S| > 1$, $\lambda > 0$, and instability. Thus, while a constant population exists for $3 < R \leq 4$, the slightest offset leads to a solution that moves rapidly away from $X_i = 1 - 1/R$ toward the oscillations predicted by Chaundy and Phillips and later explored by Metropolis, Stein, and Stein and by May.

Figure 14.6 presents a graphical version of these stability results. Here, the constant-population solutions are the points of intersection between the 45° line, $X_{\text{new}} = X_{\text{old}}$, and the logistic parabolas for $R = 2.5$, 3.0, and 3.5. The slopes of the parabolas at these points are $S = -0.5$, -1.0, and -1.5, indicating stability for the constant population at $R = 2.5$, marginal stability at $R = 3.0$, and instability at $R = 3.5$. Knowing about cobweb plots, you can see from the insets of Fig. 14.6 how an offset solution spirals inward toward the constant population at $R = 2.5$ and spirals outward from the constant population at $R = 3.5$.

Now let's look at stability in the oscillatory regime. Combining Eqs. (14.7) and (14.9), we obtain a general formula for the Liapunov exponent of the logistic map,

$$\lambda = \lim_{N \to \infty} \frac{1}{N} \sum_{i=1}^{N} \ln |R(1 - 2X_i)|. \tag{14.11}$$

The computation required for a single exponent easily fits on a modern programmable calculator: just use Eq. (14.1) to compute a sequence of iterates and for each iterate add its contribution, $\ln |R(1 - 2X_i)|$, to λ. Figure 14.7 shows λ as a function of R together with a scatter plot of the population at each R. In the range $3.5 \leq R \leq 4.0$, we find a mixture

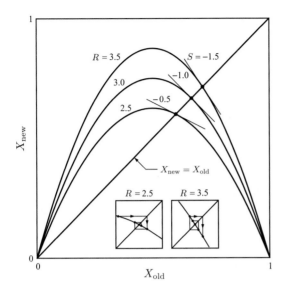

Fig. 14.6 A graphical analysis of the stability of the constant-population solutions of the logistic map for $R = 2.5$, 3.0, and 3.5. For each R, the constant solution is at the intersection of the logistic parabola with the 45° line, $X_{new} = X_{old}$, and the slopes S are as indicated. Insets show cobweb plots for trajectories beginning near the constant solutions at $R = 2.5$ and 3.5.

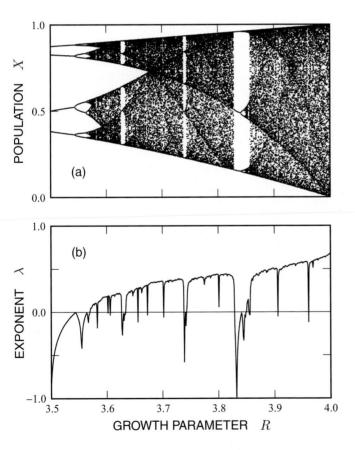

Fig. 14.7 Scatter plot of the steady-state population (a) and Liapunov exponent (b) as a function of the growth parameter R in the logistic map. In (a), 100 iterates are plotted for each R, and in (b) λ is averaged over 10^5 iterations.

of periodic and chaotic oscillations. For the periodic solutions, λ is negative, indicating stability, while for the chaotic solutions λ is positive, indicating sensitivity to initial conditions. The periodic solutions have periodicities including 2, 3, 4, 5, 6, 7, 8, 10, 12, 16, 18, 20, 24, and 32 iterations, to list a few.

Chaos predominates beyond the end of the period-doubling cascade at $R = 3.5700$, interrupted only occasionally by periodicity. Just above $R = 3.5700$, the chaotic solution is restricted to a narrow range of population values, but with increasing R the range broadens, and at $R = 4$ the chaotic solution spans the full range of populations between 0 and 1. The positive exponents of these solutions assure us that, just as in continuous-time systems, chaotic behavior in the logistic map is characterized both by apparent randomness and by the butterfly effect. The logistic map thus passes the acid test for chaotic behavior.

14.3   Stretching and folding

Wth the logistic map we can see for the first time two actions, often described as stretching and folding, that are common to all chaotic systems. Stretching refers simply to the exponential divergence of neighboring trajectories, and in the logistic map it's associated with the steep portions of the parabola, the regions where the slope has a magnitude greater than 1. Stretching leads to a positive Liapunov exponent and the now familiar butterfly effect. On the other hand, we haven't previously discussed the action of folding. It's plainly evident in Fig. 14.1, however, in the fact that the same output population follows from both X and $1 - X$. The parabolic shape of the logistic map effectively folds the range of input populations back on itself to insure that the output always falls in the interval between 0 and 1. Folding prevents the population from heading off to infinity and makes the logistic map globally stable. Thus, the logistic parabola combines stretching, the source of local instability, and folding, the source of global stability, in one remarkably simple package.

Stretching and folding were first described in Borel's paper of 1913, and they are clearly at the heart of chaos. We'll have more to say about these actions later in this chapter and in Chapter 16, where we discuss continuous-time systems.

14.4   Ulam and von Neumann—Random numbers

As noted in Chapter 11 and also by May, Ulam and von Neumann were apparently the first to appreciate the randomness inherent in the logistic map for $R = 4$. In 1947 they suggested that the map $X_{\text{new}} = 4X_{\text{old}}(1 - X_{\text{old}})$ might be a useful source of random numbers for simulating nuclear explosions and other complex phenomena. Let's take a closer look at this map and see why Ulam and von Neumann were intrigued.

As it happens, the $R = 4$ map is mathematically especially simple, in spite of being chaotic. Its simplicity first became apparent in the work of the English mathematician John Herschel (1792–1871), the son of astronomer William Herschel (1738–1822). In 1814, Herschel showed that the ith iterate of the map can be expressed as,

$$X_i = \tfrac{1}{2}[1 - \cos(2^i\theta)], \qquad (14.12)$$

where

$$\theta = \cos^{-1}(1 - 2X_0). \qquad (14.13)$$

Here $\cos^{-1}$ denotes the arc cosine function, a button found on most scientific calculators. These equations express an astonishing result. Instead of iterating the $R = 4$ map i times, the ith iterate X_i can be obtained directly from the initial population X_0 by evaluating θ using Eq. (14.13) and plugging the result into Eq. (14.12). Try it and see for yourself!

Although we know from the outset that X_i can be predicted from X_0 by iterating the map, it is nonetheless surprising that we can jump straight to any X_i without calculating all of the intermediate populations. Calculated either way, X_i is equally sensitive to small offsets in X_0, but the determinism expressed by Eqs. (14.12) and (14.13) is so simple and direct that we can only marvel at the random character of the successive populations. Herschel was either unaware of this randomness or not sufficiently impressed to remark on it.

Another simple outcome for $R = 4$ is the Liapunov exponent. Equations (14.12) and (14.13) combined with elementary calculus lead directly to the result $\lambda = \ln 2$. That is, a small offset increases on average by a factor of $e^\lambda = e^{\ln 2} = 2$ with each iteration of the map. This exponent implies strongly chaotic motion, with a perturbation doubling on average each time the map is iterated.

In seeking a source of random numbers, Ulam and von Neumann wanted a simple algorithm producing a sequence of numbers uniformly distributed between 0 and 1, all mutually independent. At first the $R = 4$ map seemed promising because they found an exact result for the probability distribution of the iterates. In particular, the probability P of obtaining an iterate between X_i and $X_i + \delta$ is

$$P(X_i, X_i + \delta) = \frac{\delta}{\pi\sqrt{X_i(1 - X_i)}} \qquad (R = 4,\ \delta \to 0), \qquad (14.14)$$

in the limit of small δ. This distribution is far from uniform, but it is mathematically exact. Figure 14.8(a) plots the distribution of 10^6 iterates X_i and compares the result with Eq. (14.14). Clearly, the

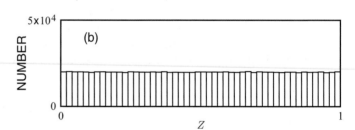

Fig. 14.8 (a) The distribution of 10^6 iterates X_i of the $R = 4$ logistic map tabulated in 50 population bins between 0 and 1. A dashed line shows the prediction of Eq. (14.14). (b) The distribution of the same iterates retabulated as $Z_i = \frac{1}{\pi} \cos^{-1}(1 - 2X_i)$.

distribution favors populations near 0 and 1. However, if we rescale all of the X_i according to

$$Z_i = \frac{1}{\pi} \cos^{-1}(1 - 2X_i), \tag{14.15}$$

then we obtain a uniform distribution, as shown in Fig. 14.8(b). Thus, Ulam and von Neumann might seem to have discovered an ideal random number generator.

But there is a fly in the ointment: successive iterates of the $R = 4$ map are not independent of one another. As seen in Fig. 14.3(d) at $i = 11$ and 28, a population near 0 is always followed by a string of regularly increasing populations. Thus X_i and X_{i+1} are strongly correlated, and the unadorned logistic map doesn't generate mutually independent random numbers. As is typical of chaotic systems, however, correlations between well separated iterates tend to be negligible. Thus, it's possible to obtain a useful random sequence from the logistic map if we include only every nth iterate in the sequence. When n is large enough, say 10 or more, then the included numbers are nearly independent, and the sequence becomes an excellent source of randomness.

Why are well-separated iterates, say X_i and X_{i+10}, only weakly correlated? The answer goes back to the Liapunov exponent. After 10 iterations, an offset in X_i grows on average by a factor of $e^{10\lambda} = e^{10 \ln 2} = 2^{10} = 1024$. That is, the first digit of X_{i+10} depends strongly on the fourth digit of X_i. Thus, while X_i completely determines X_{i+10}, the gross magnitude of X_{i+10} depends on subtle details of X_i, and the two appear to be independent. Furthermore, the larger the Liapunov exponent, the greater the degree of independence between successive iterates.

During the era of Ulam and von Neumann, the logistic map was never used as a random-number generator. In those days computer time was expensive, and more efficient maps with larger Liapunov exponents were soon discovered and put to use in simulating nuclear explosions.

14.5 Chaos explodes

When May began to study the logistic map in the early 1970s, chaos was not a recognized field of study. Although both Lorenz and Smale eventually realized that their work owed a debt to Poincaré, initially neither fully understood this heritage, and neither was aware of the other's work. It was as if Maxwell, Poincaré, Hadamard, Borel, Birkhoff, Lorenz, Smale, and Ueda were scientific fireflies with flashes of insight too far apart in time and space to attract the attention of the others. Living in separate eras and publishing in different specialties, these early explorers worked in isolation and were largely ignored by the greater scientific community.

The situation changed suddenly and dramatically in the mid-1970s. Perhaps it was the advent of cheap computing that made the difference, opening up the previously impenetrable domain of nonlinear systems. Or perhaps chaos was simply an idea whose time had arrived. In any case, scientists encountering chaos in diverse fields began to become aware of each other's work. For example, James Yorke, the mathematician at the University of Maryland who coined the term "chaos," met Robert May at Princeton. This meeting led May to use the term "chaos" (duly attributed to Yorke) in a 1974 paper, a year before Yorke himself used it in print.

In fact, May and Yorke purposely began a campaign to bring interested parties together and raise the awareness of the scientific world to the facts of chaos. It was Yorke, for example, who introduced Lorenz to Smale by sending him a copy of Lornez's 1963 paper. And in 1976, May published a review article in *Nature*, with the intention of making the basics of chaos accessible to scientists in general. May wrote that "we would all be better off if more people realized that simple nonlinear systems do not necessarily possess simple dynamical properties," and he advocated that students "be introduced to, say, [the logistic equation] early in their mathematical education."

Whatever the spark that ignited the explosion, by the end of the 1970s, word of chaos had spread throughout most of the scientific community, including my laboratory in Boulder, and many people were hopping on the chaos bandwagon. In the ensuing melee, scientists searched high and low for chaotic phenomena and found it almost everywhere they looked in the nonlinear realm. Previously inexplicable data, moldering in forgotten files, suddenly became explicable and could be written up for publication. Scientific journals began to overflow with observations of chaotic phenomena and new insights into the nature of chaos. And no field was immune: whether you worked in mechanics, fluid flow, electronics, photonics, chemistry, ecology, physiology, or economics, there were papers on chaos to be reckoned with.

In the 1980s, the excitement over chaos began to spill from professional journals into the popular press. Thus, an article on chaos by Douglas Hofstadter appeared in the November 1981 issue of *Scientific American*, one by Judith Hooper appeared in the June 1983 issue of *Omni* magazine, and thereafter popular accounts appeared with increasing frequency. But it was probably James Gleick's marvelous book, *Chaos: Making a New Science*, that made "chaos" into a household word. Published in 1987, *Chaos* is an absorbing account, touching on the lives and work of many of the scientists who founded the field, from Poincaré to Lorenz, Smale, and beyond. An international best seller, *Chaos* captured the excitement of discovery and the mystery of a phenomenon that was simultaneously predictable and random.

However, as sometimes happens with newsworthy science, the importance of chaos was often exaggerated in the popular press. Let's set the record straight. Chaos was not a breakthrough comparable to relativity or quantum mechanics. There were no new basic equations, just the tried and true dynamics of Newton. Indeed, the fundamentals of chaos weren't even new when chaos exploded, having been discovered a century earlier by Maxwell, Poincaré, and Hadamard. Instead of a revolution, the chaos explosion was a sudden awakening of the larger scientific community to an idea that had been gathering dust in the archives of science. Even though a few mathematical dynamicists knew of Poincaré's work, even though thermodynamics professors knew that atoms in a gas are easily perturbed, and even though computer scientists routinely used iterated maps to generate random numbers, the significance and widespread occurrence of chaos had escaped almost everyone's attention. Chaos exploded when the entire world suddenly sat up and took notice.

We might argue that the chaos explosion broke a spell cast by generations of dynamics professors. Having been taught that regular equations lead to regular solutions, scientists hypnotically proceeded to apply mathematical techniques that assumed regularity. These techniques are well suited to linear systems but have limited power in the nonlinear domain and are ultimately misleading. As we first saw with Lorenz, the spell was finally broken when solutions were obtained by numerical computations that follow the motion simplistically, tiny time step by tiny time step. With the assumption of regularity thrown out the window, chaotic motion at last had a chance to express itself, and scientists were freed from a spell that had bound them for centuries. The chaos explosion celebrated a strange and wonderful product of this newly found freedom.

14.6   Shift map—Bare chaos

With the logistic map we explored chaos in a simple system, but another iterated map, called the shift map, strips chaos down to its bare essentials. In this section, we'll use the shift map to explore chaos in one of its most rudimentary forms.

The shift map is made up of two linear segments separated by a discontinuity, as shown in Fig. 14.9. If Y_i is the value of Y at time i, then the map is defined by,

$$Y_{i+1} = \begin{cases} 2Y_i, & 0 \le Y_i < 1/2 \\ 2Y_i - 1, & 1/2 \le Y_i < 1 \end{cases}. \qquad (14.16)$$

Even though the separate pieces are linear, the sudden jump at $Y = 1/2$ makes the shift map highly nonlinear and a candidate for chaotic behavior. Moreover, the slope of the map is 2 at every point except $Y = 1/2$, so the Liapunov exponent is exactly $\ln 2$ for all trajectories, and every trajectory is unstable.

Figure 14.10 shows trajectories of the shift map for two specially selected initial values. For $Y_0 = 1/7$, we find periodic motion that repeats after three iterations (a), while $Y_0 = \pi - 3$ leads to apparently random motion (b). The qualitative difference between these trajectories results directly from a qualitative difference in the numbers $1/7$ and $\pi - 3$. To understand why, we'll take a mini safari into number theory, then return to the shift map.

Any mathematician will immediately tell you that $1/7$ and $\pi - 3$ are fundamentally different because $1/7$ is rational and $\pi - 3$ is irrational. That is, $1/7$ is the ratio of two integers, while $\pi - 3$ cannot be expressed as such a ratio. For the uninitiated, this statement may seem altogether strange, so let's take a closer look at what it means for a number to be irrational.

Historically, $\sqrt{2}$ was the first number known to be irrational, a fact discovered by the ancient Greeks. The proof is indirect and proceeds by first assuming that $\sqrt{2}$ is rational, then showing that this assumption leads to a contradiction. Thus, let's suppose that there are integers m

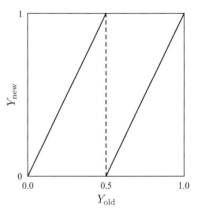

Fig. 14.9 The shift map.

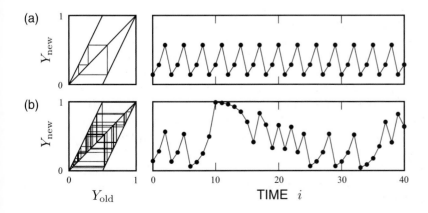

Fig. 14.10 The transient behavior of the shift map for (a) $Y_0 = 1/7$ and (b) $Y_0 = \pi - 3$. In each case, a cobweb plot and a time series record the first 40 iterations.

[1]Can you see why? Remember that according to the fundamental theorem of arithmetic any integer can be written as a unique product of prime factors.

```
      0.1 4 2 8 5 7 1 . . .
7 | 1.0 0 0 0 0 0 0
    7
    ⬚3⬚ 0
    2 8
      ⬚2⬚ 0
      1 4
        ⬚6⬚ 0
        5 6
          ⬚4⬚ 0
          3 5
            ⬚5⬚ 0
            4 9
              ⬚1⬚ 0
                7
```

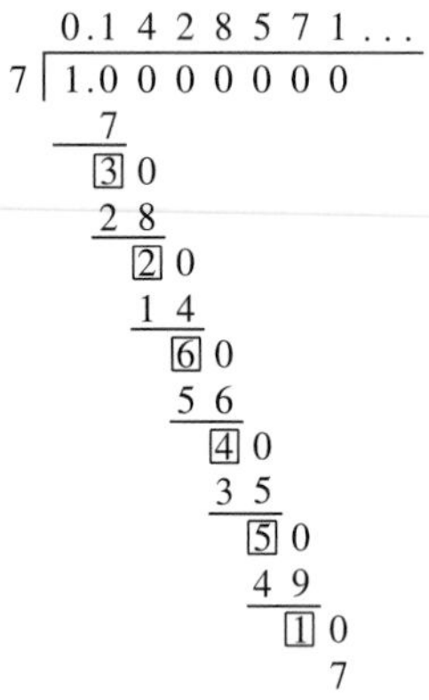

Fig. 14.11 The decimal expansion of 1/7.

and n such that $\sqrt{2} = m/n$. Without loss of generality, we can further suppose that m and n have no factors in common. Now, $2 = m^2/n^2$ or $m^2 = 2n^2$, and we know that m^2 includes 2 as a factor. But if m^2 has 2 as a factor, then m itself must have 2 as a factor.[1] Thus, we can write $m = 2p$, where p is another integer. With this substitution, we have $m^2 = (2p)^2 = 4p^2 = 2n^2$ or $n^2 = 2p^2$. Thus, we are led to the conclusion that n also has 2 as a factor. Since this contradicts our assumption that m and n have no factors in common, our assumption must be in error, and $\sqrt{2}$ cannot in fact be expressed as a ratio of integers. That is, $\sqrt{2}$ is irrational. This conclusion is no less amazing today than it was to the ancient Greeks.

The number π is also irrational, although the proof isn't so simple. The Swiss mathematician Johann Lambert (1728–1777) gave the first rigorous proof in 1767, but here we'll just accept that π is irrational. It immediately follows that the fractional part of π, namely $\pi - 3$, is also irrational. Can you see why?

In order to understand the shift map, we further note that the decimal expansion of any rational number ends in a finite sequence of digits that repeats over and over without end. Conversely, any decimal expansion ending in a repeating sequence of digits represents a rational number. Consider, for example, the decimal expansion of 1/7 by long division, as shown in Fig. 14.11. Here, division was halted at the point where the digit 1 was obtained for the second time and the process had begun to repeat itself. Carrying the division further will quickly convince you that $1/7 = 0.142857142857142857\ldots$. That is, the decimal expansion of 1/7 ends in the repeating sequence of digits 142857.

How can we be sure that the decimal expansion of any rational number m/n will end in a repeating sequence? The reason is that when dividing any integer by n the remainder can only be one of the n numbers $0, 1, \ldots, n-1$. In the long division example of Fig. 14.11, the remainder at each step is displayed in a box. In the first step, for example, 7 goes into 10 once with a remainder of 3, and in the second step 7 goes into 30 four times with a remainder of 2. If the remainder were ever 0, then the division would terminate and the expansion would end with repeated 0's. In the case of 1/7, however, the process of long division results in remainders of 3, 2, 6, 4, 5, and 1, after which the remainders and the digits of the expansion necessarily repeat. In general, dividing m by n will either end in zeros or in a repeating sequence that can be no longer than $n-1$ digits. Dividing 1 by 7 yields the longest possible sequence of digits, but the repeating sequence is typically shorter than $n-1$. For example, $1/9 = 0.11111\ldots$.

How can we be sure that every decimal expansion that ends in repeating digits represents a rational number? To answer this question, let's see how the decimal expansion of 1/7 can be converted back to a ratio of whole numbers. Because $A = 0.142857142857\ldots$ has a repeating sequence of six digits, we can perform the conversion by first multiplying A by 10^6 to obtain $10^6 A = 142857.142857142857$ and then subtracting A. This leads to

$$142857.142857142857\ldots$$
$$-0.142857142857\ldots$$
$$\overline{142857.000000000000\ldots}$$

or $10^6 A - A = 142,857$ and

$$A = \frac{142,857}{999,999} = \frac{1}{7}. \tag{14.17}$$

Thus, we arrive back at $1/7$ as expected. The trick here was to arrange a cancellation between the identical tails of the two repeating decimals, and this can done no matter how long the repeating sequence is. Thus, all repeating decimals are rational numbers in disguise.

Now let's return to the shift map. As we see from Eq. (14.16), a principal action of the map is the multiplication of Y_i by a factor of 2. While this action is cumbersome when Y_i is expressed in decimal notation, it reduces simply to moving the "decimal" point if Y_i is written in the binary system. As a refresher, recall that a decimal fraction between 0 and 1 can be written in as

$$Y = 0\,.\,D_1 D_2 D_3 \ldots$$
$$= \frac{D_1}{10} + \frac{D_2}{10^2} + \frac{D_3}{10^3} + \cdots, \tag{14.18}$$

whereas in binary notation the same number is

$$Y = 0\,.\,B_1 B_2 B_3 \ldots$$
$$= \frac{B_1}{2} + \frac{B_2}{2^2} + \frac{B_3}{2^3} + \cdots. \tag{14.19}$$

Here the decimal digits $D_1, D_2, D_3, \ldots$ range between 0 and 9, while the binary digits $B_1, B_2, B_3, \ldots$ are either 0 or 1. In Eq. (14.19) B_1, B_2, and B_3 represent the halves, fourths, and eighths digits, and the "decimal" point is called the "radix" point, as appropriate for a nondecimal number system. In binary, multiplication by 2 yields

$$2Y = B_1 + \frac{B_2}{2} + \frac{B_3}{2^2} + \cdots$$
$$= B_1\,.\,B_2 B_3 \ldots. \tag{14.20}$$

Thus, as advertised, multiplication by 2 in the binary system shifts the radix point one place to the right, just as multiplication by 10 has the same action in the decimal system. Now that we're working in binary, the name "shift map" begins to make sense.

In addition to multiplying Y_i by 2, the shift map requires us to subtract 1 if Y_i is greater than or equal to $1/2$. Subtracting 1 keeps Y_{i+1} between 0 and 1, and in general we can say that Y_{i+1} is the fractional part of $2Y_i$. When Y_i is expressed in binary, Y_{i+1} is thus obtained by shifting the radix point of Y_i one place to the right then discarding any 1 that appears in the units column. For example, if we convert $Y_0 = 1/7$ to binary, then $Y_0 = 0.001001001\ldots$, where the trailing digit sequence 001 repeats forever, and the first iterates of the shift map are

$$Y_0 = 0.001001001001001001001001001\ldots$$
$$Y_1 = 0.010010010010010010010010010\ldots$$
$$Y_2 = 0.100100100100100100100100100\ldots$$
$$Y_3 = 0.001001001001001001001001001\ldots$$

In binary, the arithmetic of the shift map is thus remarkably simple: with each iteration the digits march one step to the left and the left-most digit evaporates when it enters the ones column. The binary view also explains why the map for $Y_0 = 1/7$ has a period of three iterations. The periodicity results because $1/7$ expressed in binary ends in a repeating sequence three digits long, so that $Y_3 = Y_0$. More generally, because rational numbers expressed in binary always end in a repeating sequence, all rational values of Y_0 eventually lead to periodic behavior in the shift map.

On the other hand, $\pi - 3$ is irrational, and its binary expansion does not end in repeating digits but continues forever with apparently random digits. In the shift map, the initial value $Y_0 = \pi - 3$ thus leads to the kind of random behavior expected in a chaotic system. The first iterates are

$$Y_0 = 0.0010010000111111011010101100\ldots$$
$$Y_1 = 0.0100100001111110110101011000\ldots$$
$$Y_2 = 0.1001000011111101101010101010001\ldots \quad (14.21)$$
$$Y_3 = 0.0010000111111011010101100010\ldots$$
$$Y_4 = 0.0100001111110110101011000100\ldots$$

As long as we presume to use numbers of infinite precision, the shift map for $Y_0 = \pi - 3$ will never repeat but behave in a suitably chaotic fashion, as suggested by Fig. 14.10(b).

14.7   Origin of randomness

Now that we know how the shift map works, let's take a closer look at its mechanism to see what we can learn about chaos in general. In particular, does the shift map provide any clues to the origin of randomness?

First we observe that the shift map, like the logistic map, incorporates both stretching and folding. In the shift map, these actions are perhaps as simple as possible: multiplication by 2 and taking the fractional part. Multiplication by 2 leads to a positive Liapunov exponent, and it underlies the butterfly effect. But multiplication is a linear process, so we shouldn't be surprised that it isn't sufficient to obtain persistent random motion. Taking the fractional part is the nonlinear action of the shift map, and it acts to keep Y within its bounds between 0 and 1. Without taking the fractional part, multiplying by 2 on each iteration would simply send Y off to infinity in exactly exponential fashion, without any randomness at all. Thus, taking the fractional part is essential to chaotic behavior. It discards the most significant digit of Y and allows the less significant digits to take their turn as the most significant.

Another aspect of the shift map seems entirely peculiar: rational values of Y_0 produce periodic motion while irrational values yield aperiodic, generally random motion. In more typical chaotic systems, the nature of the motion is not directly related to a simple property of the initial conditions, and we find apparently random motion without carefully selecting the starting point. Indeed, at first glance, the shift map looks like a complete farce, since the randomness that we get out is exactly the randomness built into Y_0. If we enter only a finite number of binary digits for Y_0, then the computed Y_i will drop to 0 when the non-zero digits run out. Nonetheless, the shift map has some important things to tell us about the nature of chaos.

In particular, because there are an infinite number of rational numbers between 0 and 1, the shift map has an infinite number of unstable periodic solutions. (Remember, all solutions of the shift map are unstable.) Is this a peculiarity of the shift map? No; as we'll learn in Chapters 18 and 19, infinite numbers of unstable periodic solutions are found in all chaotic systems. The shift map is only peculiar in that the initial conditions for periodic motion are so easily identified.

Even though the periodic solutions are infinite in number, the chaotic solutions are overwhelmingly more numerous. This result is easy to see in the shift map, but to do so, we must take another mini safari into strange mathematical territory, specifically the world of transfinite numbers. The trek isn't terribly long or arduous, and it leads to some interesting insights into chaos.

The mathematics of transfinite numbers originated with the German mathematician Georg Cantor (1845–1918). Cantor's fundamental idea was to compare two infinite sets of objects by searching for a one-to-one correspondence between the elements of one set and the elements of the other. If such a correspondence can be established, then we say that the transfinite numbers of elements in the two infinite sets are equal. On the other hand, if no such correspondence exists, then the transfinite numbers of elements in the two sets are different, and one set is larger than the other. When Cantor first introduced his ideas in 1874, they were immediately controversial, and many mathematicians of eminence, including Poincaré, considered them repulsive. Poincaré later remarked that transfinite numbers were "a perverse pathological illness that would one day be cured." However, David Hilbert (1862–1943), a German mathematician of equal eminence, kept his cool and famously proclaimed that "no one will expel us from the paradise that Georg Cantor has created for us."

Perhaps the simplest set with a transfinite number of elements is the set of counting numbers, $\mathbf{C} = \{1, 2, 3, 4, \ldots\}$, clearly an infinite set. To understand how strange the world of transfinite numbers really is, let's compare the counting numbers with the set of all even numbers, $\mathbf{E} = \{2, 4, 6, 8, \ldots\}$, which omits the odd numbers and is apparently half as large as $\mathbf{C}$. However, when we apply Cantor's criterion, we see that the elements of these two sets can be put in a one-to-one correspondence. That is, we can pair each counting number with its double,

$$
\begin{array}{cccc}
1 & 2 & 3 & 4 \quad \cdots \\
\updownarrow & \updownarrow & \updownarrow & \updownarrow \\
2 & 4 & 6 & 8 \quad \cdots
\end{array}
$$

without leaving out any element of either set. Thus, strange as it may seem, **C** and **E** include the same transfinite number of elements, and we say that both sets are countable infinities. If we use ∞ to represent a countable infinity, then the relation between sets **C** and **E** suggests that doubling ∞ doesn't make it any larger, or $2\infty = \infty$, a strange sort of equality that works only for transfinite numbers.[2]

Even more surprising, Cantor found a way to pair all rational numbers with the counting numbers. That is, there are no more numbers of the form m/n where m and n are counting numbers than there are counting numbers. The proof is simple. As Cantor noted, all rational numbers can be arranged in an infinite array where the numerators in the mth column are all m and the denominators in the nth row are all n. You can find any rational number m/n simply by locating the mth entry in the nth row. Cantor then found a path through the array that visits each and every rational number in turn, as indicated by the arrows in Fig. 14.12. This path establishes the following correspondence between the counting numbers and the rational numbers

$$
\begin{array}{cccc}
1 & 2 & 3 & 4 \quad \cdots \\
\updownarrow & \updownarrow & \updownarrow & \updownarrow \\
1/1 & 2/1 & 1/2 & 1/3 \quad \cdots
\end{array}
$$

and proves that the rationals are countable. Because each row of the above array includes a countable infinity of fractions and there are a countable infinity of rows, the total number of fractions is ∞^2. But Cantor's argument tells us that the result is still a countable infinity, so $\infty^2 = \infty$. This astonishing conclusion alone might have justified Poincaré's skepticism of transfinite numbers.

The last of Cantor's results of interest here concerns the continuum of real numbers between 0 and 1. Any number in this range, whether rational or irrational, can be represented by an infinite decimal expansion. Assume for a moment that the real numbers form a countable set. In this case, we can make a list that pairs a counting number with each and every real number. The list might begin as in Fig. 14.13 and should include every real number between 0 and 1. But, as Cantor realized, there exist real numbers not on such a list. In particular, consider the number $B = 0.43672\ldots$, composed of the digits surrounded by boxes in Fig. 14.13. This number agrees with A_1 in the first decimal place, with A_2 in the second decimal place, and with A_n in the nth decimal place. If we now arbitrary substitute another digit for each digit of B to obtain, say $C = 0.54783\ldots$, we have a number that differs from every number on the list in at least one decimal place. The existence of C contradicts our assumption that all real numbers appear on the list. Thus, the real numbers cannot be paired one-to-one with the counting numbers, and the number of reals is larger than a countable infinity. The real numbers

[2]In his *Two New Sciences* of 1638, Galileo similarly noted that a one-to-one correspondence exists between the positive integers and their squares ($1 \leftrightarrow 1$, $2 \leftrightarrow 4$, $3 \leftrightarrow 9$, ...) and raised the paradox of apparent equality between two infinite sets, one of which lacks an infinity of elements included in the other. While Galileo concluded that equality and inequality are not applicable to infinite quantities, he anticipated some of Cantor's ideas by more than three centuries.

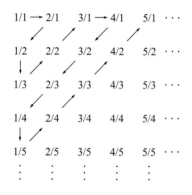

Fig. 14.12 Cantor's array of all rational numbers.

constitute an uncountable infinity, a transfinite number infinitely larger than a countable infinity.

Cantor was able to show that the transfinite number of real numbers is represented by 2^∞, where we again take ∞ to represent a countable infinity. Thus, while $2\infty = \infty$ and $\infty^2 = \infty$, the fact that the transfinite number of reals exceeds a countable infinity implies that $2^\infty > \infty$. This astounding deduction is just a first glimpse of the transfinite paradise that Cantor created.

Returning now to the shift map, we see that the periodic solutions, which correspond to rational values of Y_0, amount to a countable infinity. Removing the rational numbers from the real numbers, we are left with the irrational numbers, which like the reals are uncountable. Thus, the aperiodic solutions of the shift map, corresponding to irrational values of Y_0, are uncountable and infinitely more numerous than the periodic solutions.

What are the implications of these competing infinities in the shift map? Strangely, in spite of the superior number of aperiodic solutions, if we select Y_0 by typing a few digits into a computer, the exactly implemented shift map always yields a periodic trajectory. This results because a truncated decimal is a rational number and converts to a repeating binary expansion. For example, the decimal number $0.3000\ldots$ converts to the binary number $0.010011001\ldots$ where the digit sequence 1001 repeats infinitely. On the other hand, suppose we select Y_0 with a magic spinner that selects a number between 0 and 1 with infinite precision, such that all real numbers are equally probable. In this case, the probability of selecting a rational number is precisely 0 because the irrational numbers are infinitely more numerous. Thus, our magic spinner always selects an irrational Y_0, producing an aperiodic trajectory.

$A_1 = 0.\boxed{4}\,1\,4\,2\,1\,3\,5\,6\,2\ldots$
$A_2 = 0.7\,\boxed{3}\,2\,0\,5\,0\,8\,0\,8\ldots$
$A_3 = 0.2\,3\,\boxed{6}\,0\,6\,7\,9\,7\,7\ldots$
$A_4 = 0.6\,4\,5\,\boxed{7}\,5\,1\,3\,1\,1\ldots$
$A_5 = 0.3\,1\,6\,6\,\boxed{2}\,4\,7\,9\,0\ldots$
$\vdots$

Fig. 14.13 A list of real numbers between 0 and 1.

14.8 Mathemagic

From what we've learned so far about chaos in the shift map, you may be thinking, like Sam Sawnoff, that "The whole thing is a low put-up job on our noble credulity."[3] A map that shifts the radix point and tosses out the ones digit isn't extraordinary, but a magic spinner that selects random numbers with infinite precision is pure fiction. However, a little mathematical magic can bring us back to reality. The trick is to establish a surprising but rigorous connection between chaotic behavior in the shift map and that in the logistic map. This connection helps illuminate the ultimate origin of chaotic behavior.

Our task is to relate the sequence of numbers X_i generated by the logistic map for $R = 4$, Eq. (14.1), to a sequence of numbers Y_i generated by the shift map, Eq. (14.16). To work this mathematical hocus-pocus, we'll introduce yet another chaotic map, called the tent map, which is defined by

[3] Norman Lindsay, *The Magic Pudding*.

$$Z_{i+1} = \begin{cases} 2Z_i, & 0 \le Z_i < 1/2 \\ 2(1 - Z_i), & 1/2 \le Z_i < 1 \end{cases}, \qquad (14.22)$$

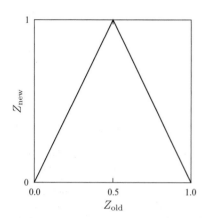

1

Z_{new}

0

0.0   0.5   1.0

Z_{old}

Fig. 14.14 The tent map.

and is plotted in Fig. 14.14. For inputs less than $1/2$, the tent map is identical to the shift map, and the output $Z_{i+1} = 2Z_i$ can be computed in binary by shifting the radix point one place to the right. For $1/2 \le Z_i < 1$, on the other hand, the tent map specifies computing $1 - Z_i$ before multiplying by 2. In binary $1 = 0.111111\ldots$ (can you see why?), so $1 - Z_i$ can be obtained from Z_i by changing all of the 1's to 0's and all of the 0's to 1's. Computer scientists say that Z_i and $1 - Z_i$ are the "one's complement" of each other. Thus, when Z_i is expressed in binary, the action of the tent map can be specified as follows. If the halves digit of Z_i is 0 (that is, $0 \le Z_i < 1/2$) then shift the radix point one digit to the right to obtain Z_{i+1}, but if the halves digit is 1 (that is, $1/2 \le Z_i < 1$) then take the one's complement of Z_i before shifting the radix point.

The tent and shift maps are so similar that their actions on a given initial number are closely related. In particular, if $Z_0 = Y_0$ then any subsequent Z_i is either equal to Y_i or its one's complement. Consider, for example, the first few binary numbers in the sequence generated by the tent map for $Z_0 = \pi - 3$.

$$\begin{aligned}
Z_0 &= 0.0010010000111111101101010100\ldots \\
Z_1 &= 0.0100100001111111011010101000\ldots \\
Z_2 &= 0.1001000011111101101010101010001\ldots \qquad (14.23) \\
Z_3 &= 0.1101111000000100101010101011101\ldots \\
Z_4 &= 0.0100001111111011010101000100\ldots
\end{aligned}$$

Because the halves digit of Z_0 is 0, we obtain Z_1 simply by shifting the radix point one digit to the right, and Z_1 is identical to Y_1 of the corresponding shift-map sequence, Eq. (14.21). Similarly, the halves digit of Z_1 is 0, so Z_2 equals Y_2. But the halves digit of Z_2 is 1, so its one's complement must be taken before shifting to obtain Z_3. As a result, Z_3 is the one's complement of Y_3. Finally, because the halves digit of Z_3 is also 1, the one's complement is taken again in computing Z_4, undoing the complement taken earlier, and Z_4 equals Y_4. Thus, Z_i and Y_i are sometimes equal and sometimes the one's complement of each other, but the two sequences are otherwise identical. Consequently, the tent map exhibits periodic motion when it's initially fed a rational number and aperiodic motion when it's fed an irrational number, just like the shift map.

Surprisingly, the tent map is also related to the $R = 4$ logistic map. In particular, for any initial X_0 of the logistic map, if we choose an initial Z_0 for the tent map given by

$$Z_0 = \frac{1}{\pi} \cos^{-1}(1 - 2X_0), \qquad (14.24)$$

then all subsequent Z_i and X_i obey the same relationship. That is,

$$Z_i = \frac{1}{\pi}\cos^{-1}(1 - 2X_i), \tag{14.25}$$

for all i. For example, if we suppose that $X_0 = 0.3$, then the corresponding Z_0 is $(1/\pi)\cos^{-1}(0.4) = 0.36901\dots$. Using Eqs. (14.1) and (14.22) to compute the next iterates for the two maps, we find $X_1 = 0.84$ and $Z_1 = 0.73802\dots$, and these values of X_1 and Z_1 satisfy Eq. (14.25), as advertised. In fact, when the initial values are chosen according to Eq. (14.24), the iterates of the two maps track each other exactly, magically maintaining the relationship of Eq. (14.25).[4] Mathematicians call maps that shadow one another like this "conjugate" maps.

Now we pull a rabbit out of a hat. Our trick is suggested by the fact that $X_0 = 3/10$ for the logistic map is shadowed in the tent map by $Z_0 = 0.36901\dots$. That is, due to the complex nature of the function $(1/\pi)\cos^{-1}(1 - 2X_0)$, a rational X_0 leads to a Z_0 that is almost certainly irrational. But if Z_0 is irrational then the tent map produces an aperiodic sequence of Z_i, and, because it shadows the tent map, the logistic map will generate X_i that are also aperiodic. Amazingly, the simple map $X_{i+1} = 4X_i(1 - X_i)$ with the simple initial value $X_0 = 3/10$ generates an infinite aperiodic sequence of numbers. When X_0 is rational, all subsequent X_i are also rational, so the logistic map generates chaos without leaving the domain of the rationals. Thus, the logistic map, like most chaotic systems, doesn't require a magic spinner to obtain chaos: almost any X_0 will do, and complexity is mathematically conjured from simplicity. This makes the logistic map a potent source of random numbers, as Ulam and von Neumann suggested so long ago.

But the fact that the logistic map with a rational X_0 produces the same kind of randomness as the shift map with an irrational Y_0, tells us something more about chaotic randomness. Because randomness in the logistic map isn't built into X_0, it must come from some property of the map itself. What property? Perhaps the only conclusion we can reach is that the actions of stretching and folding are the mechanisms that pull an infinitely long random sequence out of the integers. In this respect, chaos in the logistic map has a kinship to randomness in the digits of π, which are computed by summing an infinite series of integer ratios. In both cases, randomness emerges from a simple algorithm applied to the integers. However the logistic map is a richer source of randomness because we can begin with almost any rational X_0 and get another random sequence.

The connection between randomness and sensitivity to initial conditions, first proven by Lorenz, is also made plain by the actions of stretching and folding common to the logistic, shift, and tent maps. This is particularly clear for the shift and tent maps, where stretching is produced on each iteration when Y or Z is multiplied by 2, directly yielding exponential sensitivity. On the other hand, randomness results because multiplication by 2 shifts the digits of Y or Z toward more

[4]Of course, the relation between the logistic and tent maps isn't magic but can be proven if you remember your trigonometry. First rewrite Eq. (14.25) as $\cos(\pi Z_i) = 1 - 2X_i$. Then, given that $X_{i+1} = 4X_i(1 - X_i)$ and either $Z_{i+1} = 2Z_i$ or $Z_{i+1} = 2(1 - Z_i)$, show that $\cos(\pi Z_{i+1}) = 1 - 2X_{i+1}$. You'll need the identity $\cos(2\theta) = 2\cos^2(\theta) - 1$ and some patience, but give it a try.

significant positions, and folding ensures that each digit has its iteration as the leading digit. Thus, randomness and sensitivity are produced by the same processes, and it isn't surprising that one implies the other.

Although we like to think of chaotic systems as random number generators, it's important to understand that all of the numbers produced in an infinitely precise calculation are inherent in the equations of motion and their initial conditions. Thus, as we've said previously, the numbers produced by a chaotic mathematical system must be considered pseudorandom rather than strictly random. This fact is especially obvious for the shift map, where, in the absence of a magic spinner, randomness depends directly on our choice of Y_0, but it's equally true for the logistic map, the Lorenz equations, and the Tilt-A-Whirl.

Finally, let's turn from strictly mathematical systems to real physical systems. Suppose, for example, that the shift map is a physical system in which Y represents a quantity subject to environmental influences. In this case, the origin of randomness is not found in the infinite binary expansion of Y_0. Instead, Y is affected at some level by external gravitational or electromagnetic noise or by internal molecular noise. As a result, the value of Y beyond some apparently insignificant binary digit will change randomly. In a nonchaotic system such changes might never make an appreciable difference, but in the shift map, thanks to stretching and folding, they quickly march from insignificant bits to the most significant bit of Y. Thus, a physical chaotic system may be thought of as a mechanism for amplifying microscopic noise until it affects the macroscopic world. Again, the stretching and folding actions of the shift map are at the heart of this mechanism.

Further reading

Logistic map

- Alligood, K. T., Sauer, T. D., and Yorke, J. A., "Chaos", in *Chaos: An Introduction to Dynamical Systems* (Springer, New York, 1996) chapter 3.
- Chaundy, T. W. and Phillips, E., "The convergence of sequences defined by quadratic recurrence-formulae", *The Quarterly Journal of Mathematics, Oxford* **7**, 74–80 (1936).
- Dewdney, A. K., "Probing the strange attractions of chaos", *Scientific American* **257**(1), 108–111 (July 1987).
- Dewdney, A. K., "Leaping into Lyapunov space", *Scientific American* **265**(3), 178–180 (September 1991).
- Flake, G. W., "Nonlinear dynamics in simple maps", in *The Computational Beauty of Nature: Computer Explorations of Fractals, Chaos, Complex Systems, and Adaptation* (MIT Press, Cambridge, Massachusetts, 1998) chapter 10.
- Gleick, J., "Life's ups and downs", in *Chaos: Making a New Science* (Viking, New York, 1987) pp. 57–80.
- Herschel, J. W. F., "Consideration of various points of analysis", *Philosophical Transactions of the Royal Society of London* **104**, 440–468 (1814).
- Hofstadter, D. R., "Strange attractors: mathematical patterns poised between order and

chaos", *Scientific American* **245**(5), 22–43 (November 1981).

o Jensen, R. V., "Classical chaos", *American Scientist* **75**, 168–181 (1987).

• May, R. M., "Simple mathematical models with very complicated dynamics", *Nature* **261**, 459–467 (1976).

• Metropolis, N., Stein, M. L., and Stein, P. R., "On finite limit sets for transformations on the unit interval", *Journal of Combinatorial Theory (A)* **15**, 25–44 (1973).

o Vivaldi, F., "An experiment with mathematics", *New Scientist* **124**(1688), 46–49 (28 October 1989).

Random number generation

o Klarreich, E., "Take a chance: Scientists put randomness to work", *Science News* **166**, 362–364 (2004).

• Li, T.-Y. and Yorke, J. A., "Ergodic maps on [0, 1] and nonlinear pseudo-random number generators", *Nonlinear Analysis, Theory, Methods and Applications* **2**, 473–481 (1978).

• Phatak, S. C. and Rao, S. S., "Logistic map: A possible random-number generator", *Physical Review E* **51**, 3670–3678 (1995).

o Ulam, S. M. and von Neumann, J., "On combination of stochastic and deterministic processes", *Bulletin of the American Mathematical Society* **53** 1120 (1947).

Shift map

• Ford, J., "How random is a coin toss," *Physics Today* **36**(4), 40–47 (April 1983).

o Wolfram, S., "Iterated maps and the chaos phenomenon", in *A New Kind of Science* (Wolfram Media, Champaign, Illinois, 2002) pp. 149–155.

Transfinite numbers

o Aczel, A. D., *The Mystery of the Aleph: Mathematics, the Kabbalah, and the Search for Infinity* (Four Walls, Eight Windows, New York, 2000).

o Kaplan, R. and Kaplan, E., *The Art of the Infinite: The Pleasures of Mathematics* (Oxford University Press, Oxford, 2003).

o Maor, E., *To Infinity and Beyond: A Cultural History of the Infinite* (Princeton University Press, Princeton, New Jersey, 1987).

o Rucker, R., *Infinity and the Mind: The Science and Philosophy of the Infinite* (Birkhäuser, Boston, 1982).

Part V

Topology of motion

State space—Going with the flow

In previous chapters we examined chaotic motion as a physical and mathematical phenomenon in a variety of simple systems, exploring its random nature and its sensitivity to small perturbations. Now that we know what chaos is, it's time to look more deeply into its mathematical underpinnings and apparent paradoxes in continuous-time systems. Thus, we begin a new quest with the ultimate goal of understanding the mathematical tangle at the heart of chaos, discovered more than a century ago by Henri Poincaré. The setting of our journey will be an abstract mathematical space called "state space," where motion is viewed as a "flow." Our challenge is to understand the topology of this flow and thereby gain an overview of the kinds of motion that can occur in any given system. As you might guess, the topology of a chaotic flow is complex, and it's this complexity that Poincaré first glimpsed as a tangle.

In the next chapters, as we proceed on our quest, we'll travel historically backward in time toward Poincaré's discovery of tangles in 1889. Thus, when we reach the most profound understanding of chaos, the climax of our story, we'll arrive at the historical beginning of chaos theory, a time when chaos was a dimly recognized phenomenon. This inversion of normal events, with the deepest insight occurring before the most trivial observation, is a tribute to the genius of Poincaré.

Henri Poincaré was a child prodigy who lived up to his early promise. Born in 1854 in Nancy, where his father was a physician and professor of medicine, Henri was part of a vibrant, affluent, intellectual family. As sometimes happened in those times, at the age of 5 Henri contracted diphtheria and was unable to walk or speak for nine months. While he remained physically frail for some time, his intellectual development proceeded undeterred. Tutored by his mother, who was "très bonne, très active and très intelligente," Henri became an avid reader for whom a book, once read, was committed to memory and need not be read again. He was also playful and curious, wanting to see and understand everything around him. Even as a child, he was prone to becoming lost in thought. At school Poincaré excelled in French and Latin and placed first in most examinations.

Poincaré discovered his love of mathematics at the age of 15, while attending the high school (now the Lycée Henri Poincaré) in Nancy.

Fig. 15.1 Henri Poincaré. (AIP Emilio Segre Visual Archives.)

Owing to his excellent memory, at the lycée and later schools, Poincaré never needed to take notes in a math class. He simply integrated new knowledge immediately into his comprehensive understanding. One teacher described Poincaré as a "monster of mathematics," and before graduating from the lycée in 1873, he won first place in two national math competitions.

Electing to become an engineer, at age 19 Poincaré entered the École Polytechnique in Paris, where he placed first on the entrance exam. After two years at the Polytechnique, he studied for another three years at the École des Mines, also in Paris, and graduated as a mining engineer in 1878 at age 24. After graduation, Poincaré became a civil servant in the Corps des Mines, a career that might have lasted a lifetime. However, Poincaré had simultaneously pursued mathematics at the University of Paris under Charles Hermite (1822–1901) and, after working just eight months as a mine inspector, he was awarded his doctorate. One examiner remarked that Poincaré's doctoral thesis included enough original mathematical ideas for several theses. In December of 1879 Poincaré received an appointment to teach mathematics at the University of Caen.

For the remainder of his life, Poincaré remained officially on leave without pay from the Corps des Mines, and he retained an interest in mine safety. In 1910 he was even named inspecteur général of the Corps des Mines. This interest reflects Poincaré's enthusiasm for practical problems. During his career, he served on committees overseeing France's railroads, mail and telegraph systems, educational institutions, and the national observatory. Poincaré also served as president of the Bureau des Longitudes for three terms. Thus, while his greatest insights were mathematical, Poincaré was also a physicist and engineer, intrigued by real-world problems.

After two years at Caen, in 1881 Poincaré returned to Paris as a lecturer in mathematical analysis at the university. That year he also married Louise Poulain, by whom he would have three daughters and one son. Poincaré spent the remainder of his career at the University of Paris, becoming professor of mathematical physics and the calculus of probability in 1886. For Poincaré, mathematics had begun in earnest at Caen with the development of what are now called automorphic functions, ground breaking work that prompted his election to the Académie de Sciences in 1887. Upon returning to Paris, Poincaré started to develop his qualitative approach to dynamics that would eventually lead to the discovery of tangles and the publication of *Les Méthodes Nouvelles de la Mécanique Céleste*.

Poincaré's complete scientific works, collected and published after his death, extend to eleven volumes. They include work in both pure and applied mathematics, spanning topics from sets, groups, functions, topology, and differential equations to optics, electricity, celestial mechanics, thermodynamics, and special relativity. In addition, Poincaré is credited with originating the fields of algebraic topology and analytic functions of multiple complex variables. Sometimes called the last universalist,

Poincaré was recognized in his day as a mathematician of the highest rank, an honor that he shared only with his German counterpart, David Hilbert.

15.1   State space

In trying to understand chaotic motion, one strategy is to look closely at a simple example, and we adopted this approach in our investigation of iterated maps in Chapter 14. Another strategy is just the opposite: to move away from specific examples toward greater generality and abstraction. The latter course is exemplified by Poincaré when he sought a qualitative picture of dynamics. No matter how many specific orbits you compute, they may not reveal the full spectrum of planetary motion. However, the abstract realm of state space can provide an overview of all possible trajectories in a given system.

State space provides a second advantage. As Galileo realized when he began to study falling bodies, the ephemeral nature of motion can be frustrating. Galileo solved the problem by substituting the slower motion of a ball rolling down an incline for the speed of free-fall. Even so, motion is the epitome of transience, and generally leaves no record. By working in state space, however, Poincaré was able to effectively eliminate time from dynamics and reduce trajectories to static curves.

What is state space, and how can we study motion without considering time? In this chapter, we return to Galileo's pendulum for an answer to these and other questions. To begin, let's look again at Eqs. (5.23) and (5.24) for the motion of a pendulum. Writing these equations in a slightly different form brings us directly to the idea of state space. In particular, we see that the change in angle $\Delta\theta = \theta_n - \theta_o$ over a short time Δt can be written as

$$\Delta\theta = \tilde{v}_o \Delta t. \tag{15.1}$$

That is, the change in angle is just the present angular velocity $\tilde{v}_o$ multiplied by the time increment Δt. Similarly, the change in velocity $\Delta\tilde{v} = \tilde{v}_n - \tilde{v}_o$ is

$$\Delta\tilde{v} = -(\sin\theta_o + \rho\tilde{v}_o)\Delta t. \tag{15.2}$$

As these equations show, if we know the old angle and velocity of the pendulum, θ_o and $\tilde{v}_o$, then we can calculate $\Delta\theta$ and $\Delta\tilde{v}$ and hence the new angle and velocity, $\theta_n = \theta_o + \Delta\theta$ and $\tilde{v}_n = \tilde{v}_o + \Delta\tilde{v}$, a short time later.

In the pendulum, θ and $\tilde{v}$ are said to be "state variables" because they define the state of the system at any time and because they provide the information required to determine the state of the system at any future time. If we plot a point with coordinates $(\theta, \tilde{v})$ in the Cartesian plane, we locate the state of the pendulum in its two-dimensional "state space." Plotting several points $(\theta, \tilde{v})$ corresponding to successively later times, we obtain the pendulum's trajectory as a static curve in state

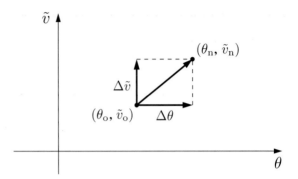

Fig. 15.2 The motion of a pendulum can be represented in the plane of angular velocity $\tilde{v}$ versus angle θ. Over a small time increment Δt, the angle changes by $\Delta \theta$ while the velocity changes by $\Delta \tilde{v}$, taking the pendulum from $(\theta_\mathrm{o}, \tilde{v}_\mathrm{o})$ to $(\theta_\mathrm{n}, \tilde{v}_\mathrm{n})$.

space. State space isn't mysterious and, as we'll discover shortly, it's very useful.

The basics of state space are shown in Fig. 15.2, where we plot the points $(\theta_\mathrm{o}, \tilde{v}_\mathrm{o})$ and $(\theta_\mathrm{n}, \tilde{v}_\mathrm{n})$ in the θ–$\tilde{v}$ plane. The increments $\Delta \theta$ and $\Delta \tilde{v}$ determine how the system moves from $(\theta_\mathrm{o}, \tilde{v}_{,\mathrm{o}})$ to $(\theta_\mathrm{n}, \tilde{v}_\mathrm{n})$ over a small time Δt. The slope of the vector connecting these points given by

$$\frac{\Delta \tilde{v}}{\Delta \theta} = -\frac{\sin \theta_\mathrm{o} + \rho \tilde{v}_\mathrm{o}}{\tilde{v}_\mathrm{o}}. \tag{15.3}$$

Remarkably, the slope depends only on the damping parameter ρ and the coordinates of the point $(\theta_\mathrm{o}, \tilde{v}_\mathrm{o})$. No matter how or when the pendulum arrives at some point $(\theta_\mathrm{o}, \tilde{v}_\mathrm{o})$ in the θ–$\tilde{v}$ plane, it always leaves by moving away in the same direction, determined by the point itself. Thus, when trajectories are plotted in state space, time effectively disappears.

Imagine, for a moment, a corn field with an arrow posted on each of the cornstalks. Beginning at a given spot, you note the direction of the arrow on a nearby cornstalk and take a step or two in the indicated direction. Arriving at another cornstalk, you find an arrow pointing in a slightly different direction and take a couple steps in the new direction. Continuing in this fashion, you slowly trace a graceful curve through the corn field. Time is of no concern: your path is determined solely by the posted arrows.

Plotting a trajectory in state space is like following arrows in a corn field. Beginning at any given point $(\theta_\mathrm{o}, \tilde{v}_\mathrm{o})$, you can use Eq. (15.3) to compute the direction in which to move, take a small step, record your new coordinates, and compute the next direction. You always know which direction to move, because the direction depends only on your present coordinates. In fact, you can get a feeling for a state space just by plotting an arrow indicating the direction of motion at selected points. In Fig. 15.3, for example, we show arrows corresponding to the motion of a damped pendulum in the θ–$\tilde{v}$ plane. As Poincaré recognized, the nice thing about this plot is that it doesn't change with time. Mathematicians like to say that Eqs. (15.1) and (15.2), define a "vector field," a space with a unique vector defined at each point.

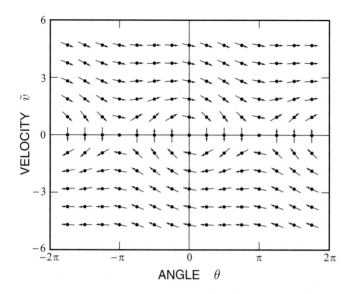

Fig. 15.3 At each point in state space, the motion of a pendulum is represented by a vector, indicating how the angle θ and velocity $\tilde{v}$ will evolve from that point. Here, vectors are drawn on a regular grid of points for a damping parameter of $\rho = 0.2$.

15.2 Attracting point

Two points in Fig. 15.3 are of special interest. No vector is drawn at either $(\theta, \tilde{v}) = (0,0)$ or $(\pi, 0)$, where $\Delta\theta$ and $\Delta\tilde{v}$ are exactly 0 according to Eqs. (15.1) and (15.2). These points are called fixed points, because motion comes to a stop when the pendulum reaches either point. We met the pendulum's fixed points in Chapter 4: $(0,0)$ corresponds to a pendulum hanging at rest, and $(\pi, 0)$ to a pendulum balanced in an upward position. These fixed points are central to understanding the topology of the pendulum's trajectories in state space.

Viewing motion as a vector field in state space is instructive, but simply plotting trajectories reveals the continuity of the motion and the significance of the fixed points. In Fig. 15.4, we show two such trajectories, beginning with the initial angles and velocities $(\theta, \tilde{v}) = (0, 5)$ and $(0, -5)$. Here, as in Fig. 15.3, we show the state space over the interval from $\theta = -2\pi$ to 2π in order to reveal the periodic nature of the angle coordinate. Thus, everything for angles between -2π and 0 is exactly repeated in the interval 0 to 2π. For the trajectory beginning at $(\theta, \tilde{v}) = (0, 5)$, the initial velocity is large enough that the pendulum completes three full revolutions in the forward direction before friction slows the pendulum and reduces it to oscillating about $\theta = 0$. (To follow this motion in Fig. 15.4, it may help to cover up the left half of the figure and recall that $\theta = 0$ is equivalent to $\theta = 2\pi$, so that a trajectory exiting the plot at $\theta = 2\pi$ is continued at $\theta = 0$.) Similarly, the trajectory beginning at $(\theta, \tilde{v}) = (0, -5)$ makes three rotations in the reverse direction before it begins to oscillate about $\theta = 0$. Comparing Figs. 15.4 and 15.3 confirms that the trajectories follow the arrows of the vector field.

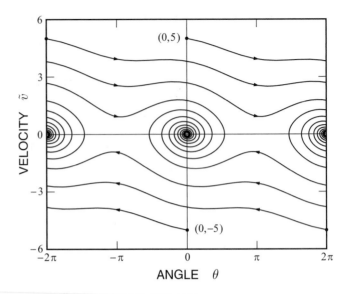

Fig. 15.4 State-space trajectories of a damped pendulum for the initial angles and velocities $(\theta, \tilde{v}) = (0, \pm 5)$. The damping parameter is $\rho = 0.2$.

The two trajectories plotted in Fig. 15.4 both end at $(\theta, \tilde{v}) = (0, 0)$, one of the fixed points of the pendulum. Indeed, trajectories beginning at almost any point in the state space end at $(0, 0)$, and for this reason $(0, 0)$ is called an "attractor" of the pendulum. Saying that $(0, 0)$ is an attractor is equivalent to saying that, almost no matter what kind of initial kick you give a pendulum, it ends up pointed downward at rest. What else could we expect? It would be devilishly difficult to give a pendulum just the right kick so that it stops at the upward balance point. Thus, like water draining from a sink, most trajectories of the pendulum swirl around and around in state space, gradually approaching the fixed point $(0, 0)$, as if it had some magic power of attraction. Experiment 20 of the Dynamics Lab allows you to plot pendulum trajectories in state space and check out the power of the attractor at $(0, 0)$.

An important property of state-space trajectories is evident in Fig. 15.4: trajectories never cross. As shown here, two trajectories can be closely interleaved and approach the same attracting point, but they never intersect. The reason is simple. At any given point in state space, the direction of motion is uniquely specified by Eq. (15.3), so there is one and only one way in which to proceed from the given point. A fork in the path is not allowed. As a result, trajectories plotted in state space tend to look like lines of flow in a fluid, and motion depicted in state space is often described as a "flow."

As Poincaré realized, the beautiful thing about state space is that every possible motion of a system is represented by the flow. If we added a few more trajectories to Fig. 15.4, we could confidently predict the qualitative motion for any given initial angle and velocity. Moreover, once the topology of a flow is understood, it is easy to prove that certain types of motion are either possible or impossible. This is the vision that Poincaré revealed in his qualitative theory of dynamics.

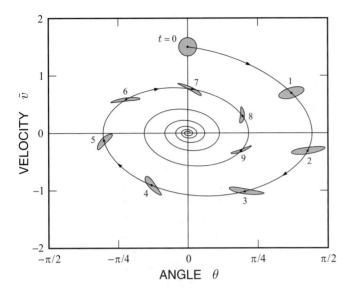

Fig. 15.5 A circular patch of dye thrown into state space at $(\theta, \tilde{v}) = (0, 1.5)$ is followed as it spirals into the pendulum's attracting point at $(0, 0)$. The patch is shown at times $t = 0, 1, 2, \ldots, 9$ as it shrinks and deforms. The damping parameter is $\rho = 0.2$.

15.3   Contracting flow

In view of the analogy between state-space trajectories and fluid flow, it's hard to resist tossing some dye into state space, just to see what happens. A numerical experiment of this kind is shown in Fig. 15.5. Here we consider all of the trajectories having initial angles and velocities within the circular region centered on $(\theta, \tilde{v}) = (0, 1.5)$. This circle is our patch of dye at $t = 0$. Following the patch in time, we find that at $t = 1$ the flow has transformed it into an approximate ellipse of slightly smaller area, and that it elongates and shrinks further as time jumps to $t = 2, 3, 4, \ldots$.

The most important conclusion of this experiment is that areas in state space contract as they move along the flow. Thus, our patch of state space shrinks as if it were part of a compressible fluid like air rather than an incompressible fluid like water. The contraction of areas in state space reflects the fact that our pendulum model includes friction and energy is gradually dissipated. In fact, we know that all of the trajectories in the initial circular patch will end up at the attracting point $(\theta, \tilde{v}) = (0, 0)$, because friction inevitably brings the pendulum to rest. Thus, the initial circular patch approaches a single point with zero area.

For the pendulum, a more complete analysis reveals that areas in state space shrink exactly in proportion to $e^{-\rho t}$, where ρ is the damping parameter. Given $\rho = 0.2$ and comparing areas to the circle at $t = 0$, we find that the ellipse at $t = 1$ has an area of $e^{-0.2} = 0.82$ or 82%, and the zucchini at $t = 3$ has an area of $e^{-0.6} = 0.55$ or 55%. In other systems, the evolution of areas in state space is typically more complex in form, but the general rule is that friction leads to contracting flows in state space.

On the other hand, in the absence of friction the damping parameter of the pendulum is $\rho = 0$, and our formula tells us that areas evolve as $e^{-0 \times t} = 1$. That is, when energy isn't dissipated by friction, the area of a patch of state space will remain constant (although it may change shape). This is a general rule: in frictionless systems, where energy is conserved, a patch of state space preserves its area as it moves with the flow, like part of an incompressible fluid. Called Liouville's theorem after the French mathematician Joseph Liouville (1809–1882), this rule implies qualitative differences in the topology of flows in systems with and without friction. In particular, systems without friction, called conservative systems because they conserve energy, cannot have attracting points or orbits, because neighboring trajectories cannot coalesce into regions of zero area. In the remainder of this book, however, we'll be primarily concerned with dissipative systems, and attractors of various sorts will be defining features of the flow.

15.4   Basin of attraction

Another perspective on the attracting point at $(\theta, \tilde{v}) = (0,0)$ is gained by attaching a counter that keeps track of the pendulum's total number of rotations. If the angle is always computed from the pendulum's physical position and the rotation count, then its final resting place might be at $\theta = 0$, 2π, -2π, or any multiple of 2π, depending on the initial angle and velocity. Thus, with a rotation counter, the pendulum can be viewed as having an infinite number of attracting points, at $(\theta, \tilde{v}) = (0,0)$, $(2\pi, 0)$, $(-2\pi, 0)$, etc. In terms of the washboard analog introduced in Fig. 6.3, these attracting points correspond to a BB resting at the bottom of one of washboards's infinite number of valleys. In either picture, any given set of initial conditions leads to a specific attractor. For example, the initial conditions $(\theta, \tilde{v}) = (0, 5)$ and $(0, -5)$ considered in Fig. 15.4 lead to the attractors at $(6\pi, 0)$ and $(-6\pi, 0)$, respectively.

What sets of initial angle and velocity lead to the attractor at $(\theta, \tilde{v}) = (0,0)$? To find out, we consider a grid of small boxes covering the state space and compute a trajectory using the initial angle and velocity at the center of each box. If the trajectory leads to the attractor at $(0,0)$, then most of the box is probably part of the attractor's "basin of attraction," the region of state space drained by the attractor in the flow analogy. In Fig. 15.6, for example, we display the approximate basin of the attractor at $(0, 0)$ by drawing all the boxes that lead to this attractor. As expected, the basin includes boxes in the immediate neighborhood of $(0,0)$, where trajectories simply spiral into the attractor. However, the basin also includes points corresponding to large negative angles and positive velocities, and for these points the pendulum completes one or more full revolutions before reaching the attractor. Thus, the initial conditions $(\theta, \tilde{v}) = (-2\pi, 3)$ lead to the attractor at $(0,0)$ rather than $(-2\pi, 0)$ because the pendulum must lose some of its initial velocity before it settles down to oscillatory motion. Similarly, the initial conditions $(2\pi, -3)$ yield a trajectory that approaches $(0, 0)$ after completing a full negative rotation.

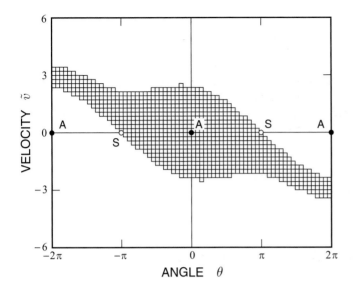

Fig. 15.6 Approximate basin of attraction for the pendulum attractor at $(\theta, \tilde{v}) = (0,0)$. Here, state space is broken into a grid of small squares, and a given square is plotted if the angle and velocity at its center initiate motion leading to the attractor at $(0,0)$. Attracting points are labeled A and other fixed points are labeled S. The damping parameter is $\rho = 0.2$.

Although not shown in Fig. 15.6, the basins of the attractors at $(2\pi, 0)$ and $(-2\pi, 0)$ are exactly similar to that for $(0,0)$ but displaced in angle by $\pm 2\pi$. Each attractor has a separate basin of attraction, and the basins tell us a lot about the topology of the state-space flow. When the basins of attraction are known, we can predict where the trajectory from any given starting point will end.

15.5 Saddle

So far, you might think that the fixed point at $(\theta, \tilde{v}) = (\pi, 0)$ is irrelevant to the topology of a pendulum's motion. This fixed point represents a balanced pendulum and appears to have nothing to do with movement. As we'll see shortly, however, the balance point and four special trajectories associated with it largely fix the topology of the flow. The balance point is also important because it incorporates both stability and instability and can be viewed as the origin of these contradictory features in chaotic motion. We have much to learn from the balance point.

In the context of state space, the fixed point at $(\pi, 0)$ is classified as a "saddle," for reasons soon to be evident. The best way to see how a saddle affects the flow is simply to plot a few trajectories in its neighborhood. Figure 15.7(a) shows how four such trajectories first approach and then recede from the saddle point. Trajectory a begins at $(\pi/2, 1.7)$ with enough velocity that the pendulum passes through the balance point and keeps going. Trajectory b begins at $(\pi/2, 1.5)$ with a slightly lower velocity, and the pendulum doesn't quite reach the top before it stops and begins to fall backward. Similarly, trajectories

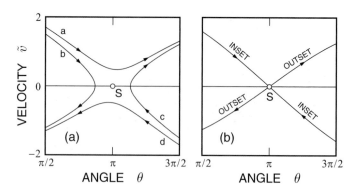

Fig. 15.7 State-space trajectories near the saddle point S at $(\theta, \tilde{v}) = (\pi, 0)$ for a pendulum with $\rho = 0.2$. Frame (a) shows trajectories defined by the initial conditions (a) $(\pi/2, 1.7)$, (b) $(\pi/2, 1.5)$, (c) $(3\pi/2, -1.5)$, and (d) $(3\pi/2, -1.7)$. Frame (b) shows the two inset and the two outset trajectories associated with the saddle point.

c and *d* begin at $\theta = 3\pi/2$ with negative velocities that are insufficient and sufficient, respectively, for the pendulum to reach $\theta = \pi$. As these trajectories suggest, the pendulum likes to avoid its saddle point.

If we fine tune the initial velocity, we expect to find a value between those for trajectories *a* and *b* which takes the pendulum to $\theta = \pi$ and no further: the pendulum just comes to a stop at the balance point. This trajectory is one of two incoming trajectories associated with the saddle. The second is similar but approaches the balance point from angles greater than π. Together, these two special trajectories, plotted in Fig. 15.7(b), form the "inset" of the saddle—the set of all trajectories that lead to the saddle. The inset trajectories are exceptional because they do not end at an attractor and thus aren't part of a basin of attraction. They are also exceptional because they reveal a vestige of stability associated with the saddle: they approach the balance point even though it is highly unstable.

Two additional trajectories associated with the saddle are outgoing trajectories, and, taken together, form the saddle's "outset." The outset trajectories, also shown in Fig. 15.7(b), are easy to compute since they result when the balanced pendulum receives the slightest touch. Starting the pendulum at $\theta = \pi$ with a slight negative velocity yields one outset trajectory, and starting it with a slight positive velocity yields the other. Although truncated in Fig. 15.7(b), these trajectories end on the attractors at $(\theta, \tilde{v}) = (0, 0)$ and $(2\pi, 0)$. The outset trajectories reveal the essential instability of the saddle point.

Experiment 21 of the Dynamics Lab allows you to animate the inset and outset trajectories of the pendulum's saddle point.

The relation between saddles, attractors, insets, and outsets is perhaps best understood by analogy with the mountain topography shown in Fig. 15.8. Here, the saddle S is analogous to a mountain pass and the surrounding topography has the shape of a horse's saddle. Like a saddle that bends upward toward the horse's head and rear and bends downward toward either flank, the terrain surrounding a pass curves upward along the ridge lines and downward toward the valleys on either side. The inset trajectories are like trails that lead down to the pass along

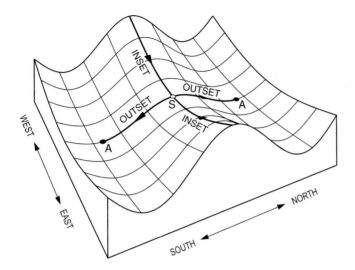

Fig. 15.8 A saddle S in state space is analogous to a pass in the mountains, and attractors A are like the bottoms of valleys. Following an inset trajectory takes you to the saddle as if you are walking down a ridge, and following an outset trajectory takes you from the saddle to the valley bottom.

the ridge lines, while the outset trajectories are like trails that lead from the pass to the attractors A at the valley bottoms. Altogether, motion in state space near a saddle point is a little like that of water flowing down the contours of a mountain pass.

Pursuing the mountain analogy further, we can say that the inset is like a continental divide. Just as water on one side of a divide flows to one ocean while water on the other side flows to another, trajectories on opposite sides of an inset flow toward different attractors. In other words, the inset of a saddle point defines the boundary between basins of attraction. For example, Fig. 15.9 shows the inset and outset trajectories of the saddle points plotted over a large portion of the pendulum's state space. Here, the area between the insets of the saddle points at $(\theta, \tilde{v}) = (-\pi, 0)$ and $(\pi, 0)$ is marked by hatching, and comparison with Fig. 15.6 reveals that this area coincides with the basin of the attractor at $(0, 0)$. Insets really do define the boundaries of the basins of attraction. Indeed, simply by plotting the insets and outsets of the saddle points as in Fig. 15.9, we obtain an excellent overview of the topology of the entire state-space flow.

Until now, we have used the word "topology" as if it were synonymous with "geometry," but it's time to distinguish between the two. Topology is often called rubber-sheet geometry, because it deals more with the way different parts of an object are connected than with the object's exact shape. Two objects are topologically equivalent if one can be deformed into the other without cutting or gluing. For example, a round ball and a cubical block are equivalent because a clay model of one can be deformed into the other. There's a standard joke that a topologist can't tell her donut from her coffee mug. Why? Because the mug's handle gives it a hole like that of the donut, and the mug's coffee reservoir is just a large dent that could easily be put in one side of the donut with a little stretching. The donut and mug, like the ball and the

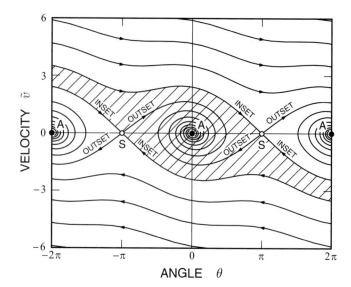

Fig. 15.9 The insets and outsets of the saddle points for a pendulum with $\rho = 0.2$. Attractors and saddles are indicated by A and S. The basin of the attractor at $(\theta, \tilde{v}) = (0, 0)$ is hatched.

cube, are topologically equivalent. But a ball and a donut have different topologies. A donut can be transformed into ball only by cutting the ring to eliminate the hole, and a ball, once it's been rolled into a cylinder, can be transformed into a donut only by gluing the ends of the cylinder together.

Returning to the damped pendulum, it isn't hard to see that changing the damping parameter will change the geometry of the flow without changing its topology. Increasing the damping will make trajectories approach the attractors more rapidly, but the attractors and saddles will remain at the same points. As we'll see next, however, the topology is changed completely if we introduce a steady driving torque.

15.6   Limit cycle

With friction acting on a pendulum, no matter how big the initial kick, its unforced motion will eventually come to a stop. For Poincaré, who wanted to discover all possible types of motion, it was important to look beyond fixed points. Are there other possibilities? Yes, as we learned in Chapter 6, if we apply a torque to overcome friction, then a pendulum can keep rotating forever. In this case, the pendulum settles into exactly periodic motion called a limit cycle, which gives us a new kind of attractor and a different topology for the flow in state space.

To get a closer look at a limit cycle, we consider the pendulum shown in Fig. 6.1. Here, air blown through a straw pushes against a paddle wheel attached to the pendulum's shaft. In our new style, the equations of motion for this case, Eqs. (6.3) and (6.4), are

$$\Delta\theta = \tilde{v}\Delta t, \tag{15.4}$$

$$\Delta\tilde{v} = (\tilde{\tau}_0 - \sin\theta - \rho\tilde{v})\Delta t, \tag{15.5}$$

where $\tilde{\tau}_0$ is the ratio of torque applied by the straw to the maximum gravitational torque (which occurs when the pendulum is at $\theta = \pm\pi/2$). In Eqs. (15.4) and (15.5), we have dropped the subscript "o" (for "old") from θ and $\tilde{v}$, with the understanding that the current values of θ and $\tilde{v}$ are always used to predict the changes, $\Delta\theta$ and $\Delta\tilde{v}$, over the next interval in time Δt. In state space, as in life, "No matter where you go, there you are." Just check your coordinates, and you'll know where to go next.

Armed with Eqs. (15.4) and (15.5), we can investigate the motion of a pendulum with a constant applied torque. First, we ask whether the equations allow any fixed points. That is, are there values of θ and $\tilde{v}$ for which $\Delta\theta = \Delta\tilde{v} = 0$? Solving these equations yields $\tilde{v} = 0$ and

$$\sin\theta = \tilde{\tau}_0. \tag{15.6}$$

Because $|\sin\theta|$ is never greater than 1, Eq. (15.6) has a solution only if $|\tilde{\tau}_0| \le 1$. This makes sense because the weight of the bob can balance the applied torque only if τ_0 doesn't exceed the maximum gravitational torque. When $|\tilde{\tau}_0| \le 1$ we can express the solutions of Eq. (15.6) using the arcsine function $(\sin^{-1})$,

$$\theta = \begin{cases} \sin^{-1}\tilde{\tau}_0 \\ \pi - \sin^{-1}\tilde{\tau}_0 \end{cases}, \tag{15.7}$$

where we assume that the arcsine function returns an angle between $-\pi/2$ and $\pi/2$. Thus, if $\tilde{\tau}_0 = 0.5$ then Eq. (15.7) predicts fixed points at $(\theta, \tilde{v}) = (\pi/6, 0)$ and $(5\pi/6, 0)$. As in the absence of an applied torque, one of these fixed points is a stable attractor and the other is a saddle point. They are plotted in state space as A1 and S in Fig. 15.10. Physically, the attractor A1 corresponds to the situation in which the applied torque is counterbalanced by the weight of the bob.

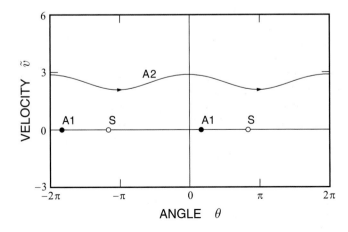

Fig. 15.10 With a constant applied torque, a damped pendulum can have a limit cycle and two types of fixed points. Here, we locate the fixed-point attractor A1, the fixed-point saddle S, and the limit-cycle attractor A2 in state space. The applied torque is $\tilde{\tau}_0 = 0.5$ and the damping is $\rho = 0.2$.

But fixed points are old hat. The new feature produced by the applied torque is a limit cycle. If we give the pendulum a kick to get it started, the applied torque can keep it rotating forever. Solving Eqs. (15.4) and (15.5) on a computer for $\tilde{\tau}_0 = 0.5$ and $\rho = 0.2$, we find that the rotation quickly settles into the repeating pattern identified as attractor A2 in Fig. 15.10. In the steady-state motion represented by this limit cycle, the angle of the pendulum increases continuously with only a slight variation in velocity as the bob cycles from its low point to its high point and back again. As intuition suggests, the bob moves slowest after climbing to its high point and fastest after falling to its low point.

Like the stable fixed point, the limit cycle is an attractor. If the initial kick starts the pendulum rotating too rapidly, then friction, which is proportional to velocity, will gradually slow the pendulum until its trajectory merges with the limit cycle. If the initial kick is small but enough to start the pendulum rotating, then its speed will gradually build until the trajectory again merges with the limit cycle. Thus, the limit cycle is expected to have its own basin of attraction, separate from that of the stable fixed point.

You can explore the rotating and stationary attractors of the pendulum in Experiment 22 of the Dynamics Lab. Alternatively, Fig. 15.11 presents a static picture of the flow as revealed by the inset and outset of the saddle point. If the pendulum is given a nudge from the saddle point toward positive angles, then it accelerates toward the limit cycle A2 along an outset trajectory, aided by the applied torque. If instead it's given an nudge toward negative angles, then the pendulum falls backward toward $\theta = 0$ along the other outset trajectory, but its motion is slowed by the applied torque, and it spirals into the stationary attractor A1. As usual, the inset trajectories define the boundaries between the basins of attraction. In Fig. 15.11, the hatched area is the basin of the

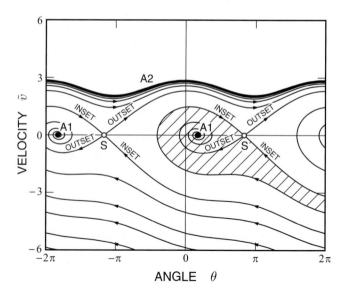

Fig. 15.11 The flow of a damped pendulum with a constant applied torque is illustrated by the inset and outset trajectories of the saddle point S. The hatched area bounded by the inset is the basin of the fixed point A1 at $(\theta, \tilde{v}) = (\pi/6, 0)$. As in Fig. 15.10, the applied torque is $\tilde{\tau}_0 = 0.5$ and the damping is $\rho = 0.2$.

stationary attractor at $(\theta, \tilde{v}) = (\pi/6, 0)$, and similar basins are associated with the other A1 attractors. The remainder of state space is the basin of the rotating A2 attractor. As without an applied torque, the inset and outset of the saddle point give a reasonably complete picture of the state-space flow.

15.7 Poincaré–Bendixson theorem

Are there yet more kinds of attractors? Amazingly, Poincaré was able to prove that, as long as there are just two state variables, then the only possible attractors are fixed points and limit cycles. With two state variables, the state space has two dimensions (θ and $\tilde{v}$ in the case of a pendulum) and state-space trajectories can be plotted in a plane (the pendulum's θ–$\tilde{v}$ plane). Thus, given that the state space is two dimensional, we immediately know that only two types of stable, long-term motion are possible—fixed points and limit cycles—regardless of what mechanical contraption we might consider. What a nifty theorem! It's exactly the kind of general result that Poincaré had hoped to prove by investigating motion in state space, where all of the trajectories are visible at once.

As seen in Fig. 15.11, the limit cycle A2 is a wavy line that extends forever in angle. On the other hand, if we recall that $\theta = \pi$ and $-\pi$ correspond to the same position of the pendulum, then it makes sense to wrap the figure around a cylinder, joining the plot along the lines $\theta = \pi$ and $\theta = -\pi$. The result, shown in Fig. 15.12(a), reveals the true nature of the limit cycle: it's a loop in state space. The A2 trajectory connects smoothly with itself at $\theta = \pm\pi$. An alternative view of this attractor is shown using polar coordinates in Fig. 15.12(b). Here, the pendulum angle is plotted around a circle and the velocity is plotted radially. This plot preserves the circular topology of the limit cycle while eliminating the apparent three-dimensional aspect of the cylinder in Fig. 15.12(a). The polar plot also has the nice feature that an imaginary line drawn from a given point to the origin represents the angle of the pendulum's shaft, with $\theta = 0$ and π corresponding to down and up. We'll find such polar plots useful in the next chapter.

Now that we know limit cycles are loops, let's reconsider Poincaré's idea that attractors must be either fixed points or limit cycles in a two-dimensional state space. Known formally as the Poincaré–Bendixson[1] theorem, this restriction on attracting trajectories is illuminated by a simple experiment requiring only a pencil and a sheet of paper.

In this experiment, the paper represents our two-dimensional state space, and the pencil is used to draw a single trajectory. As the experimenter, you get to draw whatever trajectory you desire with three restrictions: your trajectory isn't allowed to leave the paper, it isn't allowed to cross itself, and it isn't allowed to pass arbitrarily close to itself when headed in the opposite direction. That is, you can't draw trajectories that head off to infinity or ones that aren't specified by a smooth vector field. Your challenge is to find a trajectory that satisfies

[1]Ivar Bendixson (1861–1936) was a contemporary of Poincaré who extended his work on the topology of motion.

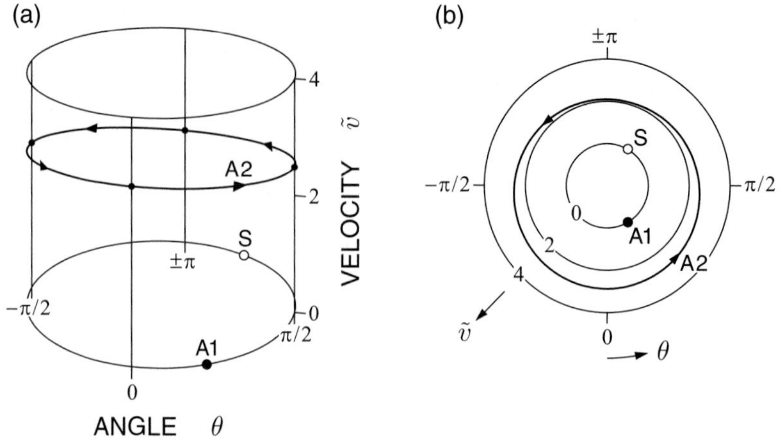

(a)

(b)

Fig. 15.12 When the pendulum's state space is wrapped around a cylinder, the trajectory of the limit cycle forms a closed loop. As in Figs. 15.10 and 15.11, the applied torque is $\tilde{\tau}_0 = 0.5$ and the damping is $\rho = 0.2$.

these constraints while forever exploring new territory. A little experimentation will probably convince you that this task is simply impossible. After a while, you'll find that your trajectory is forced either to approach a closed loop or simply come to an end. Thus, as Poincaré and Bendixson discovered, the trajectories specified by a two-dimensional vector field inevitably approach either a limit cycle or a fixed point. There just isn't room in a two-dimensional state space for more complicated attractors.

Galileo and Newton would not have been surprised by the Poincaré–Bendixson theorem. They created the science of dynamics to explain regular motion, and limit cycles are just the sort of behavior they expected to observe. Strangely, however, allowing just one more state-space dimension changes everything. As we'll see in the next chapter, a three-dimensional state space has plenty of room for chaotic motion: trajectories, described by a smooth vector field, that follow a random sequence of curves without repeating.

Further reading

Poincaré

o James, I., "Henri Poncaré (1854–1912)", in *Remarkable Mathematicians: From Euler to von Neumann* (Cambridge University Press, Cambridge, 2002) pp. 237–245.

o O'Shea, D., "Henri Poincaré", in *The Poincaré Conjecture* (Walker, New York, 2007) pp. 111–121.

o Szpiro, G. G., "The forensic engineer", in *Poincaré's Prize: The Hundred-Year Quest to Solve One of Math's Greatest Puzzles* (Dutton, New York, 2007) chapter 3.

o Yandell, B. H., "How famous can a function theorist be?", in *The Honors Class: Hilbert's Problems and Their Solvers* (A. K. Peters, Natick, Massachusetts, 2002) pp. 297–329.

Topology of motion

- Abraham, R. H. and Shaw, C. D., *Dynamics: The Geometry of Behavior*, 2nd edition (Addison-Wesley, Reading, 1992).

- Jackson, E. A., *Perspectives of Nonlinear Dynamics*, Volumes 1 and 2 (Cambridge University Press, Cambridge, 1991).

- Nolte, D. D., "The tangled tale of phase space", *Physics Today* **63**(4), 33–38 (April 2010).

16

Strange attractor

Enough of wandering through two-dimensional Iowa corn fields. It's time to don your scuba gear and plunge into the three-dimensional world of a Caribbean lagoon. There's a new degree of freedom here—you can move up or down as well as east or west and north or south—and strange new territory is suddenly open to exploration.

While perhaps not as exciting as a first scuba dive, in this chapter we begin to explore the possibilities inherent in a three-dimensional state space, including flows that represent chaotic motion. Chaos may seem almost ordinary by now, but, as we'll learn here and in the following chapters, the geometry of a chaotic flow is mind boggling in its complexity.

As a first example of a 3-D state space, consider the Lorenz equations, Eqs. (10.1)–(10.3), that model fluid convection. In our new notation, they are,

$$\Delta x = \sigma(y - x)\Delta t, \tag{16.1}$$

$$\Delta y = (x(r - z) - y)\Delta t, \tag{16.2}$$

$$\Delta z = (xy - bz)\Delta t. \tag{16.3}$$

The variables x, y, and z are the state variables of this system, and they are the coordinates of the 3-D state space. According to Eqs. (16.1)–(16.3), the position (x, y, z) records the present state of the convection cell and provides the information required to calculate where the system will move next in x–y–z space. Thus, if we imagined a 2-D state space as a corn field with an arrow on each stalk pointing the way, the Lorenz system would be a lagoon with a regular grid of buoys tethered at various positions and depths, each indicating where to swim next.

By comparison, the weather model in which Lorenz first discovered chaos involved 12 variables, and its motion can only be plotted in a 12-dimensional state space. We should be thankful that chaos can be visualized without understanding the topology of 12-D flows, a truly impossible mission. Even if you've never scuba dived, you have a good intuition of 3-D geometry and a chance of understanding a chaotic flow in three dimensions. As far as flows are concerned, however, the Poincaré–Bendixson theorem assures us that three is the absolute minimum number of dimensions allowing chaos. In two dimensions the only possibility for continued motion in a finite region of state space is a closed loop.

Chaos can occur in a smaller number of dimensions, however, if we consider systems with discontinuous time, like the iterated maps in

Chapter 14. The logistic and shift maps are examples of systems with
1-D state spaces that allow chaos. But to follow Poincaré's path, we'll
stick with continuous time and chaos in three dimensions.

Like Lorenz's convection cell, the Tilt-A-Whirl and the periodically
driven pendulum also exemplify chaos in three dimensions. However,
the equations don't immediately fall into the proper form. Consider, for
example, a pendulum driven by a torque having both a constant term
$\tilde{\tau}_0$ and an oscillatory term $\tilde{\tau}_1 \sin(\omega t)$. In this case, Eqs. (6.3) and (6.4)
become

$$\Delta\theta = \tilde{v}\Delta t, \tag{16.4}$$

$$\Delta\tilde{v} = (\tilde{\tau}_0 + \tilde{\tau}_1 \sin(\omega t) - \sin\theta - \rho\tilde{v})\Delta t. \tag{16.5}$$

The problem here is that $\Delta\theta$ and $\Delta\tilde{v}$ depend not only on the pendulum
angle θ and the velocity $\tilde{v}$, but also on the time t. So, if θ and $\tilde{v}$ are
the only state variables, we don't know where to go next because we
don't know the time. The trick is to introduce a third state variable,
$\phi = \omega t$, the phase of the oscillatory torque, that keeps track of time.
The equations of motion for the pendulum then become,

$$\Delta\phi = \omega\Delta t, \tag{16.6}$$

$$\Delta\theta = \tilde{v}\Delta t, \tag{16.7}$$

$$\Delta\tilde{v} = (\tilde{\tau}_0 + \tilde{\tau}_1 \sin\phi - \sin\theta - \rho\tilde{v})\Delta t. \tag{16.8}$$

Now we have three state variables and a 3-D state space, as required
for chaos, and time doesn't appear explicitly in the equations of motion.
The same trick can be applied to show that the Tilt-A-Whirl has a 3-
D state space and is also among the simplest-possible flows to exhibit
chaos.

16.1 Poincaré section

Without further ado, let's look at some trajectories of the driven pen-
dulum in its 3-D state space, that is, ϕ–θ–$\tilde{v}$ space. This space is a little
strange because it's periodic in both the phase ϕ of the oscillatory drive
and the pendulum angle θ. That is, incrementing either ϕ or θ by 2π
leaves the system unchanged, with the phase at the same point in the
drive cycle and the pendulum at the same angle.

To reflect these periodicities, we adopt the coordinate system for ϕ–θ–
$\tilde{v}$ space shown in Fig. 16.1(a) Here ϕ increases by 2π as we move around
the dashed circle in the horizontal plane. At any given ϕ, we plot θ and
$\tilde{v}$ on a vertical bull's-eye diagram like that in Fig. 15.12(b). In Fig. 16.1,
bull's-eyes are drawn for the phase angles $\phi = 0, \pm\pi/2$, and $\pm\pi$. In each
bull's-eye, the pendulum angle is the angle from the downward direction,
and the velocity is plotted in the radial direction, with $\tilde{v} = -5$ at the
center of the bull's-eye and $\tilde{v} = 5$ at the outer edge.

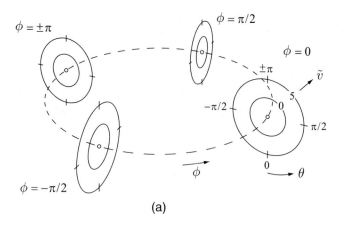

(a)

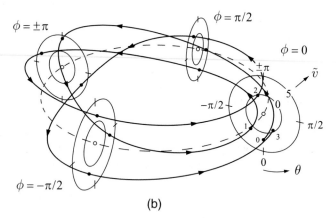

(b)

Fig. 16.1 (a) The state space of a driven pendulum. (b) The state-space trajectory of a pendulum with parameters $\rho = 0.2$, $\omega = 3$, $\tilde{\tau}_0 = 0$, and $\tilde{\tau}_1 = 5$ and initial conditions $(\phi, \theta, \tilde{v}) = (0, 0, 1.9)$. The trajectory is plotted over three drive cycles.

The trajectory shown in Fig. 16.1(b) begins with the arbitrarily chosen initial state vector $(\phi, \theta, \tilde{v}) = (0, 0, 1.9)$ and extends for three drive cycles. The initial point is the black dot labeled "0" on the $\phi = 0$ bull's-eye, located exactly below the center of the eye ($\theta = 0$) and at a radius corresponding to $\tilde{v} = 1.9$. Over three drive cycles, ϕ completes three circuits of the dashed circle, and the trajectory is marked by a black dot at each point where it intersects one of the bull's-eyes. The trajectory returns to the $\phi = 0$ bull's-eye three times, as indicated by the points labeled "1", "2", and "3".

With some study, you can get an idea of the 3-D nature of the state-space trajectory in Fig. 16.1(b). You may notice that the trajectory crosses itself at several points, apparently contradicting our dictum that state-space trajectories never intersect. Of course, the problem is that we have represented a 3-D curve by projecting it onto a 2-D piece of paper. If the trajectory were instead formed from wire in three dimensions, it would be clear that the wire never crosses itself. Figure 16.1(b) only shows the shadow of this wire frame as if it were held in front of a projector, and it bears the same relation as a shadow rabbit does to the hands that make the shadow.

The difficulty of visualizing a 3-D trajectory from its 2-D projection prompted Poincaré to look instead at a cross-section, now called a Poincaré section. For the driven pendulum, a Poincaré section is easily constructed by considering the points at which the trajectory crosses the $\phi = 0$ plane. Thus, the four points, "0–3", in the $\phi = 0$ bull's-eye constitute a Poincaré section of the trajectory in Fig. 16.1(b). These four points don't tell us nearly as much as the full trajectory, but as the trajectory is extended and more points are plotted, the Poincaré section becomes much more useful than the shadow of a tangled mass of wire.

For example, consider the Poincaré section in Fig. 16.2 obtained by continuing the trajectory of Fig. 16.1(b). Here, the θ–$\tilde{v}$ plane at $\phi = 0$ is shown in rectangular rather than polar coordinates, but the points labeled "0–3" are the same as in Fig. 16.1(b). The succeeding points bounce around somewhat arbitrarily, but gradually converge to the circled point labeled "A". As you might expect, A represents an attractor, a limit cycle that appears as a point in the Poincaré section.

The cyclic nature of this attractor is revealed in Fig. 16.3(a), where we return to the full 3-D state space. Here I've drawn a line from the center of each bull's-eye plot to the point at which the trajectory pierces the bull's-eye. These lines give us miniature representations of the pendulum at $\phi = 0$, $\pm\pi/2$, and $\pm\pi$ that correctly reveal the angle of the pendulum from vertical (although the length of the "pendulum" is determined by its velocity). Following the pendulum around one cycle reveals that it is simply swinging back and forth like a clock pendulum. It hangs nearly straight down at $\phi = 0$ and $\phi = \pm\pi$ and reaches its maximum displacement to one side or the other at roughly $\phi = \pm\pi/2$. The orbit repeats itself on every cycle of applied torque, as ϕ advances by 2π.

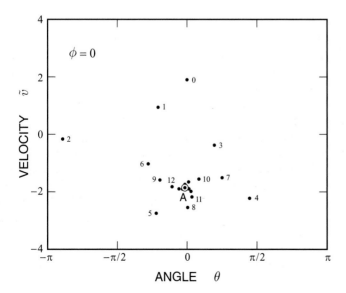

Fig. 16.2 The Poincaré section of the trajectory in Fig. 16.1(b) extended to an indefinite number of drive cycles. Each point plots the angle θ and velocity $\tilde{v}$ of the pendulum as the phase of the drive torque passes through $\phi = 0$. The first dozen points are labeled in order of appearance. The circled point A represents the limit cycle to which the trajectory converges.

Experiment 23 of the Dynamics Lab allows you to further explore the attracting nature of this orbit through a 2-D projection of its state-space trajectory and through its Poincaré section.

16.2   Saddle orbit

We can think of the attracting orbit of Fig. 16.3(a) as evolving from the attracting point of the free pendulum as the amplitude of the oscillatory torque is increased from zero. This observation leads us to ask if the unstable balance point of the pendulum similarly evolves from a static saddle into an unstable periodic orbit with the application of an oscillatory torque. The answer is yes, and the resulting "saddle orbit" introduces a new topology into our state-space that is essential to the existence of chaos. Indeed, as we'll learn in Chapter 19, saddle orbits are at the root of Poincaré's topological tangle. Understanding saddle orbits now will pay back handsomely when the time comes to peer into the heart of chaotic flows.

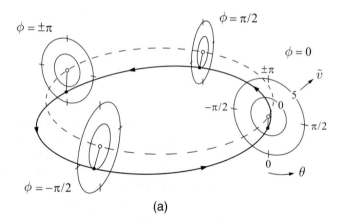

(a)

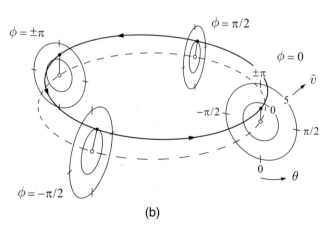

Fig. 16.3 (a) The state-space trajectory of the attracting orbit A in Fig. 16.2. (b) The state-space trajectory of the unstable saddle orbit. In both (a) and (b) the parameters are $\rho = 0.2$, $\omega = 3$, $\tilde{\tau}_0 = 0$, and $\tilde{\tau}_1 = 5$.

(b)

The saddle orbit itself is not especially mysterious. As shown in Fig. 16.3(b), this orbit is much like the attracting orbit of the driven pendulum except that the pendulum oscillates around its balance point, $\theta = \pi$, rather than its equilibrium point, $\theta = 0$. The motion is a little like that of a tightrope walker who has begun to lose his balance and sways from side to side just before falling off the rope. Of course, the saddle orbit is highly unstable, and even the slightest disturbance will upset the periodic motion shown in Fig. 16.3(b).

One of the trickiest aspects of saddle orbits is finding one. The only clue we have to go on is that after one drive cycle, beginning at say $\phi = 0$ and ending at $\phi = 2\pi$, the trajectory must return to the values of θ and $\tilde{v}$ at which it began. This completes the loop and insures the trajectory will continue in the same orbit forever. But because a saddle orbit is unstable, the only way to discover one is by trial and error. We simply choose initial values of ϕ and $\tilde{v}$, compute the trajectory over one drive cycle, and see if we come back to the initial ϕ and $\tilde{v}$. If not, we adjust the initial values and try again. Fortunately, a few such trials will indicate how to make better guesses for the initial values, and the whole process can be automated. Thus, using simple search strategies, a computer can locate saddle orbits quite quickly.

For the 2-D state space of the last chapter, we found that the topology was largely determined by the inset and outset trajectories associated with the saddle points. Something similar is true in three dimensions, but the saddle points are replaced by saddle orbits, and the insets and outsets of saddle orbits are surfaces rather than individual trajectories. This sounds complicated, but an example will help explain the geometry. The overall picture is presented in Fig. 16.4(a), which replots half of the saddle orbit in Fig. 16.3(b). Note that the saddle orbit never strays far from the angle $\theta = \pi$, the pendulum's balance point in the absence of an applied torque. The inset and outset are depicted here as surfaces that intersect the saddle orbit.

By definition, the inset is the set of all trajectories that approach the saddle orbit more and more closely as time goes on. This convergence is illustrated in Fig. 16.4(b), where we plot a few trajectories chosen from the inset. Similarly, the outset is defined as the set of all trajectories that begin arbitrarily close to the saddle orbit and diverge from it as time goes on. A few examples of outset trajectories are shown in Fig. 16.4(c). As the inset and outset demonstrate, saddle orbits include elements of both stability, reflected by the inset, and instability, reflected by the outset. Of course, this combination of stability and instability is also one of the principal paradoxes of chaotic motion.

How can we discover trajectories on the inset or outset of a saddle orbit? Outset trajectories are the easiest to find. Simply begin with initial conditions very very close to the saddle orbit and see where they lead as time proceeds. Even if the initial point isn't exactly on the outset, the trajectory will rapidly approach it with great accuracy. To find an inset trajectory, begin again with a point very very close to the saddle orbit but use the equations of motion with a negative Δt to go backward

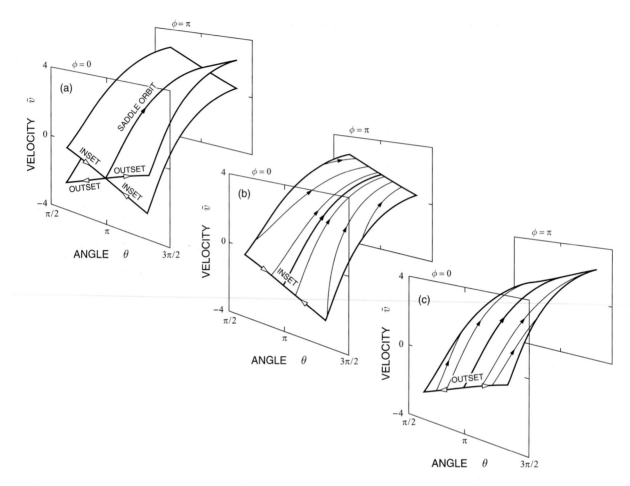

Fig. 16.4 Half of the saddle orbit shown in Fig. 16.3(b) is plotted here between the planes $\phi = 0$ and $\phi = \pi$. Frame (a) shows this section of the orbit together with portions of the associated inset and outset surfaces. Frame (b) focuses on the inset and shows how several trajectories of the inset approach the saddle orbit. Similarly, frame (c) shows how several trajectories of the outset diverge from the saddle orbit. Open arrows in the $\phi = 0$ plane indicate the trend of trajectories in the inset and outset to converge to or diverge from the saddle orbit.

in time. Moving backward will take you farther and farther from the saddle orbit, but if you then switch to a positive Δt, you'll return to where you started, very very close to the saddle. Thus, you will have discovered an inset trajectory. If this logic seems confusing, go back to Experiment 19 of the Dynamics Lab, and you'll see how it works for a saddle point.

Figure 16.4 reveals only a narrow strip of the inset and outset surfaces near the saddle orbit. To see more, we switch to the Poincaré section plotted in Fig. 16.5 for the $\phi = 0$ plane. Here the saddle orbit appears as the point S, and the inset and outset form a cross through this point, just as in the $\phi = 0$ plane of Fig. 16.4(a). Overall, Fig. 16.5 looks remarkably like Fig. 15.9 for the free pendulum, but perhaps on drugs. However, it's

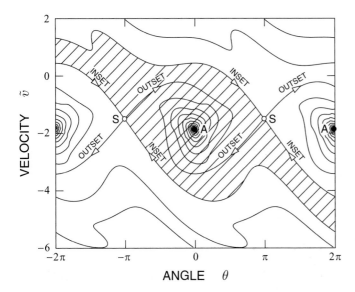

Fig. 16.5 Poincaré section at $\phi = 0$ of the saddle orbit (S) and its inset and outset for a driven pendulum with $\rho = 0.2$, $\omega = 3$, $\tilde{\tau}_0 = 0$, and $\tilde{\tau}_1 = 5$. Point A near the center is an attracting orbit, and the basin of this attractor is hatched.

crucial to understand the difference between these figures. Figure 15.9 is for a 2-D state space and shows all there is to show, while Fig. 16.5 is a Poincaré section of a 3-D state space and hides an entire dimension. Thus, while the point S in Fig. 15.9 is all there is to the saddle point, the point S in Fig. 16.5 is just one point of a saddle orbit that extends into the hidden ϕ dimension.

Likewise, while the directed inset and outset lines in Fig. 15.9 represent individual trajectories, those in Fig. 16.5 are the cross-sections of surfaces composed of a continuum of trajectories and are not themselves trajectories. Thus, the arrows in Fig. 16.5 are intended to indicate only that the trajectories of the inset approach the saddle and that those of the outset diverge from the saddle. To emphasize this difference, open arrows are drawn in Fig. 16.5, in contrast to the filled arrows of Fig. 15.9. Plotting the Poincaré sections of the inset and outset actually requires computing thousands of trajectories to establish where the inset and outset surfaces cross the $\phi = 0$ plane.

Nonetheless, the role of the inset surface in Fig. 16.5 is like of that the inset trajectories in Fig. 15.9 in that it defines the boundary between basins of attraction. Thinking again in terms of a particle on a washboard, the point A near $\theta = 0$ represents oscillatory motion within the central valley of the washboard. In this case, the hatched area identifies values of θ and $\tilde{v}$ at $\phi = 0$ that lead to this attracting orbit. Of course, the basin of attraction is actually three dimensional, extending through the full 2π extent of the ϕ dimension, but we'll be content to see just the cross-section of Fig 16.5. All in all, this section reveals almost everything we might want to know about the 3-D flow of a pendulum driven by a small oscillatory torque. Still, it's important to remember that Poincaré sections always hide one dimension of state space.

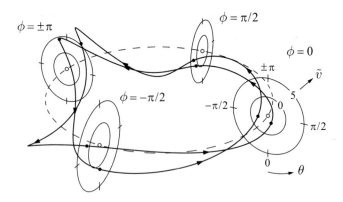

Fig. 16.6 The state-space trajectory of a stable periodic orbit of the driven pendulum in which the pendulum advances by two revolutions during two cycles of the driving torque. The parameters are $\rho = 0.1$, $\omega = 1.4$, $\tilde{\tau}_0 = 0$, and $\tilde{\tau}_1 = 4.9$.

16.3   Period doubling

Before proceeding to chaotic flows, we pause for a brief look at period doubling in our 3-D state space. Figure 16.6 shows an attracting orbit in which the pendulum advances by two revolutions over two drive cycles. This is the same solution we plotted in Fig. 7.3(b), where the net rotation on alternate cycles is slightly less and slightly more than one revolution. As discussed previously, period doubling is a stepping stone to chaos, because it breaks the symmetry implied by the exact repetition of the drive torque from cycle to cycle.

Carefully following the trajectory in Fig. 16.6 reveals that it begins at one point in the $\phi = 0$ plane, comes back to a second point after one drive cycle, and finally returns to the first point after another drive cycle. That is, the Poincaré section consists of two points that are visited on alternate drive cycles. From this, we can predict that the Poincaré section of a chaotic solution, which never repeats, will consist of an infinite number of points. Let's see how this works.

16.4   Strange attractor

At last, we show in Fig. 16.7 the Poincaré section of the trajectory of a chaotic pendulum. This portrait of chaos was generated by first allowing the pendulum to relax to steady-state motion, then plotting the points $(\theta, \tilde{v})$ at the beginning (phase $\phi = 0$) of the next 10,000 drive cycles. The first 50 points are shown as open circles and numbered sequentially.

With Fig. 16.7, the complexity of a chaotic flow begins to become apparent. First, it's important to understand that the points plotted here represent an attracting set. No matter what initial values are chosen for θ and $\tilde{v}$, the trajectory quickly moves to the intricate filigree of points in Fig. 16.7 and gradually fills out the pattern as more and more points are plotted. This filigree clearly represents an attractor of some kind, and in 1971 the mathematical physicists David Ruelle (Belgian, born 1935) and Floris Takens (Dutch, born 1940) coined the term "strange attractor" to

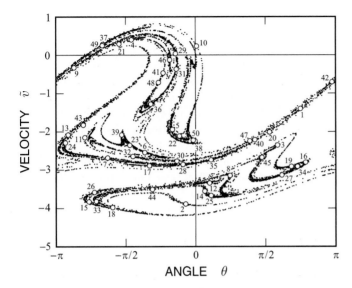

Fig. 16.7 A Poincaré section at $\phi = 0$ of the chaotic pendulum. One point $(\theta, \tilde{v})$ is plotted at the beginning of each of 10,000 successive drive cycles. The first 50 points, indicated by open circles, are numbered sequentially. The parameters are $\rho = 0.1$, $\omega = 0.8$, $\tilde{\tau}_0 = 0$, and $\tilde{\tau}_1 = 1.6$.

describe the attracting sets associated with chaotic motion. Compared to a fixed point or periodic orbit, the attracting set of Fig. 16.7 is indeed strange.

The 10,000 points plotted in Fig. 16.7 only suggest the strange attractor that they represent. Because chaotic motion never repeats itself, the attracting set must include an infinity of points. Thus, any finite number of points doesn't completely capture a strange attractor. The particular trajectory shown in Fig. 16.7, if continued forever, would contribute a countable infinity of points to the attractor. However, there is a continuum of possible initial conditions leading to similar trajectories, so the attractor must include an uncountable infinity of points. Nevertheless, Fig. 16.7 is a good representation of the attractor in that additional points merely fill out the pattern established here.

Another important thing to understand about Fig. 16.7 is that the points are added in a haphazard fashion as the pendulum follows its equation of motion. If you take time to follow the sequence of the first 50 points plotted here, you'll discover that there's no obvious pattern. The points jump from one part of the attractor to another as if chosen at random rather than computed with mathematical precision, each new point from the last. Of course, we've met this random aspect of chaos previously, but here we see that, contrary to expectations, a beautiful, complex pattern emerges from chaos as successive points are plotted.

We learned in Chapter 7 that chaotic trajectories of the pendulum are strongly correlated over a few drive cycles, so we know that chaos isn't completely random. But the pattern apparent in Fig. 16.7 demonstrates another kind of order. The strange attractor represents chaotic motion over the longest possible time scales, and the motion never strays from the attractor, no matter how far we extend our computation. This long-term localization is an apparent paradox, however, because the butterfly effect prevents us from accurately predicting the state of the pendulum far into the future.

How can the computed motion of a pendulum remain forever on a strange attractor, when we can't even accurately predict its state-space coordinates after a few drive cycles? A detailed answer to this question will be given in the next section, but for now suffice it to say that two nearby trajectories can rapidly diverge from each other while both remain on the attractor. That is, the butterfly effect is real and powerful, but it's constrained to the attractor. In meteorological terms, a butterfly can shift the time or place of a tornado but it can't produce something outside the range of natural possibilities. The flap of a butterfly's wings doesn't budge the weather system from its attracting set, but it does shift its trajectory within the attractor, changing the order in which different parts are visited.

The geometric shapes taken by strange attractors of the pendulum often remind me of the ocean waves depicted in Japanese art. In fact, the attractor in Fig. 16.7 bears a striking resemblance to a famous woodcut by the nineteenth century artist Katsushika Hokusai (1760–1849). Called *The Great Wave off Kanagawa*, Hokusai's picture (Fig. 16.8) shares several features in common with our attractor. First and most obvious, both pictures include a spectacular breaking wave. Second, given the affinity of fluids for chaotic motion established by Lorenz, we can argue that Hokusai's great wave is almost certainly also a product of chaos. Finally, both pictures evince similar structures found in a range of sizes: a giant wave, smaller waves, and fingers of water ending in yet smaller fingers. As we'll discuss in the next chapter, such self-similarity over a range of scales is a hallmark of fractal geometry, and both the attractor and the great wave are fractals.

Fig. 16.8 *The Great Wave off Kanagawa*, nineteenth century Japanese woodcut by Katsushika Hokusai. (By permission of the U. S. Library of Congress.)

However, while Figs. 16.7 and 16.8 are both rooted in chaotic motion, they have entirely different origins. *The Great Wave* portrays a single instant in the history of an infinite-dimensional dynamic system (the ocean) as a scene from the real world. In contrast, the strange attractor represents the total time history of a three-dimensional dynamic system in an abstract state space, far removed from the simple pendulum that is its source. Thus, the similarity of these pictures is an intriguing coincidence rather than a cosmic connection.

To see the construction of a strange attractor in animation, give Experiment 24 of the Dynamics Lab a whirl.

16.5   Chaotic flow

An attractor, strange or otherwise, is simply the set of points in state space that a damped or dissipative system inevitably approaches when initial conditions are chosen within a basin of attraction. As a set of points, an attractor is entirely static and provides no obvious clue to the nature of the flow in its neighborhood. Of course, the direction of flow at a given point in state space is easily determined by evaluating the equations of motion at the point.

But when we restrict our view of state space to the plane of a Poincaré section, the direction of flow is always out of the plane, and it tells us very little about where a point will reappear when its trajectory next returns to the section. Is there any way to see how points move within the section on successive returns? Yes, but only in special cases. To see how, reconsider the Poincaré section shown in Fig. 16.5. Here the point S represents a periodic saddle orbit, which returns to the same point after each cycle of drive torque. More important, all points on the inset and outset of S move either closer to or farther from the saddle orbit with successive drive cycles. Thus, a point on the Poincaré section of the inset will remain on the inset after a drive cycle but move closer to S. Similarly, a point on the outset will remain on the outset but step farther and farther from S with each drive cycle.

When viewed in the Poincaré section, the points of the inset and outset are like strings of Christmas lights programmed to give the illusion of motion by flashing on and off in sequence. A sequence of illuminated lights on the inset string moves toward the saddle, and an illuminated sequence on the outset string moves away from the saddle. Thus, the inset and outset give us a sense of the flow near a saddle orbit in the Poincaré section.

If you're wondering what all of this has to do with strange attractors, recall from Chapter 14 that there are typically an infinite number of unstable periodic solutions associated with chaotic motion. While we've previously ignored these solutions in the driven pendulum, they're not difficult to locate, and we shouldn't be surprised to discover that they're embedded in the strange attractor. An unstable orbit in the middle of chaotic instability fits right in.

These ideas come together in Fig. 16.9(a), where we redraw the strange attractor of Fig. 16.7 using 100,000 points. Superimposed on this more complete image of the attractor, we plot the locations of 33 periodic saddle orbits. Five of the orbits have a period of one drive cycle, and the pendulum advances by 0, ±1, or ±2 revolutions on each cycle. There are also 14 orbits having a period of two cycles (each contributing two points to the Poincaré section), for which the pendulum advances by a total of 0, ±1, ±2, ±3, or ±4 revolutions over the period. These solutions are just a sampling of the infinite number of saddle orbits of various periods embedded in the strange attractor.

For each saddle orbit in Fig. 16.9(a), we show the orientation of the inset and outset surfaces by inward and outward directed arrows. Each inward arrow indicates a line along which points move closer to the saddle each time they return to the section after following their trajectory for one period of the saddle orbit. Similarly, an outward arrow shows the direction in which points move away from the saddle after one

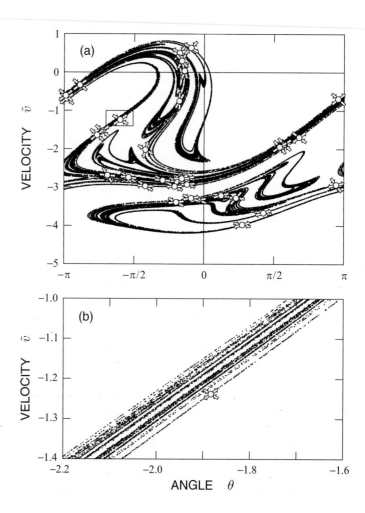

Fig. 16.9 (a) The Poincaré section of the strange attractor in Fig. 16.7 as it appears when 10^5 drive cycles are included. The locations of 33 saddle orbits are plotted as open circles, and the orientation of the associated inset and outset surfaces of each saddle are indicated by inward and outward directed arrows. (b) A close-up of the area outlined by the box in (a), including a saddle orbit for which the pendulum advances by two revolutions in two drive cycles. The points of the attractor in (b) were accumulated over 10^6 cycles.

period. Thus, as always, the inset and outset reveal elements of both stability and instability associated with a saddle orbit.

However, because the saddle orbits in Fig. 16.9(a) are imbedded in a strange attractor, their insets and outsets also tell us about the flow near the attractor. In particular, we note that the outset of each saddle orbit is aligned with the laminar structure of the attractor. Thus, trajectories that diverge from the saddle orbit remain on the strange attractor. In contrast, the inset is oriented crosswise to the laminations, indicating that trajectories not on the attractor move toward it. These relationships are seen more clearly in Fig. 16.9(b), which presents a magnified view of the boxed region of Fig. 16.9(a), containing a single saddle orbit.

The flows near the saddle orbits are conveniently analyzed, but they are also characteristic of the entire strange attractor. Trajectories not on the attractor move quickly towards it, indicating the global stability of the attracting set. In contrast, neighboring trajectories within the attractor move rapidly apart while remaining on the attracting set, revealing the local instability of chaotic motion and the origin of the butterfly effect.

This picture helps resolve the paradox of motion that is globally stable and locally unstable at the same time. The global stability of a strange attractor makes it difficult to budge a system from its attracting set. At the same time, motion within the attractor is highly unstable, and the slightest perturbation will quickly shift a trajectory to a part of the attractor that it otherwise might not have visited for a long time. Thus, with the aid of state space, we begin to see chaotic motion in a new and revealing light. The tiny arrows in Fig. 16.9 give us deep insight into the nature of chaos.

16.6 Stretching and folding

We can also learn about chaotic flows by tossing some dye into state space and watching what happens as it moves with the flow. For this experiment, we turn to a somewhat simpler chaotic attractor than that shown in Fig. 16.9. Motion on the latter attractor is complicated by the fact, discussed in Chapter 8, that the pendulum executes a random walk, allowing its angle to range over an interval much greater than 2π. For our present purpose, it's useful to consider a more restricted form of chaos in which the pendulum is confined to a narrow range of angles.

Results for our chosen example are shown in Fig. 16.10. Here we follow all of the trajectories defined by initial conditions $(\theta, \tilde{v})$ chosen within a circular patch of state space at $t = 0$ and $\phi = 0$. What happens to this patch as time passes? After one drive cycle of duration $t_{P1} = 2\pi/\omega$, the patch returns to the $\phi = 0$ plane, but now it's much smaller in area and shaped like a banana. After a second drive cycle ($t = 2t_{P1}$), it returns as a shoelace, and after two additional cycles ($t = 4t_{P1}$) it has morphed into a piece of spaghetti with a fold at one end. Finally, after six cycles, our patch is yet longer and thinner and sports a couple more folds. At

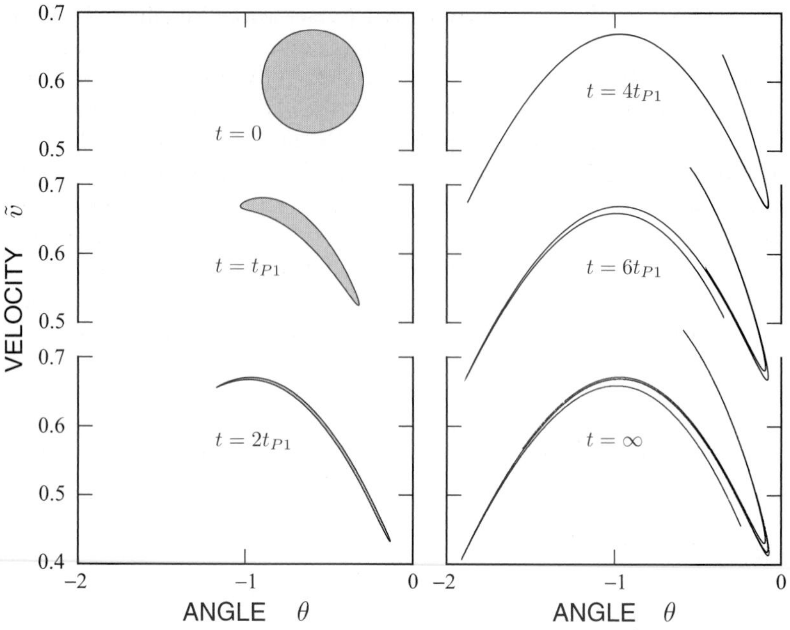

Fig. 16.10 The evolution of a circular patch in the state space of a driven pendulum as it evolves through successive drive cycles within the basin of a strange attractor. The patch is shown in Poincaré section at $t = 0$ and after 1, 2, 4, and 6 cycles of the drive. For comparison, the attracting set at $t = \infty$ is also plotted, based on points accumulated over 20,000 cycles. The pendulum parameters are $\rho = 0.2$, $\omega = 1$, $\tilde{\tau}_0 = 1.885$, and $\tilde{\tau}_1 = 10.2$.

this point, the patch has become a good representation of the strange attractor that it approaches as time goes to infinity.

There are a number of lessons to be learned from watching a circular patch of state space morph into a strange attractor. First, we note that the area of the patch decreases as it evolves. In the last chapter, we learned that for a free pendulum the area of such a patch decreases with time exactly as $e^{-\rho t}$, where ρ is the damping parameter. Perhaps surprisingly, the same is true of a driven pendulum. In particular, after each drive cycle our patch will be smaller by a factor of $e^{-\rho t_{P1}} = e^{-2\pi\rho/\omega}$, or 0.285 for the parameters in Fig. 16.10. Thus, the banana's area is 28.5% of the original circle, while the areas of the shoelace and spaghetti strand are $(0.285)^2 = 8.1\%$ and $(0.285)^4 = 0.66\%$ of the circle. The inevitable conclusion is that the strange attractor itself has no area at all, since $(0.285)^\infty = 0$. The attractor includes an uncountable infinity of points, but, like a line, its area is zero.

The flow acts on the patch by stretching and folding it, the same actions noted earlier in the shift map. With each drive cycle, the patch is further elongated and, when it becomes long enough, the patch develops folds at the ends. If you've ever kneaded bread dough, this action might sound familiar. You first press on the dough to stretch it out, then fold it over to return it to the shape of a loaf, and repeat the process. The same thing happens when taffy is "pulled" to make candy. To give it the right texture, the taffy is repeatedly pulled into a long string then folded back on itself, ready to be pulled again. In a state-space flow, the process of stretching moves neighboring trajectories farther and farther apart at an exponential rate, giving rise to the butterfly effect. And the process of folding prevents the trajectories from heading off to infinity, thereby confining the motion to a finite region of state space. Stretching and folding are thus essential ingredients to chaotic motion. They were first described by Borel in his 1913 paper on molecular chaos.

Figure 16.10 doesn't follow the transformation of the circular patch beyond six drive cycles, but the stretching and folding continue in principle forever. In the strange attractor, we end up with a finely layered structure having a infinite number of laminations. Only a few layers are visible in Fig. 16.10, but in the next chapter, we'll explore in detail other examples of this layered structure and learn why it gives strange attractors their fractal geometry.

Further reading

Strange attractor

- Flake, G. W., "Strange attractors", in *The Computational Beauty of Nature: Computer Explorations of Fractals, Chaos, Complex Systems, and Adaptation* (MIT Press, Cambridge, Massachusetts, 1998) chapter 11.

- Gleick, J., "Strange attractors", in *Chaos: Making a New Science* (Viking, New York, 1987) pp. 119–153.

- Ruelle, D., "Strange attractors", *The Mathematical Intelligencer* **2**, 126–137 (1980).

- Ruelle, D., "Turbulence: Strange attractors", in *Chance and Chaos* (Princeton University Press, Princeton, 1991) chapter 10.

- Stewart, I., "Strange attractors", in *Does God Play Dice? The New Mathematics of Chaos*, 2nd edition (Blackwell, Malden, Massachusetts, 2002) chapter 6.

17 Fractal geometry

Strange attractors have a fractal geometry with a dimension that is typically fractional. What precisely does this mean? In ordinary geometry, the dimension of an object is related to the number of coordinate axes required to specify a position within it. A point on a line is specified by its distance x from an origin marked on the line, so a line is one dimensional. Similarly, a point in a plane is specified by the coordinates, x and y, measured along two axes, and a plane is two dimensional. On the other hand, in this chapter we'll show that the Poincaré section of a certain strange attractor has a fractal dimension of approximately 1.45. Does this mean that 1.45 axes are required to specify a point on the attractor?

Don't worry if you can't imagine what ghostly form 45/100 of a coordinate axis might assume. As we'll see, fractal dimensions aren't related to the number of axes required to specify a point but to the way in which the structure of an intricate object scales with magnification. Fractional dimensions sound stranger than they actually are.

We have already glimpsed the intricate structure of a strange attractor in Fig. 16.9, where frame (b) magnifies a portion of the attractor in frame (a), revealing further structure on a smaller scale. What looks like a broad line in (a) is seen to be a dozen or more closely spaced lines in (b). And this is just the tip of the iceberg. As we'll discover, no matter how much a strange attractor is magnified, there is always more structure to be observed at smaller scales. This persistence of structure at all scales is precisely what makes any object a fractal.

17.1 Mathematical monster

Fractals made their first appearance in mathematics in the nineteenth century, when mathematicians probed the foundations of the calculus. The methods invented by Newton and Leibniz for finding the area under a curve or a line tangent to a curve required working with infinitesimals, quantities smaller than any finite size. Although these methods generally worked well, not all mathematicians were satisfied with their fundamental rigour. This skepticism led to the invention of devilish curves, just to see what could go wrong with the calculus in extreme cases. Initially, Poincaré scorned such curves as a "gallery of monsters" more important to philosophy than mathematics, but later admitted in *Les Méthodes*

Nouvelles de la Mécanique Céleste that his dynamical tangles gave rise to just such monsters.

In 1872 the German mathematician Karl Weierstrass (1815–1897) constructed the prototypical monster curve. His curve is everywhere continuous (making no finite jump at any point) but so infinitely convoluted that it doesn't have a proper tangent anywhere. This curve can be drawn without lifting your pencil from the paper, but it requires a very shaky hand. In devising his monster, Weierstrass wanted only to warn that the calculus cannot be applied to every curve, but in the process he discovered one of the first fractals in mathematics.

Following Weierstrass, other continuous curves without tangents were soon discovered. One of the simplest was introduced in 1904 by the Swedish mathematician Helge von Koch (1870–1924). Called the Koch snowflake, its construction is shown in Fig. 17.1. The snowflake is created by an infinite process that begins with the equilateral triangle in Fig. 17.1(a). In the first step, each side of the triangle is modified by removing the middle third and replacing it with a vee made of two segments each 1/3 the length of the original side. This creates an equilateral bump on each side of the triangle and yields the six-pointed star of Fig. 17.1(b). The following steps are all similar to the first and simply add a 1/3 size equilateral bump to each side of the previous curve, as in Figs. 17.1(c) and (d), creating an ever more prickly snowflake.

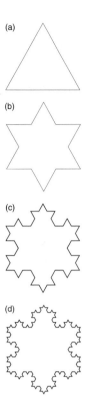

Fig. 17.1 Construction of the Koch snowflake.

The Koch snowflake may look innocent, but the limiting curve of the construction process is infinitely intricate and has surprising properties. For example, the initial triangle lacks a tangent only at the three vertices, where the curve suddenly changes direction and no parallel line can be drawn. However, when the construction of equilateral bumps continues to completion, the length of a side goes to zero, and the curve changes direction at every point. Thus, as advertised, the curve is continuous but lacks a tangent everywhere.

Another strange property of the snowflake is suggested by a fragment from Jonathan Swift's 1733 work *On Poetry: A Rhapsody.*

So, Nat'ralists observe, a flea
Hath smaller fleas that on him prey,
And these have smaller fleas to bite 'em,
And so proceed *ad infinitum.*

Mathematicians say that the snowflake is self-similar, because a side from the nth step with all its added equilateral bumps upon bumps is identical to a side from the next step and all its added bumps, except that the latter is 1/3 as big. Each time we magnify the snowflake by a factor of three, we find the same pattern, just as each of Swift's fleas has its own smaller fleas.

Equally strange, the length of the snowflake curve is infinite. If we assume that the original triangle has sides of length 1, then its perimeter is 3 units. With each step in the construction process, one side is replaced by four sides each 1/3 as long, so after n steps the snowflake has 3×4^n sides, each of which is $1/3^n$ units long. The perimeter of the snowflake

(a)

(b)

(c)

(d)

Fig. 17.2 Construction of the Cantor dust.

(e)

is thus $3 \times (4/3)^n$ units after n steps. That is, the length of the curve increases exponentially with n and exceeds any finite bound as n goes to infinity. Despite its apparent innocence, the snowflake really is a monster. As we'll see shortly, it's also a fractal.

Another mathematical monster was introduced in 1873 by Georg Cantor, who we met earlier in our discussion of transfinite numbers. Like the snowflake, Cantor's monster is constructed by an infinite process, but one in which segments are removed without replacement. The result is known as a Cantor dust. The process begins with a line segment, say 1 unit in length, as in Fig. 17.2(a). In the first step of construction, the middle third of the segment is removed to create two segments, each $1/3$ unit long, Fig. 17.2(b). Further steps remove the middle third of the segments created by previous steps, as in Figs. 17.2(c) through (e), and the process is repeated infinitely.

With each step in construction, the number of segments in the dust increases by a factor of 2 and their size is reduced by a factor of $1/3$. Thus, after n steps, the dust consists of 2^n segments each of length $1/3^n$, and the total length of all the segments is $(2/3)^n$. The segment length and the total length decrease exponentially with n, and when the construction is complete we are left with an infinite number of segments that have been reduced to mere points. No wonder the resulting monster is called a dust. As with the snowflake, we'll see that the Cantor dust is a fractal.

17.2   Hausdorff—Fractal dimension

The defining property of fractal geometry is a type of dimension first introduced in 1918 by the German mathematician Felix Hausdorff (1868–1942). Hausdorff came to his idea for the dimension of a geometric object by considering how many small disks or boxes of uniform size are required to cover the object. He realized that the required number N will increase as the size ϵ of the disk or box is reduced, but the way in which N scales with ϵ will depend on the dimension of the object in question. Turning this observation around led Hausdorff to a definition of dimension.

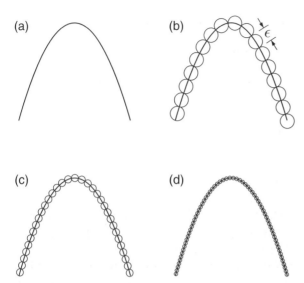

Fig. 17.3 The Hausdorff dimension of a parabola is determined by counting the number of disks of diameter ϵ required to cover the parabola.

Two examples will help explain Hausdorff's idea. Consider first the parabola in Fig. 17.3(a). If we use disks of diameter ϵ to cover this parabola, then the number required will go up as ϵ is reduced. Thus for relative diameters of 1, 1/2, and 1/4, we find in Figs. 17.3 (b)–(d) that 17, 33, and 66 disks are required to cover the parabola. That is, halving the size of the disk roughly doubles the number required. Of course, this makes perfect sense because for a curve of length ℓ, we expect N to be approximately ℓ/ϵ.

Now consider the heart-shaped area, bounded by a curve known as a cardioid, shown in Fig. 17.4(a). For simplicity, we'll use square boxes of side ϵ to cover the heart. As shown in Figs. 17.4(b)–(d), the number of

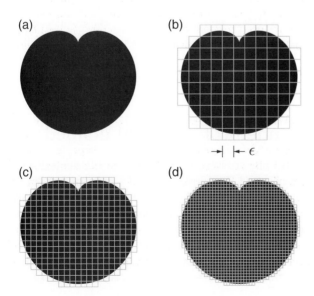

Fig. 17.4 The Hausdorff dimension of the area bounded by a cardioid is determined by counting the number of squares of side ϵ required to cover it.

boxes required for full coverage increases from 94 to 340 to 1,290 as the relative size of the boxes decreases from 1 to 1/2 to 1/4. In this case, N increases by roughly a factor of 4 each time the box size is halved. Again, this makes sense because to cover a heart of area A, we need approximately A/ϵ^2 boxes of area ϵ^2.

Because N is proportional to $1/\epsilon^1$ for a one-dimensional parabola and proportional to $1/\epsilon^2$ for a two-dimensional heart, we might argue that an object is D-dimensional if N scales as $1/\epsilon^D$ in the limit of small ϵ. This is exactly Hausdorff's idea, his new definition of dimension. As we've seen, for simple objects like the parabola and the heart, this definition reduces to our conventional notion of dimension and D proves to be an integer. But for complex objects like the Koch snowflake and the Cantor dust, N doesn't grow simply in proportion to $1/\epsilon$ or $1/\epsilon^2$, and that's where the fun begins.

How can we use Hausdorff's idea to calculate the dimension of an arbitrary object? Suppose C is the constant of proportionality between N and $1/\epsilon^D$, such that $N = C/\epsilon^D$. With a little help from logarithms, we can transform this equation into one that's more helpful. Taking the logarithm of both sides yields,

$$\ln(N) = \ln(C/\epsilon^D)$$
$$= \ln(C) + \ln[(1/\epsilon)^D]$$
$$= \ln(C) + D\ln(1/\epsilon), \qquad (17.1)$$

where we have have expanded the right-hand side using the identities $\ln(xy) = \ln(x) + \ln(y)$ and $\ln(x^y) = y\ln(x)$.[1] Equation (17.1) suggests an easy way to determine D. It says that if we plot $\ln(N)$ as a function of $\ln(1/\epsilon)$ the result should be a straight line with a slope equal to D.

Let's see how this slope method of determining dimension works for the parabola and the heart. Figure 17.5 plots $\ln(N)$ for several values of ϵ beginning with that shown in Figs. 17.3(b) and 17.4(b) and going down to ϵ's just 1/128 as large. In both cases, the result is very nearly a straight line, but for the parabola the slope is 1, while for the heart it's 2. Thus, we really can read the dimension from a plot of $\ln(N)$ versus $\ln(1/\epsilon)$. Later we'll use this technique to find the dimension of a strange attractor.

[1] Both of these identities follow from the definition of the natural logarithm, $x = e^{\ln(x)}$, and the exponential identities in Chapter 10. For example, we can write xy either as $(e^{\ln(x)})(e^{\ln(y)}) = e^{\ln(x)+\ln(y)}$ or as $e^{\ln(xy)}$. Equating the exponents in the last two expressions yields the first identity, $\ln(xy) = \ln(x) + \ln(y)$. Can you use a similar approach to show that $\ln(x^y) = y\ln(x)$?

17.3   Mandelbrot—Fractal defined

Although fractal objects were introduced into mathematics during the nineteenth century and Hausdorff provided a definition of fractal dimension in 1918, the word "fractal" wasn't coined until 1975. The term was introduced by the French-American mathematician Benoît Mandelbrot (born 1924) in his book *Les Objets Fractals: Forme, Hasard et Dimension*, a pioneering work that gathered together bits and pieces of mathematics from the archives of the previous century to create a new discipline: fractal geometry. Previously, the monsters of Cantor and

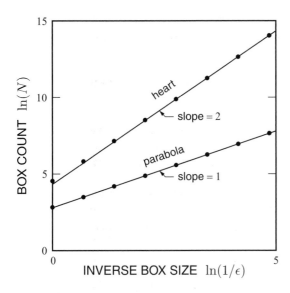

Fig. 17.5 A plot of $\ln(N)$ as a function of $\ln(1/\epsilon)$ for the parabola and heart in Figs. 17.3 and 17.4. The disk and box sizes ϵ in Figs. 17.3(b) and 17.4(b) are taken to be 1.

Koch and the dimension of Hausdorff had been considered mere mathematical curiosities. But Mandelbrot realized that kindred geometric forms appear all around us in the shapes of trees, clouds, and coastlines and that their geometry deserves to be studied in its own right. He coined the word "fractal" from the Latin adjective "fractus," meaning broken or fragmentary, and thereby gave his new discipline its name. After Mandelbrot, the wide applicability of fractal geometry could no longer be ignored.

In 1977 Mandelbrot formally defined a fractal as an object having a fractal dimension D that exceeds its topological dimension D_T. Here we take the topological dimension to coincide with the intuitive notion of dimension suggested by Euclid. That is, an object is of dimension 0 if it has neither length, width, nor height, of dimension 1 if it has length but not width or height, of dimension 2 if it has length and width but not height, and of dimension 3 if it has length, width and height. While this definition of D_T isn't rigorous, some examples will help clarify the essential idea.

Consider first the parabola and the heart of Figs. 17.3 and 17.4. The parabola has a length but no width, so its topological dimension is 1. Earlier we argued that the parabola has a fractal dimension of 1, so $D = D_T$, and the parabola isn't a fractal according to Mandelbrot. Similarly, the heart has both length and width, so its topological dimension also matches its fractal dimension of 2, and the heart isn't fractal.

The cases of the Cantor dust and the Koch snowflake are entirely different. As argued previously, when the construction of a Cantor dust is complete, we're left with a countable infinity of points without a length to be found. Thus, the topological dimension of the dust is $D_T = 0$. But what is the dust's fractal dimension? Suppose we use discs to cover the line segments remaining after each step of the construction process, as

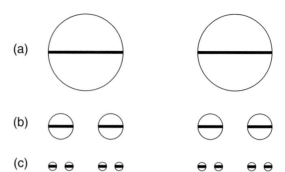

Fig. 17.6 The fractal dimension of the Cantor dust can be calculated by covering the line segments at each stage of construction by a set of disks with diameters matching the lengths of the segments.

shown in Fig. 17.6. After the nth step of the process we need $N = 2^n$ disks of diameter $\epsilon = 1/3^n$ to cover all of the segments. Taking the logarithm of these equations yields,

$$\ln(N) = \ln(2^n) = n \ln 2, \tag{17.2}$$

$$\ln(1/\epsilon) = \ln(3^n) = n \ln 3, \tag{17.3}$$

and if these equations are combined to eliminate n, we obtain,

$$\ln(N) = \left(\frac{\ln 2}{\ln 3}\right) \ln(1/\epsilon). \tag{17.4}$$

According to Eq. (17.1), the above relation between $\ln(N)$ and $\ln(1/\epsilon)$ is exactly that expected for an object with fractal dimension,

$$D = \frac{\ln 2}{\ln 3} = 0.6309\ldots. \tag{17.5}$$

Thus, the Cantor dust fits Mandelbrot's definition of a fractal: its fractal dimension $D = 0.6309\ldots$ is greater than its topological dimension $D_T = 0$.

A dimension of $0.6309\ldots$ now begins to make sense. The Cantor dust is no more than a countable infinity of points with a topological dimension 0. But, the closer we look, the more closely spaced points we find, almost as if we were approaching a continuum of points. Thus, it isn't completely crazy to assign the dust a fractal dimension between that of a point and that of a line.

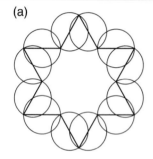

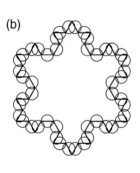

Fig. 17.7 The fractal dimension of the Koch snowflake can be calculated by covering the line segments at each stage of construction by a set of disks with diameters matching the lengths of the segments.

By contrast, the topological dimension of the Koch snowflake is $D_T = 1$, since the snowflake remains a continuous curve no matter how many kinks it ultimately develops. Its fractal dimension can be computed just as for the Cantor dust, by covering the line segments at each stage of construction with disks, as shown in Fig. 17.7. After the nth step, we need $N = 3 \times 4^n$ disks of diameter $\epsilon = 1/3^n$ to cover the entire snowflake. Taking the logarithm of these equations yields,

$$\ln(N) = \ln(3 \times 4^n) = \ln 3 + n \ln 4, \tag{17.6}$$

$$\ln(1/\epsilon) = \ln(3^n) = n \ln 3, \tag{17.7}$$

or

$$\ln(N) = \ln 3 + \left(\frac{\ln 4}{\ln 3}\right)\ln(1/\epsilon). \qquad (17.8)$$

Again, comparison with Eq. (17.1) reveals that the fractal dimension of the snowflake is,

$$D = \frac{\ln 4}{\ln 3} = 1.2618\ldots. \qquad (17.9)$$

Thus, $D > D_T$ for the snowflake, and it too is a fractal.

The fractal dimension $1.2618\ldots$ also makes sense. While no more than a closed curve, the snowflake packs an infinite length into a finite area without occupying any area at all. Thus, its intricate structure is plausibly described by a fractal dimension between 1 and 2.

17.4   Physical fractals

In searching the literature for fractals, Mandelbrot discovered a notable example in a paper by the English mathematician Lewis Richardson (1881–1953). Richardson observed that the measured length of an island's coastline depends on whether one applies a long or short yardstick. If you jump from point to point along the coast in kilometer steps, your tally for the coast length will be less than if your steps are only 100 meters. The shorter yardstick yields a greater total because it follows the nooks and crannies of the coast more closely. When Richardson plotted the logarithm of the coast length against the logarithm of the yardstick length, he found a straight line. As the reader can surely guess and Mandelbrot immediately recognized, this linear relation is exactly that expected for a fractal. Indeed, Richardson's data for the west coast of Britain implies a fractal dimension of $D = 1.24$, very close to that of the Koch snowflake.

The coast of Britain is just one example of the many fractals found in the physical world. However, it typifies the differences between real-world fractals and mathematical fractals such as the Koch snowflake and Cantor dust. First, a coastline lacks the exact self-similarity of the snowflake when viewed at different scales. Magnifying a coastline does not reveal an exact copy in miniature, but the magnified view still has the feel of a coastline. A coastline is self-similar at different scales only in a statistical sense, so it's called a random fractal. Second, a coastline maintains self-similarity over only a limited range of scales. While the Koch snowflake was constructed to have the same features at arbitrary magnification, examining a coastline with say an atomic-scale yardstick is unlikely to reveal the same kinds of structures that we see at macroscopic scales. Physical fractals typically display statistical self-similarity over scales differing by just a few factors of 10. Nonetheless, as Mandelbrot first observed, it's extremely useful to recognize the fractal properties of natural shapes.

In addition to coastlines, physical fractals are found in the topology of mountain ranges, the branching structures of trees, river systems, and blood vessels, cracks in dried mud, and vortices in turbulent flow. Indeed, fractal structures are so prevalent in the natural world, it's hard to believe that the mathematical form they hold in common went unrecognized until the work of Mandelbrot. Now we not only analyze nature in terms of fractal dimensions but use random fractals to add realism to computer generated landscapes.

17.5   Fractal attractor

We now turn from real-world fractals to mathematical fractals that are even more abstract than those of Cantor and Koch. They are the strange attractors of chaotic motion recorded in state space, mere ghosts of positions and velocities once visited by a dynamic system. The fractal properties of such attracting sets were first noted by Lorenz in his 1963 paper on convection. For the present we'll view the strange attractor simply as a geometric object.

As an example, consider the chaotic attractor of the driven pendulum shown in Fig. 17.8(a) by its Poincaré section. As previously, we plot here the angle and angular velocity $(\theta, \tilde{v})$ of the pendulum after each of many successive drive cycles. This yields a two-dimensional cross-section, composed of individual points, of the three-dimensional attractor: a slice taken at the drive phase $\phi = 0$. The 20,000 points in Fig. 17.8(a) provide a reasonably complete picture of the attractor's cross-section.

As with the coast of Britain, we can get a feel for the fractal nature of our strange attractor by examining it at smaller scales. For example, frame (b) of Fig. 17.8 magnifies the rectangular region outlined in frame (a). Here we see that each of the four broad lines in frame (a) is resolved into a few more lines. To fill out this picture, we include additional points in the magnified region obtained by extending the calculation to 500,000 drive cycles.

The additional structure discovered in frame (b) gives a hint of the fractal nature of the strange attractor. But if it's truly fractal, further magnification should reveal yet more detailed structure. Sure enough, when the small rectangle in frame (b) is magnified to obtain frame (c), we discover that the three broad lines in (b) are each resolved into several lines. Revealing these details requires extending the calculation to 20 million drive cycles. Finally, by extending the calculation to one billion drive cycles, we can magnify the small rectangle in frame (c) to reveal the additional structure in frame (d). With frame (d), we have magnified the attractor in (a) by a factor of about 3,000 and further detailed structure is still being discovered.

The intricate detail of the strange attractor in Fig. 17.8 certainly suggests that it's a fractal, but we should compare its topological and fractal dimensions to be certain. In this case, however, the topological dimension isn't immediately obvious. In Fig. 17.8, we explored the

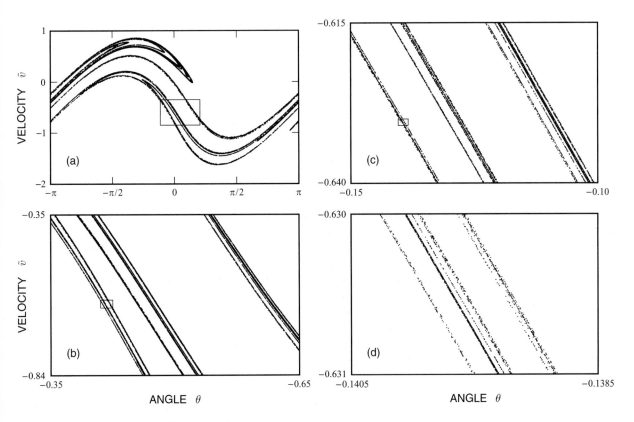

Fig. 17.8 The Poincaré section of a strange attractor of the driven pendulum for $\rho = 0.2$, $\omega = 1$, $\tilde{\tau}_0 = 1.9$, and $\tilde{\tau}_1 = 10.2$. Frame (a) plots the angle and angular velocity of the pendulum $(\theta, \tilde{v})$ at phase $\phi = 0$ after each of 2×10^4 successive drive cycles and shows the entire attractor. Frame (b) magnifies the rectangular region in frame (a), including the points obtained after 5×10^5 successive drive cycles. Similarly, frame (c) magnifies the rectangle in frame (b) after 2×10^7 drive cycles, and frame (d) magnifies the rectangle in frame (c) after 10^9 drive cycles.

attractor in cross-section by plotting the points from a single trajectory after it has relaxed to the attractor. Since this trajectory yields only a countable infinity of points on the attractor, we might assume that the section doesn't include a continuum of points and has a topological dimension of $D_T = 0$. But, by definition the attractor includes all of the points of the uncountable trajectories having initial conditions chosen within its basin of attraction. Thus, the Poincaré section includes an uncountable infinity of points and its topological dimension could be $D_T = 1$ or even 2.

In fact, the Poincaré section of our attractor has a topological dimension of $D_T = 1$. This conclusion becomes clear if we think back to the experiment shown in Fig. 16.10, where we followed a circular patch within the basin of attraction as it relaxed to the attractor. In the limit of large times, the initial area morphed into a line of zero width but infinite length, as it was repeatedly stretched and folded by the dynamics of the driven pendulum. Thus, the basic topology of the Poincaré section is that of a line, and its topological dimension is $D_T = 1$.

To determine the fractal dimension of our attractor, we return to the method of plotting $\ln(N)$ versus $\ln(1/\epsilon)$ described earlier. Here we imagine overlaying the Poincaré section of Fig. 17.8(a) with various rectangular grids like those shown for the heart in Fig. 17.4. For a grid with a given box size ϵ, we then count the number of boxes that include part of the attractor. Plotting $\ln(n)$ versus $\ln(1/\epsilon)$ as in Fig. 17.9, we obtain a nearly linear relation, and the slope of this line tells us that the fractal dimension of the Poincaré section is $D = 1.45$. Thus, $D > D_T$, and the strange attractor really is a fractal.

While we are now accustomed to two-dimensional Poincaré sections, let's return for a moment to the full three-dimensional state space of the driven pendulum. With the state variable ϕ added back, the state space becomes θ–$\tilde{v}$–ϕ space, and trajectories appear as curves rather than points. Moreover, the curves representing the strange attractor in the Poincaré section appear as surfaces in the full state space. This adds 1 to both the topological and fractal dimensions of the strange attractor. Thus the sets of roughly parallel lines in Fig. 17.8 are the cross-sections of stacked planes in the three-dimensional view. I like this view because it reminds me of those delicious flaky pastries made from paper-thin sheets of phyllo dough. Our chaotic attractor is really a pastry with a countable infinity of layers and has a fractal dimension of $D = 2.45$. If it were edible, the fractal attractor might be infinitely delicious.

Let's round out our picture of the driven pendulum's attractors in three-dimensional state space. First, recall that a periodic attractor is a simple closed loop in this space, as shown in Figs. 16.3(a) and 16.6. Periodic trajectories have topological and fractal dimensions of $D_T = D = 1$ and aren't fractals. Although we haven't previously discussed them, the driven pendulum also allows what are called almost periodic solutions, which lack the exponential sensitivity of chaotic solutions but

Fig. 17.9 The fractal dimension of the strange attractor in Fig. 17.8 is determined by counting the number of boxes N in a rectangular grid overlaying the Poincaré section that include part of the attractor. The count was determined for five grid sizes: the largest grid spanning the attractor from $-\pi$ to π with $2^8 = 256$ boxes and the smallest spanning the attractor with $2^{12} = 4,096$ boxes. Box sizes ϵ are normalized to the largest box considered, and counts are for a Poincaré section including 10^8 points.

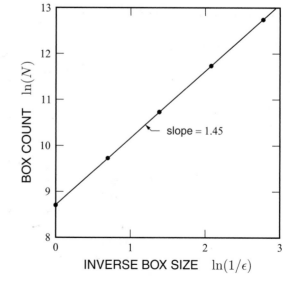

have trajectories that never quite connect back on themselves. An almost periodic solution has a Liapunov exponent of zero and an attracting set that is a simple surface with $D_T = D = 2$. In contrast, a chaotic solution has a fractal attractor, with a topological dimension of $D_T = 2$ and a fractal dimension in the range $2 < D < 3$ and consists of a complex structure of closely interleaved surfaces.

The fact that a chaotic attractor of the driven pendulum has a fractal dimension $D > 2 = D_T$ is precisely what makes it a fractal. However, the inequality $D < 3$ is also an important defining property. The fractal dimension can never be as large as 3 because, as discussed in Chapter 15, the state-space flow of a system with friction always contracts to a region of zero volume. Indeed, this contraction is the property of the flow that gives rise to attracting sets in the first place. In a sense, it is fractal geometry that opens up the possibility of chaotic motion by allowing attractors with dimensions in the range $2 < D < 3$. The fractal dimension of 2.45 found here provides the room needed for a trajectory to go beyond almost periodic motion into the realm of randomness, even though it is constrained to a region of state space having no volume. A fractal attractor is thus an essential element of chaos in the driven pendulum and not an incidental by-product.

Further reading

Fractals

- Barnsley, M., *Fractals Everywhere*, 2nd edition (Academic, Boston, 1993).

- Clarke, A. C., editor, *The Colours of Infinity: The Beauty and Power of Fractals* (Clear Books, Bath, 2004).

- Edgar, G. A., editor, *Classics on Fractals* (Addison-Wesley, Reading Massachusetts, 1993).

- Gardner, M., "Mandelbrot's fractals", in *Penrose Tiles to Trapdoor Ciphers* (W. H. Freeman, New York, 1989) chapter 3.

- Jürgens, H., Peitgen, H.-O., and Saupe, D., "The language of fractals", *Scientific American* **263**(2), 60–67 (August 1990).

- Mandelbrot, B. B., *The Fractal Geometry of Nature* (W. H. Freeman, San Francisco, 1982).

- Peitgen, H.-O. and Richter, P. H., *The Beauty of Fractals: Images of Complex Dynamical Systems* (Springer–Verlag, Berlin, 1986).

- Peterson, I., "Snowflake curves", in *Islands of Truth: A Mathematical Mystery Cruise* (W. H. Freeman, New York, 1990) chapter 4.

- Pickover, C. A., *Computers, Pattern, Chaos and Beauty* (St. Martin's Press, New York, 1990).

- Pickover, C. A., editor, *Fractal Horizons: The Future Use of Fractals* (St. Martin's Press, New York, 1996).

- Stewart, I., "Gallery of monsters", in *The Magical Maze: Seeing the World Through Mathematical Eyes* (Wiley, New York, 1997) chapter 8.

18 Stephen Smale— Horseshoe map

Stephen Smale arrived in Rio de Janeiro with his wife and two small children in January of 1960. He was ready to begin work on dynamical systems at the Instituto de Matemática Pura e Aplicada (IMPA), supported by a postdoctoral fellowship from the U.S. National Science Foundation (NSF). Smale had completed his doctorate in 1956 at the University of Michigan, writing his thesis on a problem in topology. Two years later at the University of Chicago he had stunned the mathematical community by proving that a sphere can be "everted" or turned inside out in a smooth fashion, a possibility that many mathematicians had considered unlikely. Why was a rising talent in topology taking up dynamical systems? For Smale, it would be just one of several times during his career when he demonstrated how ideas from one field of mathematics can benefit another.

Smale was attracted to IMPA by one of its founders, the Brazilian Mauricio Peixoto (born 1921), who was an expert in dynamical systems. Rio proved an ideal setting for Smale, with Peixoto at hand to fill in gaps in his knowledge of dynamics and sparkling beaches to act as inspiration. Mornings typically found Smale at Leme beach, scribbling down ideas on a pad of paper and going for an occasional swim in the ocean. Afternoons were spent at IMPA, reading in the library and consulting with Peixoto and others. Indeed, Rio proved so conducive to mathematics that during his six-month stay Smale produced two seminal results, one in dynamics and another in topology. His work on dynamics focused on extending the kind of catalog of possibilities given by the Poincaré–Bendixson theorem to higher dimensional flows, and it led to the discovery of "a horseshoe on the beaches of Rio," as he would later put it.

Fig. 18.1 Stephen Smale. (Courtesy of Stephen Smale.)

18.1   Horseshoe map

As you'll recall from Chapter 15, the Poincaré–Bendixson theorem tells us that stable steady-state motion within a bounded region of a two-dimensional state space is restricted to two simple possibilities: fixed points and periodic orbits. We now know that three-dimensional flows allow additional possibilities, including chaotic motion, but in 1960 Smale knew nothing of the sort. In fact, he had already written a paper including a conjecture that implied, in modern terms, that chaos doesn't exist. But soon after his paper appeared in January of 1960, Smale received a letter from Norman Levinson setting him straight. Levinson pointed out that his 1949 paper, which identified one of Poincaré's tangles in a driven oscillator, gave a counterexample to Smale's conjecture.

Taken aback, Smale set to work translating Levinson's algebraic arguments into his own geometric terms. In the process, he proved to himself that Levinson was correct and discovered a simple iterated map, the horseshoe map, illustrating the consequences.

By viewing a dynamical system as a flow in state space, Poincaré had raised the level of abstraction and gained the advantage of seeing all possible trajectories at once. Smale raised the level of abstraction yet again by forgetting about the dynamical origins of the flow and allowing himself to simply imagine different flow topologies. To be sure that his imaginary flows mimicked dynamical systems, they needed to be smooth, so that neighboring points remained neighbors in the short term, but otherwise the imagination could run free. No wonder Smale's office in Rio could be the beach.

In contrast to the one-dimensional iterated maps of Chapter 14, the horseshoe map is two-dimensional and describes a flow in three-dimensional state space. While two-dimensional maps might sound like unfamiliar territory, the driven pendulum provides a perfect example if we think in terms of its Poincaré section. In Figs. 16.1, 16.3, and 16.6, we plotted trajectories of the driven pendulum in the full three-dimensional θ–$\tilde{v}$–ϕ state space. These trajectories give us a taste of the three-dimensional flow. However, we have often found it convenient to plot only the points $(\theta, \tilde{v})$ at the end of each drive cycle, where the phase is $\phi = 0$, creating a Poincaré section of the trajectory. In effect, the points of the section are successive iterates of a two-dimensional map in θ–$\tilde{v}$ space. That is, if we start with values of pendulum angle and velocity $(\theta_0, \tilde{v}_0)$ at the beginning of a drive cycle, we can solve the equations of motion to obtain the values $(\theta_1, \tilde{v}_1)$ at the end of the drive cycle. But $(\theta_1, \tilde{v}_1)$ are also the state variables at the beginning of the next cycle, so the process can be repeated over and over to obtain $(\theta_i, \tilde{v}_i)$ after any number of drive cycles. In this picture, time comes in discrete lumps (drive cycles), and the equations of motion define $(\theta_{i+1}, \tilde{v}_{i+1})$ in terms of $(\theta_i, \tilde{v}_i)$, giving us a two-dimensional iterated map.

Smale found it convenient to think of his horseshoe map not in terms of its action on individual points but its effect on an entire region of state space. This strategy gives a more immediate picture of the map's action, and we previously applied it to the driven pendulum in Fig. 16.10. There we showed how a circular area of θ–$\tilde{v}$ space changed after 1, 2, 4, and 6 drive cycles. Of course, Fig. 16.10 was obtained by strictly following the pendulum's equations of motion, while Smale was free to dream up the action of his map.

The horseshoe map is defined by its action on a square region of state space, as shown in Fig. 18.2. Here, rather than follow individual trajectories through a time step, we see what happens to all the points

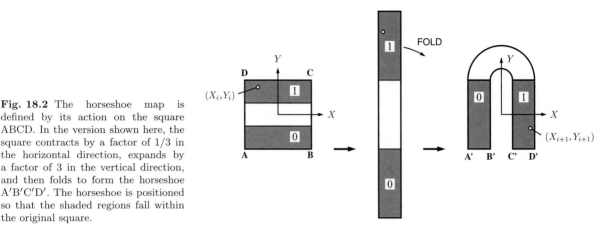

Fig. 18.2 The horseshoe map is defined by its action on the square ABCD. In the version shown here, the square contracts by a factor of 1/3 in the horizontal direction, expands by a factor of 3 in the vertical direction, and then folds to form the horseshoe A′B′C′D′. The horseshoe is positioned so that the shaded regions fall within the original square.

within the square ABCD at once. In particular, the square first contracts in the X direction and expands in the Y direction, then folds to create the horseshoe A′B′C′D′ which is finally positioned with its ends within the original bounds of the square, as shown. Thus, one iteration maps point A to point A′, B to B′, etc., and more generally the point (X_i, Y_i) is mapped to (X_{i+1}, Y_{i+1}). Smale didn't worry about what happens between times i and $i + 1$, although we can imagine that a point follows a smooth trajectory in the 3-D state space between these times. In Fig. 18.2, all of the points in the shaded regions labeled "0" and "1" of the original square map back into the square, while points in the white region become the horseshoe bend and fall outside. All in all, the horseshoe map is very simple, although it incorporates the stretching and folding actions that we have come to associate with chaotic motion. The contraction and expansion are linear operations, while the folding is highly nonlinear, so the map has at least some of the elements needed for chaos.

In drawing Fig. 18.2, we have chosen a particular instance of the horseshoe map in which the square contracts in the X direction by a factor of 1/3 and expands in the Y direction by a factor of 3. If we further assume that ABCD is a unit square centered at the origin ($|X| \leq 1/2$ and $|Y| \leq 1/2$), then we can describe its action on the shaded regions by a simple set of equations.

$$X_{i+1} = \begin{cases} (X_i - 1)/3, & Y_i \leq -1/6 \quad (\text{area0, unrotated}) \\ (1 - X_i)/3, & Y_i \geq 1/6 \quad (\text{area1, rotated}) \end{cases} \tag{18.1}$$

$$Y_{i+1} = \begin{cases} 3Y_i + 1, & Y_i \leq -1/6 \quad (\text{area0, unrotated}) \\ 1 - 3Y_i, & Y_i \geq 1/6 \quad (\text{area1, rotated}). \end{cases} \tag{18.2}$$

The nonlinearity in these equations results from the difference in treatment between the shaded areas 0 and 1. Area 0 is simply scaled horizontally and vertically then translated to its final location, while area 1 is scaled and then rotated by 180° before being translated. We will refer to area 1 as the rotated region and area 0 as the unrotated region. Note that shaded areas 0 and 1 and the unshaded area between them all have a width of 1/3, both before and after the mapping. By adopting a particular form for the horseshoe map, we can obtain quantitative results, although Smale found that qualitative arguments were sufficient for proving his theorems.

Unlike the one-dimensional maps of Chapter 14, the horseshoe map is invertible. That is, given any point (X_i, Y_i), we can work backwards to find the preceding point (X_{i-1}, Y_{i-1}). The effect of reversing the sequence of steps in Fig. 18.2 is captured by the equations,

$$X_{i-1} = \begin{cases} 3X_i + 1, & X_i \leq -1/6 \quad (\text{area 0, unrotated}) \\ 1 - 3X_i, & X_i \geq 1/6 \quad (\text{area 1, rotated}) \end{cases} \quad (18.3)$$

$$Y_{i-1} = \begin{cases} (Y_i - 1)/3, & X_i \leq -1/6 \quad (\text{area 0, unrotated}) \\ (1 - Y_i)/3, & X_i \geq 1/6 \quad (\text{area 1, rotated}) \end{cases} \quad (18.4)$$

where we now associate (X_i, Y_i) with the shaded areas of the final square and (X_{i-1}, Y_{i-1}) with the shaded areas of the initial square. The invertibility of the horseshoe map gives it kinship to the reversible dynamics of systems governed by Newton's equations.

18.2 Invariant set

It usually makes sense to explore a dynamical system by choosing some arbitrary initial conditions and simply observing the motion that results. However, the horseshoe map requires another approach. As Smale realized, the key to understanding the horseshoe is to search out some very special initial conditions, specifically those leading to trajectories that remain within the unit square for all time, both past and future. According to this strategy, we seek initial conditions (X_0, Y_0) for which the iterates (X_i, Y_i) remain within the unit square for times $-\infty < i < \infty$. Moreover, we want to make a complete catalog of these trajectories by tabulating all such initial conditions. The points (X_0, Y_0) in this catalog are called the invariant set of the horseshoe map, a set usually designated Λ. As Smale discovered on the beaches of Rio, Λ is an infinite set that includes initial conditions corresponding to chaotic motion.

We can begin to narrow down the points (X_0, Y_0) in Λ by carefully inspecting Fig. 18.2. In particular, note that the unshaded stripe $|Y| < 1/6$ of the initial unit square becomes the horseshoe bend, so this entire area leaves the square after a single forward iteration. Thus, no points (X_0, Y_0) in the stripe $|Y_0| < 1/6$ are included in Λ. Similarly, the unshaded stripe $|X| < 1/6$ of the final square in Fig. 18.2 does not fall in the initial square when the mapping is reversed. Because points are included in Λ only if the trajectory remains in the unit square for all

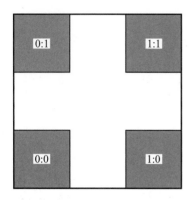

Fig. 18.3 From the properties of one forward and one reverse iteration of the horseshoe map, the map's invariant set Λ is restricted to the shaded areas shown here within the unit square. Each area is labeled by a symbol sequence indicating whether the preceding and following iterations involved rotation (1) or not (0). For example, points within the upper-left area, labeled "0:1", were not rotated on the preceding iteration but will be on the next iteration.

time, the stripe $|X_0| < 1/6$ is also excluded from the invariant set. With both of these stripes eliminated, all of the points in Λ must lie in one of the four shaded regions shown in Fig. 18.3. We don't yet know which points are in Λ, but we've narrowed the possibilities.

The shaded areas in Fig. 18.3 are labeled by what are called symbol sequences, which tell us something about the trajectories associated with the points in each area. For example, the sequence "0:1" in the upper left indicates that the points in this area were not rotated in the previous iteration but will be rotated in the next iteration, as can be verified by inspecting Fig. 18.2. That is, symbols to the left of the colon tell us about the past trajectory and symbols to the right tell about the future. The colon is a bookmark that separates the past from the future.

To learn more about the invariant set, we can extend our analysis of the map to two iterations and look for initial points (X_0, Y_0) that remain within the unit square over this longer period. Figure 18.4 shows how the unit square evolves when it's contracted, expanded, and folded twice in succession. As before, we have shaded the areas of the initial square that remain within the unit square at the end. Each initial shaded area is labeled by a symbol sequence indicating whether or not it was rotated in each of the two iterations. For example, the top horizontal stripe is labeled ":10" because it is rotated on the first iteration but not the second. Similarly, the final vertical stripes are labeled according to whether they were rotated in the two preceding iterations. Thus, the stripe at the far right labeled "01:" was rotated on the previous iteration but not on the one before that. These labels can be verified by following each stripe as it proceeds either forwards or backwards through the two iterations depicted in Fig. 18.4.

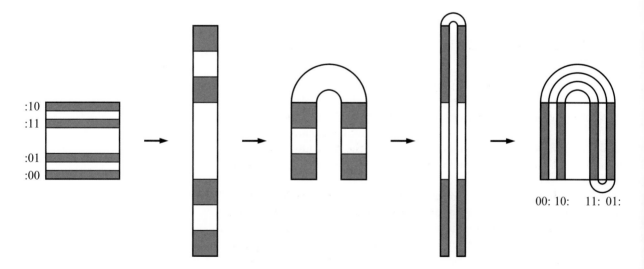

Fig. 18.4 Two successive iterations of the horseshoe map. Shading shows areas of the initial square that remain within the unit square at the end.

Again, for a point (X_0, Y_0) to be a member of Λ, both its past and future iterates must lie within the unit square, so the invariant set is restricted to the intersections between the initial horizontal stripes and the final vertical stripes in Fig. 18.4. This closer approximation to Λ is shown in Fig. 18.5. Each area of intersection is labeled with the symbol sequence that applies to the points within. Thus, a point within area 01:10 at top right was rotated in the previous iteration but not the one before that, and it will be rotated in the next iteration but not the one after that.

Of course, after two iterations, we still can't say exactly which points are included in Λ, but perhaps the pattern has become clear. Considering one forward or reverse iteration allowed us to eliminate from Λ the vertical and horizontal middle-third stripes of the unit square, leaving four squares, each $1/3$ unit in size (Fig. 18.3). Then, considering two forward or reverse iterations further eliminated horizontal and vertical middle-third stripes from these smaller squares, leaving 16 squares, each $1/9$ unit in size (Fig. 18.5). If this pattern continues with additional iterations, as indeed it does, we see that it creates a kind of two-dimensional Cantor dust. That is, Λ is nothing more than a dust of points.

Just for practice, let's calculate the fractal dimension of Λ. From the arguments above, after n iterations of the map are taken into account, we are left with $N = 4^n$ squares each of size $\epsilon = 1/3^n$. Thus, we have

$$\ln(N) = \ln(4^n) = n \ln 4, \tag{18.5}$$

$$\ln(1/\epsilon) = \ln(3^n) = n \ln 3, \tag{18.6}$$

and eliminating n yields,

$$\ln(N) = \left(\frac{\ln 4}{\ln 3}\right) \ln(1/\epsilon). \tag{18.7}$$

Comparing this result with Eq. (17.1), we find that the fractal dimension of the invariant set is

$$D = \frac{\ln 4}{\ln 3} = 1.2618\ldots, \tag{18.8}$$

or exactly the same as the Koch snowflake. For being no more than a dust of points, Λ has a surprisingly large fractal dimension, larger than that of a line. From its construction, we can see that the number of trajectories in the invariant set is at least a countable infinity, but its fractal dimension suggests that Λ is even larger.

Given the fractal nature of the invariant set, one is tempted to suppose that it represents a chaotic attractor. However, Λ isn't an attracting set. Because the points in Λ are all distinct from one another, the slightest offset from any initial point (X_0, Y_0) in Λ generally yields a trajectory that leaves the unit square. In this regard, chaotic motion in the horseshoe map is like that of Hadamard's geodesics on a surface of negative curvature rather than that of Lorenz's convection cell, the driven pendulum, or the Tilt-A-Whirl, where motion is attracted to

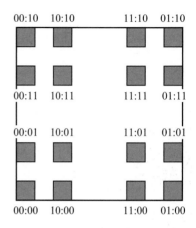

Fig. 18.5 Areas of the unit square that remain within the square after two forward iterations and two reverse iterations of the horseshoe map.

chaotic trajectories. Simply put, invariant motion in Smale's horseshoe map is dynamically fragile and can be destroyed by the slightest noise.

18.3   Symbolic dynamics

We have introduced symbol sequences such as 01:10 to describe how the trajectories within a small area of state space wend their way through the horseshoe map, both forward and backward in time. However, as Smale realized, extending such sequences to infinity in both directions provides a way to specify a particular trajectory. That is, a bi-infinite symbol sequence like ... 101:011 ... is an alternative to the initial conditions (X_0, Y_0) when it comes to identifying trajectories. In addition, some aspects of horseshoe dynamics are simplified when expressed in terms of such sequences. Indeed, the technique of "symbolic dynamics" has a long history, having been introduced by Hadamard in his 1898 paper on geodesic flows and further developed by Birkhoff. To learn more, let's take a look at how symbolic dynamics applies to the horseshoe map.

The relation between symbol sequences and initial conditions is most easily explored through periodic trajectories. Suppose for example that a trajectory of the horseshoe map returns to the same point (X_0, Y_0) within the unit square after every iteration. In this case, the trajectory must be either rotated or not rotated on every iteration. That is, the symbol sequence must be either repeating 1's or 0's, namely ... 111:$\overline{1}$11 ... or ... 000:$\overline{0}$00 .... Here a bar is used to indicate the shortest sequence of repeating symbols, $\overline{1}$ and $\overline{0}$.

Given the symbol sequences for these period-1 trajectories, how can we find the initial conditions? Because the period is one iteration, we must have $X_1 = X_0$ and $Y_1 = Y_0$, and these equations are sufficient to determine X_0 and Y_0. If the trajectory is rotated on each iteration, then from Eqs. (18.1) and (18.2) we have,

$$X_0 = X_1 = (1 - X_0)/3, \quad \text{(rotated)} \tag{18.9}$$

$$Y_0 = Y_1 = 1 - 3Y_0, \quad \text{(rotated)} \tag{18.10}$$

which yields $X_0 = Y_0 = 1/4$. Thus, the trajectories specified by the initial condition $(X_0, Y_0) = (1/4, 1/4)$ and the symbol sequence ... 111:$\overline{1}$11 ... are equivalent.

$$(1/4, 1/4) \quad \Longleftrightarrow \quad \ldots 111{:}\overline{1}11 \ldots \tag{18.11}$$

This equivalence makes some sense because the point $(1/4, 1/4)$ falls in area 11:11 of Fig. 18.5. On the other hand, if the trajectory is not rotated on each iteration, then Eqs. (18.1) and (18.2) yield,

$$X_0 = X_1 = (X_0 - 1)/3, \quad \text{(unrotated)} \tag{18.12}$$

$$Y_0 = Y_1 = 3Y_0 + 1, \quad \text{(unrotated)} \tag{18.13}$$

in which case $X_0 = Y_0 = -1/2$, and we have the equivalence,

$$(-1/2, -1/2) \quad \Longleftrightarrow \quad \ldots 000{:}\overline{000}\ldots . \qquad (18.14)$$

The point $(-1/2, -1/2)$ is located at the lower-left corner of the unit square and falls in area 00:00 of Fig. 18.5, as expected.

As a slightly more complicated example, consider the symbol sequence $\ldots 101{:}\overline{101}\ldots$, based on the repeating sequence $\overline{101}$. This trajectory has a period of three iterations, and its initial conditions can be found from the equations $X_3 = X_0$ and $Y_3 = Y_0$, which guarantee that the system returns to its starting point after three iterations. The sequence $\overline{101}$ implies that the first and last iterations involve rotation while the second doesn't. Applying the appropriate forms of Eq. (18.1) yields,

$$X_1 = (1 - X_0)/3, \quad \text{(rotated)} \qquad (18.15)$$

$$X_2 = (X_1 - 1)/3, \quad \text{(unrotated)} \qquad (18.16)$$

$$X_3 = (1 - X_2)/3, \quad \text{(rotated)} \qquad (18.17)$$

which, when combined with $X_3 = X_0$, can be solved to obtain $X_0 = 11/26$. A similar set of equations in Y leads to $Y_0 = 11/26$, so that the initial conditions for this period-3 trajectory are $(X_0, Y_0) = (11/26, 11/26)$. Thus, we have the equivalence,

$$(11/26, 11/26) \quad \Longleftrightarrow \quad \ldots 101{:}\overline{101}\ldots . \qquad (18.18)$$

Of course, our period-3 trajectory can equally well be specified by (X_1, Y_1) or (X_2, Y_2), if we simply wait one or two iterations before announcing the start.

An overview of the period-3 orbit is shown in Fig. 18.6. Here, the shaded squares are regions of the map that remain within the unit square after three forward and three reverse iterations, and any trajectory of the invariant set must remain within these squares. Our period-3 orbit begins at the point $(X_0, Y_0) = (11/26, 11/26)$ in the shaded square 101:101 near the upper-right corner of the unit square. Of course, it's no coincidence that the symbol sequence of this square matches the central symbols of the bi-infinite sequence $\ldots 101{:}\overline{101}\ldots$ corresponding to (X_0, Y_0). Only points within this square can give rise to the partial sequence 101:101. Upon iteration, the map jumps from (X_0, Y_0) to $(X_1, Y_1) = (5/26, -7/26)$ and then to $(X_2, Y_2) = (-7/26, 5/26)$ before beginning to repeat.

Let's consider in detail what happens in the jump from (X_0, Y_0) to (X_1, Y_1). The symbol sequence corresponding to (X_1, Y_1) is obtained from the original $\ldots 101{:}\overline{101}\ldots$ by shifting the colon to the right by one symbol, yielding $\ldots 101\overline{1}{:}\overline{01}\ldots$ or $\ldots 011{:}\overline{011}\ldots$. This shift makes sense because the iteration (symbol 1) to the right of the colon in $\ldots 101{:}\overline{101}\ldots$ has been completed, and at (X_1, Y_1) it's part of the past instead of the future. In fact, a forward iteration always moves the colon one symbol

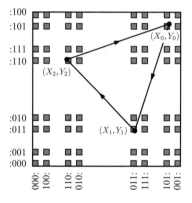

Fig. 18.6 A period-3 solution from the invariant set Λ of the horseshoe map. This solution hops from $(X_0, Y_0) = (11/26, 11/26)$ to $(X_1, Y_1) = (5/26, -7/26)$ and then $(X_2, Y_2) = (-7/26, 5/26)$ before beginning to repeat. Shaded squares are regions of the map that remain within the unit square after three forward and three reverse iterations.

to the right and a reverse iteration moves it one symbol to the left. In any case, the state of the system at (X_1, Y_1) is specified equivalently by,

$$(5/26, -7/26) \quad \Longleftrightarrow \quad \dots 011\!:\!\overline{011}\dots. \qquad (18.19)$$

Similarly, after a second iteration we find the equivalence,

$$(-7/26, 5/26) \quad \Longleftrightarrow \quad \dots 110\!:\!\overline{110}\dots, \qquad (18.20)$$

and one more iteration brings us back to (X_0, Y_0). The period-3 solution thus reveals the initial conditions and symbol sequences of three related trajectories that always remain within the unit square and are part of the invariant set Λ.

But periodic solutions are old hat, being well known even in two-dimensional flows. How does the horseshoe map go beyond two dimensions? As Smale realized, the restriction on Λ imposed by three forward and backward iterations, shown by the shaded squares in Fig. 18.6, includes all $2^6 = 64$ possible symbol sequences of length 6, from 000:000 to 111:111. More generally, the restriction imposed by n forward and reverse iterations allows all 2^{2n} possible sequences of length $2n$. Furthermore, for finite n there is a periodic solution corresponding to every such sequence. Experiment 25 the Dynamics Lab explores some of these long-period solutions. However, letting n go to infinity suggests that Λ includes every possible bi-infinite sequence. That is, we can arbitrarily choose every symbol in a bi-infinite sequence of 0's and 1's and still have a trajectory that remains within the unit square for all time, future or past. This is Smale's fundamental insight into the horseshoe map, and it tells us that motion within a bounded region of a three-dimensional flow is not restricted to fixed points and periodic orbits. That is, the invariant set Λ includes trajectories with symbol sequences that never repeat and are appropriately described as chaotic.

Smale didn't find just a few chaotic orbits within the unit square: he found infinitely many. Because every symbol in the bi-infinite sequence can be chosen arbitrarily as 0 or 1, there are as many orbits as there are binary expansions of real numbers on a line. That is, Λ includes an uncountable infinity of orbits, most of which represent completely random sequences of 0's and 1's. No wonder the fractal dimension of Λ is greater than 1—it includes symbol sequences equivalent to a continuum even if the trajectories aren't represented by a continuum of initial points (X_0, Y_0) in the X–Y plane.

Smale liked to say that the dynamics of the horseshoe map "is as unpredictable as coin-flipping." Suppose, for example, that we use the magic spinner of Chapter 14 to randomly select two real numbers between 0 and 1. If the binary expansion of one number is used as the future symbol sequence and that of the other is reversed and used as the past symbol sequence, we have a bi-infinite sequence with each symbol chosen as randomly as if we had flipped a coin to decide between 0 and 1. For this symbol sequence, successive iterations of the horseshoe map alternate randomly between rotated and unrotated, and we couldn't expect a more chaotic trajectory. Thus, although chaotic trajectories of the horseshoe are dynamically fragile and can be destroyed by the slightest noise, they are proof that steady-state motion within a bounded region of a three-dimensional flow doesn't always take the form of a fixed point or a periodic orbit. In dynamics, the possibilities in three dimensions are much richer than in two. This revelation of the horseshoe was Smale's true discovery on the beach in Rio.

18.4   3-D flow

To probe the horseshoe map more deeply, let's take a direct look at the topology of its three-dimensional flow. Of course, the map doesn't specify the trajectories between the integer times i and $i + 1$, so our look will require some imagination.

To set the scene, recall that a 3-D flow is determined by a vector field, which we previously pictured as arrows posted on a grid of buoys tethered at various positions and depths in a Caribbean lagoon. Smale assumed that the vector field of the horseshoe map is continuous, so the arrows on neighboring buoys always point in almost the same direction. Donning our scuba gear and following the arrows from buoy to buoy will lead us along a smooth trajectory in this 3-D state space of the Caribbean.

We can construct a crude picture of the flow for the horseshoe map by using the time integer i as the third state variable, as in Fig. 18.7. Here we show the unit square at four times, $i = 0$, 1, 2, and 3, with each succeeding square displaced by one unit in the i direction. On this 3-D grid, we plot three periodic trajectories drawn from the invariant set Λ: two with a period of one iteration (A and B) and the one of period three (C). A dot marks the point at which a trajectory passes through each unit square, and dashed lines connect successive dots.

These three trajectories produce a very sketchy picture of the flow, but let's see what our imagination can fill in. First, the dashed lines are only an approximation and should be replaced by smooth curves. However, it's clear that trajectory C encircles trajectory B as it goes from $i = 0$ to 3. That is, trajectories B and C are topologically entwined,

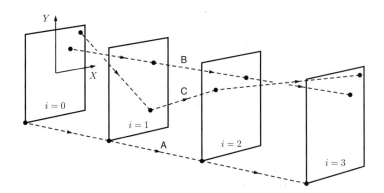

Fig. 18.7 Interpolated trajectories of the horseshoe map in X–Y–i space. Each trajectory is marked by a solid dot at the point where it passes through the unit square at times $i = 0$, 1, 2, and 3, and successive points are connected by dashed lines. Trajectories A and B are defined by the initial conditions $(X_0, Y_0) = (-1/2, -1/2)$ and $(1/4, 1/4)$ and have periods of one iteration. Trajectory C has initial conditions $(X_0, Y_0) = (11/26, 11/26)$ and a period of three iterations.

an effect of the folding that occurs on each iteration of the map. So the flow is not as simple as water moving straight through a pipe. But this is an understatement since we have only begun to plot the infinity of trajectories in Λ. There are trajectories with every conceivable period and every conceivable symbol sequence that are entwined with each other in a braid of fantastic complexity. And there are trajectories that lack periodicity altogether, bouncing from point to point in the unit square as if controlled by the flip of a coin. If you can add up all of these possibilities in your mind, you'll begin to see the complexity of a chaotic flow. And yet all of this complexity results from a vector field that is continuous at every point in state space. It boggles the mind.

18.5   Structural stability

Before leaving the horseshoe, we should mention one last aspect of Smale's work. In previous chapters, we have often considered the dynamic stability or instability of the trajectories of a flow, determined by whether neighboring trajectories converge or diverge from each other. But Smale was especially interested in another type of stability, known as "structural stability," which considers whether the topology of a flow is sensitive to its defining parameters. Thinking of the horseshoe, for example, we might wonder if the chaotic orbits might disappear with a slight change in the map. If so, chaos would not be typical of similar flows and could be highly improbable in general. However, Smale was able to prove that chaos in the horseshoe map is structurally stable and is thus a robust and important phenomenon. Thus, even though chaotic motion on the horseshoe is dynamically fragile, the horseshoe itself is structurally robust. Horseshoes aren't mathematical flukes.

18.6   Protests and prizes

After returning to the U.S. in July of 1960, Smale frequently lectured on the horseshoe map, but more pressing matters delayed publication of the horseshoe until 1965. One reason for the delay was Smale's second idea from Brazil, concerning a topological conjecture put forward by Poincaré in 1904. Known simply as Poincaré's conjecture, by 1960 it had become the most famous unsolved problem in topology, and Smale was about to make a significant contribution.

Although a rigorous statement of Poincaré's conjecture requires some technical sophistication, it says roughly that in a four-dimensional space any three-dimensional object without holes is topologically equivalent to a 3-sphere, where a 3-sphere in W–X–Y–Z space is defined by $W^2 + X^2 + Y^2 + Z^2 = 1$. The conjecture was also extended to higher dimensions with the claim that any n-dimensional object without holes is topologically equivalent to an n-sphere. Mathematicians assumed that these higher-dimensional cases, with $n \geq 4$, would be increasingly difficult to prove. But Smale's insight in Rio suggested that they might in fact be easier, and in 1961 he proved the Poincaré conjecture for $n \geq 5$. Smale had leapfrogged $n = 3$ and 4 to establish the conjecture in all higher dimensions, and the mathematical community was duly impressed. After everting the sphere and proving a major part of the Poincaré conjecture, no one could doubt that Smale was an exceptional mathematician.

In the fall of 1964, Smale assumed a tenured position at the University of California at Berkeley, where he would remain for the next 30 years. In other times, Smale might have been content to pursue mathematics to the exclusion of all else, but in 1964 he was unable to ignore what he perceived as the unwarranted military intervention by the U.S. in Vietnam. Thus, while continuing to teach and pursue research, he became an active critic of the Vietnam War. Smale's anti-war activism soon led to his becoming co-chairman with Jerry Rubin of the Berkeley-centered Vietnam Day Committee. As co-chair, Smale helped organize a Vietnam War teach-in held on the Berkeley campus in May of 1965 and troop-train protests in Oakland later that year. Also in 1965, Smale received the Veblen Prize for Geometry awarded by the American Mathematical Society.

For Smale, protests and prizes would be intertwined even more closely the following summer. Smale left for Europe in May of 1966, allowing time to vacation and visit European colleagues on his way to the International Congress of Mathematics to be held in Moscow in August. In Paris he spoke out against the U.S. involvement in Vietnam at a rally called "Six Hours for Vietnam," co-organized by a mathematical friend. A few days later, another French colleague informed him that he had been chosen to receive the Fields Medal at the Moscow congress. No ordinary award, the Fields Medal is often described as the Nobel Prize of mathematics, the highest honor a mathematician can receive. After stays in Switzerland and Greece, Smale headed for Moscow. On his final flight, Smale was startled to learn from another mathematician that he had been subpoenaed to testify before the House Un-American Activities Committee (HUAC) of the U.S. Congress, in relation to his anti-war activities. In Moscow, Smale duly presented his invited lecture and accepted the Fields Medal. However, in response to a request from a North Vietnamese reporter, at the end of the congress Smale gave a press conference on the steps of Moscow University. He read a brief statement, which criticized both the U.S. for its role in Vietnam and the U.S.S.R. for its invasion of Hungary 10 years earlier, then answered questions from the press. Afterwards, Smale was whisked away by Soviet officials, briefly detained, then returned to the closing reception of the congress. The following day, Smale's abduction was reported as front-page news in the *New York Times*.

As it happened, Smale never testified before the HUAC. Its hearings proved to be a better forum for anti-war activists like Jerry Rubin than for Congress, and the hearings were concluded before Smale returned from Europe. However, Smale wasn't off the hook. While he was still in

Moscow, the Director of the NSF initiated an investigation, questioning whether Smale's NSF summer support had been used appropriately. When he arrived back in the U.S., Smale learned that Berkeley was withholding the remainder of his summer salary, pending receipt of a detailed account of his summer mathematical activities. He provided the required justification, including an aside saying that "My best-known work was done on the beaches of Rio...," but expressed resentment that the NSF was responding to Congressional pressure. Indeed, three congressmen had publicly called for the cancellation of Smale's NSF grant, all motivated by his anti-war activities, not his mathematics. Smale's summer salary was restored after a reporter for the journal *Science* revealed the NSF's action.

The following year, under further Congressional pressure, the NSF subjected a new grant proposal from Smale to special treatment. Although an NSF division director was certain that the proposal would receive "outstanding substantive reviews," he argued that Smale was unsuitable as a principal investigator. Smale would eventually receive his grant, but not before the case had again been aired in the pages of *Science* and the scientific community had expressed its outrage. Somewhat later, the science advisor to President Johnson would refer to Smale indirectly in a public address, saying,

> This blithe spirit leads mathematicians to seriously propose that the common man who pays the taxes ought to feel that mathematical creation should be supported with public funds on the beaches of Rio de Janeiro....

Perhaps the science advisor didn't know what Smale had actually discovered in Rio.

Smale continued to pursue dynamical systems until about 1970, when his attention turned to economic theory. In dynamics, the horseshoe is Smale's best known result, but it's just the tip of the iceberg, being followed by many mathematically deeper theorems. Nonetheless, as we'll see in the next chapter, the horseshoe is at the topological core of chaotic flows.

Further reading

Horseshoe map

- Diacu, F. and Holmes, P., "Symbolic dynamics", in *Celestial Encounters: The Origins of Chaos and Stability* (Princeton University Press, Princeton, 1996) pp. 51–79.
- Ott, E., "The horseshoe map and symbolic dynamics", in *Chaos in Dynamical Systems* (Cambridge University Press, Cambridge, 1993) pp. 108–114.

- Smale, S., "Diffeomorphisms with many periodic points", in *Differential and Combinatorial Topology: A Symposium in Honor of Marston Morse*, edited by S. Cairns (Princeton University Press, Princeton, 1965) pp. 63–80.

Poincaré conjecture

- O'Shea, D., *The Poincaré Conjecture: In Search of the Shape of the Universe* (Walker, New York, 2007).
- Szpiro, G., *Poincaré's Prize: The Hundred-Year Quest to Solve One of Math's Greatest Puzzles* (Dutton, New York, 2007).

Smale

- Albers, D. J., Alexanderson, G. L., and Reid, C., editors, "Steve Smale", in *More Mathematical People: Contemporary Conversations* (Harcourt, Brace, Jovanovich, Boston, 1990) pp. 304–323.
- Batterson, S., *Stephen Smale: The Mathematician Who Broke the Dimension Barrier* (American Mathematical Society, Providence, Rhode Island, 2000).
- Hirsch, M. W., Marsden, J. E., and Shub, M., editors, *From Topology to Computation: Proceedings of the Smalefest* (Springer-Verlag, New York, 1990).

- Smale, S., "How I got started in dynamical systems", in *The Mathematics of Time: Essays on Dynamical Systems, Economic Processes, and Related Topics* (Springer, New York, 1980) pp. 147–151.
- Smale, S., "On the steps of Moscow University", *The Mathematical Intelligencer* **6**(2), 21–27 (1984).
- Smale, S., "The story of the higher dimensional Poincaré conjecture (What actually happened of the beaches of Rio)", *Mathematical Intelligencer* **12**(2), 44–51 (Spring 1990).
- Smale, S., "Finding a horseshoe on the beaches of Rio", *The Mathematical Intelligencer* **20**(1), 39–44 (Winter 1998).

Structural stability

- Abraham, R. H., and Shaw, C. D., "Structural stability", in *Dynamics: The Geometry of Behavior*, 2nd edition (Addison-Wesley, Reading, 1992) chapter 12.

19 Henri Poincaré— Topological tangle

In 1885, the Swedish mathematician Gösta Mittag-Leffler (1846–1927) announced an international mathematical competition with a prize of 2,500 crowns, to be awarded in celebration of the sixtieth birthday of King Oscar II of Sweden and Norway. Competitors were invited to submit an essay on one of four unsolved problems before June 1, 1888, and the winner would be announced on January 21, 1889, the King's sixtieth birthday.

Henri Poincaré, then a professor at the University of Paris, was intrigued by one of Mittag-Leffler's problems. Just 31 when the competition was announced, Poincaré knew that winning would greatly enhance his reputation as a mathematician, and he spent the next three years preparing his entry. The problem that caught Poincaré's fancy concerned the stability of the Solar System: will the planets continue to circle the Sun *ad infinitum*, or is there a chance, according to Newton's laws of motion, that Earth or some other planet will eventually be hurled into deep space? The prospect of instability seemed unlikely, but the possibility was fascinating, given that humanity's continued existence might hang in the balance.

Poincaré realized that the stability problem demanded a level of abstraction beyond the direct solution of Newton's equations. Of course, if there were a general formula for the motion of three or more gravitationally attracting bodies, the question of stability might be solved by inspection. After two centuries of work on the three-body problem, however, it was evident that a general formula was unlikely to be found. Instead, Poincaré attacked the problem using methods that were qualitative rather than quantitative, hoping to reach a general conclusion without computing every possible orbit.

Ultimately, Poincaré's efforts were successful. As we've learned, he discovered that all possible motions of a given system can be viewed simultaneously in an abstract space, called state space, and that many general results can be deduced from the topology of the trajectories in this space. This new perspective on Newton's equations was revolutionary, and Poincaré used it in his essay to deduce that motion in a three-body system is generally stable. One of the judges of the competition wrote to Mittag-Leffler concerning Poincaré's essay, "You may tell your sovereign that this work cannot indeed be considered as furnishing the complete solution of the question proposed, but that it is nevertheless of such importance that its publication will inaugurate a new era in the history of celestial mechanics." Poincaré was duly awarded first prize in King Oscar's contest, he received the 2,500 crowns, and his winning essay was published as Volume 13 of Mittag-Leffler's journal, *Acta Mathematica*.

In the history of chaos theory, the aftermath of King Oscar's prize proved to be far more important than the contest itself. During the process of publication, Poincaré received a comment from a reviewer that led him to consider a strange possibility he had previously overlooked. Pursuing his investigation further, Poincaré realized that the topology of trajectories in state space could be infinitely complex, even in the three-body problem. It was an astonishing conclusion. Whereas one might expect that state space could be divided into a few distinct regions in which the characteristic motion is qualitatively different, Poincaré discovered an infinity of different behaviors crowded within a finite region. He would later write, "One is struck by the complexity of this picture, which I do not even attempt to draw." Today, we call Poincaré's complex structure a "tangle" and recognize it as the state-space geometry underpinning chaotic motion.

Poincaré could not let his prize-winning essay stand as originally published. While largely correct, the mistake it contained was crucial, because he now realized that the stability of a three-body system is far from assured. Moreover, the mistake was embarrassing both to Poincaré and Mittag-Leffler, given that the essay had been awarded King Oscar's prize. Thus, Mittag-Leffler acted quickly to retrieve and destroy the original printing of Volume 13 of *Acta Mathematica*, while Poincaré hastened to correct his essay, expanding it from 158 to 270 pages in the process. His work was complete by January of 1890, and the corrected Volume 13 was printed at Poincaré's expense later that year. Although the printer's bill consumed most of his prize money, Volume 13 presented to the world a great discovery, which introduced topology to celestial mechanics and cracked the lid on the Pandora's box of chaos.

19.1 Homoclinic point

In his prize-winning essay, Poincaré considered a special case of the three-body problem in which an infinitesimal mass is subject to the gravitational pull of two equal masses orbiting one another. With some additional restrictions, Poincaré found that the motion of the infinitesimal mass is equivalent to the motion of a pendulum driven by a periodic force, and it was in this pendulum system that he first discovered a state-space tangle. Poincaré's driven pendulum differs from the one we've considered throughout this book,[1] but we'll avoid switching horses in midstream and introduce the tangle using our familiar example.

Poincaré's tangle develops from a single point, which he called a homoclinic point. An example from the driven pendulum system, Eqs. (16.6)–

[1] Poincaré's pendulum model was free of friction and driven by an oscillation in one of its parameters, rather than a directly applied torque.

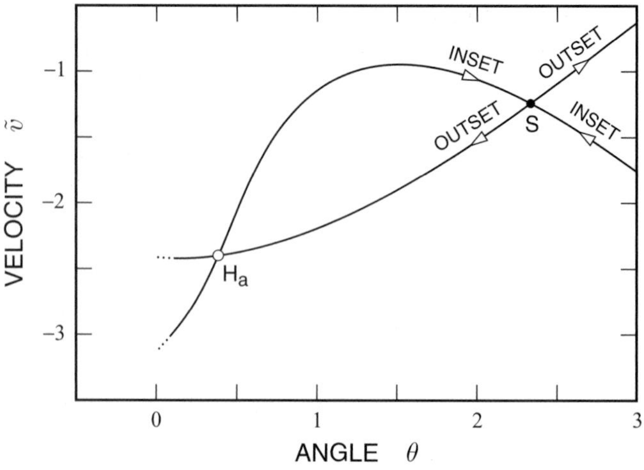

Fig. 19.1 A homoclinic point H_a in the Poincaré section of a damped driven pendulum for drive phase $\phi = 0$. The homoclinic point is defined by the intersection between the inset and outset of the saddle orbit S. The pendulum parameters are $\rho = 0.1$, $\omega = 1.6$, $\tilde{\tau}_0 = 0.6$, and $\tilde{\tau}_1 = 2.5$.

(16.8), is shown in Fig. 19.1. Here we see the Poincaré section of a saddle orbit, represented by the point S, which corresponds to simple oscillatory motion of the pendulum with the same period as the drive cycle, $t_{P1} = 2\pi/\omega$. Being a saddle orbit, however, we know that the motion is intrinsically unstable, and there is an inset and an outset associated with S, also shown in Fig. 19.1. The situation here is similar to that for the saddle orbit of Fig. 16.5, with one crucial difference. In Fig. 16.5, the inset and outset remain well separated, but in Fig, 19.1 the inset and outset cross each other at the point H_a. This intersection between the inset and outset is precisely the possibility that Poincaré failed to consider in the first version of his essay, and H_a is the homoclinic point from which his tangle develops.

Before exploring this tangle, let's set the stage with a few lines from the movie *Ghostbusters*. In the movie, three professors of parapsychology are expelled from an unnamed New York university and establish "Ghostbusters," a service offering "professional paranormal investigations and eliminations." In their first job as eliminators, the ghostbusters strap on nuclear powered proton packs, capable of generating directed particle streams, and set about trapping a lively green apparition. The following dialogue ensues.

Egon:	There's something very important I forgot to tell you.
Venkman:	What?
Egon:	Don't cross the streams.
Venkman:	Why?
Egon:	It would be bad.
Venkman:	I'm fuzzy on the good/bad thing. What do you mean by "bad?"
Egon:	Try to imagine all life as you know it stopping instantaneously and every molecule in your body exploding at the speed of light.

> Ray: Total protonic reversal.
> Venkman: Right. That's bad. Okay. Important safety tip.
>   Thanks, Egon.

Of course, in the movie's climatic scene, the ghostbusters must defy death and cross their streams to close an interdimensional portal and thwart Gozer the destructor.

Although a crossing between the inset and outset of a saddle orbit doesn't have the dramatic impact of crossed particle streams, it has revolutionized our understanding of the possibilities inherent in Newtonian dynamics. And, as we'll see in this chapter, the homoclinic point is the crux of Poincaré's window on chaos, which, if less fearsome than total protonic reversal, can still claim to be the dark side of dynamical systems. When insets and outsets cross, dynamic chaos is loosed upon the world.

At first glance, the homoclinic point H_a of the saddle orbit S in Fig. 19.1 might seem entirely innocuous. However, if we think back on the analogy between dynamic saddles and mountain passes introduced in Fig. 15.8, a homoclinic point may seem strange. This analogy is best suited to a 2-D state space in which the saddle is a fixed point. The inset is analogous to a trail that follows the ridgeline down to the pass or saddle, and the outset is a path from the saddle into a valley. So, how can the outset in the valley cross the inset on the ridge? The answer is that such crossings can't happen in a 2-D state space. As you may remember from Chapter 15, in two dimensions the inset and outset of a saddle point are simply trajectories, and we know that the trajectories of a flow can never cross. No wonder chaos requires more than two dimensions.

The situation is completely different in a 3-D state space, where the inset and outset of a saddle orbit are surfaces rather than single trajectories, and the mountain pass analogy becomes less reliable as an intuitive guide. We previously encountered the inset and outset surfaces of a saddle orbit in Fig. 16.4 for a case in which the surfaces meet only along the saddle orbit. However, when we expand the Poincaré section of Fig. 19.1 into the full θ–$\tilde{v}$–ϕ space of the driven pendulum, as in Fig. 19.2, we find that these surfaces intersect along both the saddle orbit S and the homoclinic trajectory H_a. In this 3-D view, the homoclinic point is just the spot at which the homoclinic trajectory pierces the $\phi = 0$ plane.

19.2   Homoclinic trajectory

The nature of a homoclinic trajectory, the next key to Poincaré's tangle, might at first seem self-contradictory. By definition the outset is made up of all the trajectories that diverge from the saddle orbit S, while the inset includes all of the trajectories that converge to S. The divergence and convergence of trajectories in the outset and inset are suggested by the open arrows in Figs. 19.1 and 19.2. Because a homoclinic trajectory

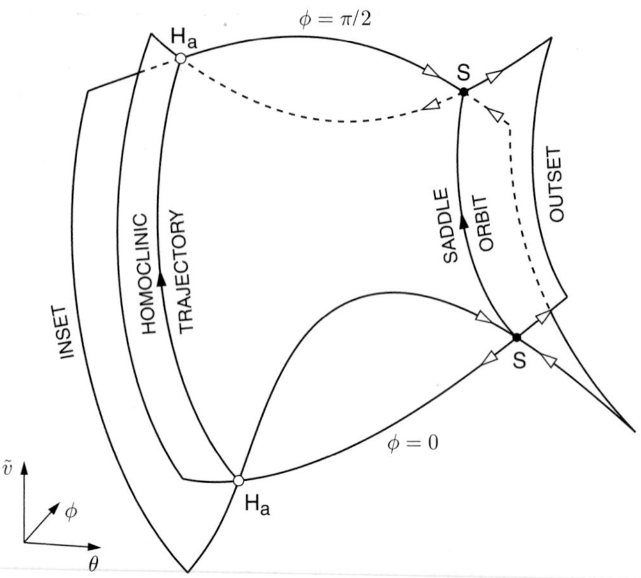

Fig. 19.2 In the full 3-D state space of the driven pendulum, the inset and outset of a saddle orbit are surfaces. For the case shown in Fig. 19.1, these surfaces intersect along both the saddle orbit S and along a homoclinic trajectory H_a. These surfaces are shown here between the two planes of constant phase, $\phi = 0$ and $\pi/2$, or for the first quarter of a drive cycle.

is in both the outset and the inset, it must both diverge from the saddle orbit and converge to it.

Impossible? No, the homoclinic trajectory H_a is exceptional, but it isn't impossible. Being a member of the outset, H_a approaches S if we track it backward in time, and, as a member of the inset, it also approaches S if we follow it forward in time. It's a little strange, but the picture is clear. If we begin on S and give the pendulum just the right nudge to start it on H_a, it will first diverge from S, then, without further interference, come right back to S. This is precisely the property that led Poincaré to call H_a a homoclinic trajectory: homo = same and clinic = inclining. H_a approaches S if we follow it either forward or backward in time.

To better understand the implications of our homoclinic trajectory, let's track it forward in time from the point labeled H_a in Fig. 19.1. To do this, we extend all of the outset trajectories to reveal where they intersect the $\phi = 0$ plane after an additional drive cycle, as shown by the Poincaré section in Fig. 19.3. In doing so, we discover that the outset now intersects the inset at two additional homoclinic points, labeled H_b^0 and H_a^{+1}. We've also relabeled H_a as H_a^0. This labeling is logical because H_a^{+1} is the point at which the homoclinic trajectory H_a hits the $\phi = 0$ plane one drive cycle after beginning at H_a^0. We might have anticipated that H_a^0 would return as another homoclinic point, because a trajectory included in both the inset and outset will always be in both. Moreover, it isn't surprising that H_a^{+1} is closer to S than H_a^0, since we know that H_a approaches S as time advances.

What about the homoclinic point H_b^0 in Fig. 19.3? This point defines a second homoclinic trajectory H_b, distinct from H_a, and H_a and H_b are called the primary homoclinic trajectories of the saddle orbit S. Clearly, the existence of one homoclinic point leads to others—actually an infinite number, as we'll soon discover.

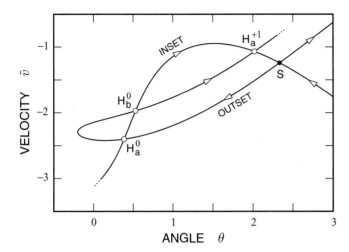

Fig. 19.3 When the outset of Fig. 19.1 is extended further away from the saddle orbit (forward in time), the Poincaré section reveals two additional intersections between the inset and outset or two additional homoclinic points.

Now let's extend the inset of Fig. 19.1 backward in time. Evolving all of the inset trajectories backward by one drive cycle, as shown in Fig. 19.4, reveals four more intersections between the inset and outset: four more homoclinic points. Two of these are labeled H_a^{-1} and H_b^{-1}, and they precede H_a^0 and H_b^0 by one drive cycle. Note that H_a^{-1} is closer to S than H_a^0, in agreement with the fact that H_a approaches S as time goes backward. The unlabeled open circles identify secondary homoclinic points, and we'll return to them later.

To help demystify homoclinic trajectories, let's plot H_a and H_b as a function of time. Figure 19.5 shows how the pendulum angle evolves in time for H_a, H_b, and S. Here the homoclinic trajectories are assumed to pass through H_a^0 and H_b^0 at time $t = 0$. As expected, H_a and H_b approach the periodic saddle orbit S both for times much less than 0 and much greater than 0. Indeed, a few drive cycles in either direction

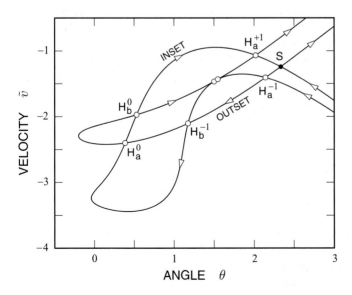

Fig. 19.4 When the inset of Fig. 19.1 is extended further away from the saddle orbit (backward in time), we find that it intersects the outset at four additional homoclinic points.

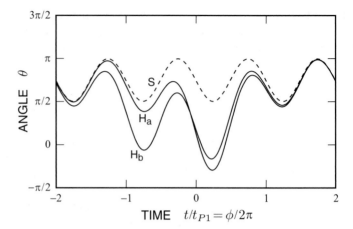

Fig. 19.5 Pendulum angle as a function of time for the homoclinic trajectories H_a and H_b (solid curves) and for the saddle orbit S (dashed curve). The system parameters are the same as in Fig. 19.1.

suffice to bring the homoclinic trajectories very close to the saddle orbit. All the same, H_a and H_b only approach S asymptotically, getting closer and closer to S in the distant past or the distant future, but never actually reaching S.

19.3 Homoclinic tangle

To further explore the mess created by the intersection of the inset and outset of S, let's extend the outset further forward in time and the inset further backward by one more drive cycle. With these additions, we obtain the Poincaré section in Fig. 19.6. Consider first the outset, which

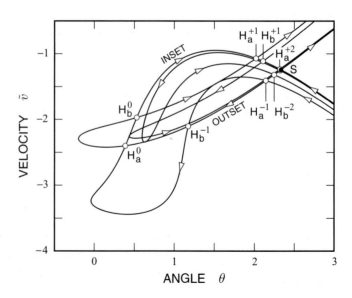

Fig. 19.6 The Poincaré section of the inset and outset of saddle orbit S extended by one drive cycle beyond those in Fig. 19.4.

was last seen headed out of frame in Fig. 19.4 after intersecting the inset at H_a^{+1}. With another drive cycle added to the outset, it doubles back on itself and intersects the inset again at H_b^{+1} as it returns to frame. Then, after another about-face, the outset heads out of frame again, intersecting the inset at H_a^{+2} before it leaves.

These folds in the outset may seem bizarre, but Poincaré saw them in his mind's eye as soon as he realized that the inset and outset could intersect. The reasoning is simple. We know that the homoclinic trajectories H_a and H_b approach S along the inset, taking a step closer with each drive cycle. But H_a and H_b always define an intersection between the inset and outset, so the outset must meet the inset each step of the way, at H_a^0, H_a^{+1}, H_a^{+2}, etc., and at H_b^0, H_b^{+1}, etc. Thus, if we extended the outset by yet another drive cycle, we would find a very thin outset loop slipped in between H_a^{+2} and S, which intersects the inset at H_b^{+2} and H_a^{+3}. Because the homoclinic trajectories only approach S and never actually arrive, the outset must include an infinite number of such loops stacked next to S, ultimately separated by infinitesimal spacings. Here we begin to see the origin of the flaky-pastry texture of chaotic attractors.

A similar repeated folding of the inset is found when we follow H_a and H_b backward in time. The inset loop added in Fig. 19.6 is a backward continuation of the curve that passes through H_a^{-1}, and it intersects the outset at H_b^{-2} and H_a^{-2}. The latter intersection is very close to S and isn't labeled. Of course, the necessary existence of H_b^{-3} and H_a^{-3} implies that another inset loop will intersect the outset at points even closer to S, and so on *ad infinitum*. That is, the inset, like the outset, takes the form of flaky pastry.

We have now come to the heart of Poincaré's tangle. With a countable infinity of inset and outset loops crossing each other in the neighborhood of the saddle orbit, we at last begin to see the full complexity of Poincaré's discovery. In his *Les Méthodes Nouvelles de la Méchanique Céleste* III, published in 1899, Poincaré describes his tangle in terms of the inset and outset curves as follows.

> When we try to represent the figure formed by these two curves and their infinitely many intersections, each corresponding to a doubly asymptotic solution, these intersections form a type of trellis, web, or net with infinitely fine mesh. Neither of the two curves must ever cut across itself again, but it must bend back upon itself in a very complex manner in order to cut across all of the meshes in the grid an infinite number of times.

Poincaré also remarks that this tangled mess of homoclinic trajectories (doubly asymptotic solutions) illustrates "the complexity of the three-body problem and of dynamics in general . . . ." However, he does not go so far as to say that the tangle might lead to chaotic motion as we understand it today. The full implication of Poincaré's tangle would be left for others to discover.

19.4   Fixed-point theorem

The thread that leads to our ultimate understanding of tangles does, however, begin with Poincaré. Near the end of his life, Poincaré began to perceive that a dynamical system with a 3-D state space might include an infinite number of periodic orbits. Indeed, he could see that a proof depended only on demonstrating a simply stated topological theorem. In December of 1911 he submitted an unfinished paper, "Sur un théorèm de géométrie," to an Italian journal, explaining to the editor that, after two years of work, "at my age, I may not be able to solve it, and the results obtained . . . seem to me too full of promise . . . that I should resign myself to sacrificing them." This was to be Poincaré's final paper, as he died in July of 1912, but his hope that it would "put researchers on a new and unexpected path" proved prescient.

Poincaré's challenge was taken up almost immediately by George Birkhoff, who was then 28, already an expert in dynamical systems, and about to become a Harvard professor. Birkhoff soon discovered a proof of Poincaré's topological theorem, which he presented to the American Mathematical Society in October of 1912. The proof was Birkhoff's first important result and it would make him world famous.

The theorem proposed by Poincaré and proved by Birkhoff falls into a class known as fixed-point theorems. Perhaps the simplest such theorem, due to the Dutch mathematician L. E. J. Brouwer, can be visualized in terms of the layer of foam on a cup of latte. When gently stirred, the foam swirls around, and we might assume that each point on the surface moves to a new point. However, Brouwer showed that if the foam remains continuous, there will always be at least one point that stays at or returns to its original position, no matter how long we choose to stir our latte. What an amazing (and delicious) proposition!

The fixed-point theorem of Poincaré and Birkhoff is similar to Brouwer's but concerns a 2-D mapping of a ring-shaped region or annulus onto itself. The situation is illustrated in Fig. 19.7. Here, points on the inner boundary of the annulus are assumed to remain on the boundary but rotate in say the clockwise direction, while those on the outer boundary rotate counterclockwise. Interior points like A can move to any other interior point A′ as long as neighboring points remain neighbors (the map is continuous). And any region B must map to a region B′ with the same area. With these restrictions, Poincaré proposed and Birkhoff proved that the map must include two fixed points like C that map back to themselves.

Of course, the significance of the Poincaré–Birkhoff fixed-point theorem is its implication of dynamical systems with an infinite number of periodic orbits. Even more to the point, in 1927 Birkhoff used the same techniques to prove "that every homoclinic motion is always in the immediate neighborhood of infinitely many periodic motions." Thus, a homoclinic point not only implies an infinity of homoclinic trajectories

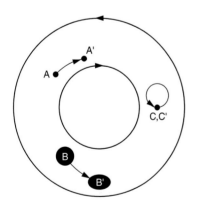

Fig. 19.7 The Poincaré–Birkhoff fixed-point theorem applies to a 2-D map that defines a one-to-one correspondence between the points of an annular region.

but an infinity of periodic orbits as well. Altogether, Poincaré and Birkhoff had established an almost unimaginable complexity associated with a homoclinic point.

19.5 Horseshoe

But didn't we also encounter an infinity of periodic orbits in the last chapter? Couldn't Birkhoff's periodic orbits be the same as those of Smale? Indeed, Smale's 1965 paper pointed out that a horseshoe map can always be found in the neighborhood of a homoclinic point. To tie everything together, we only need to find the horseshoe map hidden in the Poincaré section of Fig. 19.4.

It isn't immediately obvious, but the horseshoe in Fig. 19.4 can be seen if we relabel a few of the points and use the definition of outset and inset to track them through one drive cycle. To begin, we relabel points S, H_a^0, H_b^{-1}, and H_a^{-1} with the names A, B, C, and D to define the "rectangle" outlined by bold lines in Fig. 19.8(a). Note that two "sides" of this rectangle (AB and CD) are part of the inset, while the other two sides (BC and DA) are part of the outset.

Where does this rectangle end up after one drive cycle? The answer is revealed in Fig. 19.8(b), where the corners of the new rectangle are labeled A′, B′, C′, and D′. The logic of this mapping is simple. Since A marks the periodic orbit S, it returns as A′ at the same point in state space after the drive cycle. Because the points on side AB are all on inset trajectories, they remain on the inset but move closer to S during the drive cycle. In particular, B is actually H_a^0, so it moves to B′ at H_a^{+1}. As a result, AB is scrunched into A′B′. On the other hand, DA is on the outset, with D at H_a^{-1}, so DA expands along the outset into D′A′ with D′ at H_a^0. By similar logic, CD scrunches into C′D′ and BC expands into B′C′.

All in all, the transformation of ABCD into A′B′C′D′ should begin to remind you of last chapter's horseshoe map. Just as in the horseshoe, during one drive cycle (iteration), ABCD contracts in one direction while expanding in the other. Although ABCD unfolds rather than folding to make A′B′C′D′, the net result is the same: the initial and final rectangles overlap at each end, as indicated by the shading in Fig. 19.8(b). Moreover, this overlap pairs the ends AD and BC of the initial rectangle with the ends A′B′ and C′D′ of the final rectangle, just as in the horseshoe map of Fig. 18.2. Thus, the pendulum map of Fig. 19.8 is topologically equivalence to the Smale horseshoe.

Based on this equivalence, there should be periodic saddle orbits of the driven pendulum, similar to S but with arbitrarily long periods, that remain within the shaded region of Fig. 19.8(b). That is, we expect to find the stuff of chaos in this region.

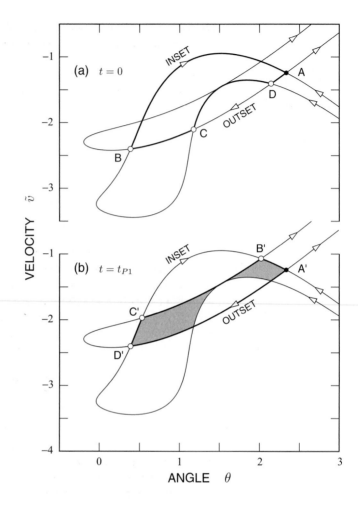

Fig. 19.8 A horseshoe in the driven pendulum. The state-space rectangle ABCD, outlined by wide lines in (a) at $t = 0$, maps into rectangle A'B'C'D' in (b) after one drive cycle ($t = t_{P1}$). The area of overlap between the initial and final rectangles is shaded. The system parameters are the same as in Fig. 19.1.

19.6   Poincaré–Birkhoff–Smale theorem

The picture of a homoclinic tangle that emerges from the work of Poincaré, Birkhoff, and Smale, sometimes called the Poincaré–Birkhoff–Smale theorem, is illustrated by the Poincaré sections of the driven pendulum in Fig. 19.9. We can't plot an infinity of points, but Figure 19.9(a) reveals 31 intersections between the inset and outset. Nine of these correspond to points on the primary homoclinic trajectories H_a and H_b, and the remaining 22 are from a few of the countable infinity of secondary homoclinic trajectories. This figure represents the tangle discovered by Poincaré in 1890.

Similarly, Fig. 19.9(b) reveals a few of the periodic orbits predicted by Birkhoff in 1927 and by Smale in 1965. From the horseshoe of Fig. 19.8, we know that all such orbits must fall within the regions of overlap between the rectangles ABCD and A'B'C'D' (shaded areas in Fig. 19.8). By considering the extensions of the inset and outset plotted in Fig. 19.6,

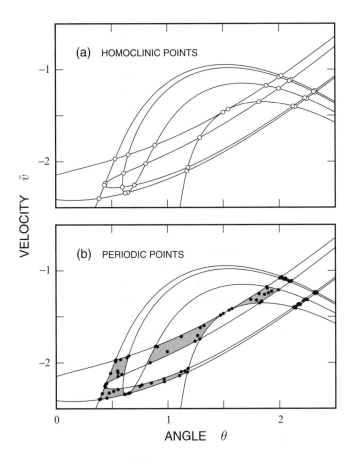

VELOCITY $\tilde{v}$

ANGLE θ

Fig. 19.9 Poincaré section of the tangle created by the homoclinic point of Fig. 19.1. Frame (a) displays 31 of the infinity of homoclinic points as open circles, and (b) displays 21 of the infinity of periodic saddle orbits as filled circles. The eight shaded regions in (b) show the areas allowed to periodic orbits based on one forward and one backward iteration of the 2-D map of Fig. 19.6. The plotted saddle orbits have periods of six or fewer drive cycles and include two orbits of period 1, one of period 2, two of period 3, three of period 4, six of period 5, and seven of period 6.

we can further restrict periodic orbits to the eight shaded areas in Fig. 19.9(b). However, there is no simple algorithm for calculating the initial conditions of the periodic orbits of the driven pendulum, as there was for the horseshoe map of the last chapter. Instead, all of the 21 solutions shown by the black dots in Fig. 19.9(b) were discovered by numerical trial and error. The periodicities of these orbits range from one to six drive cycles, so they are represented by from 1 to 6 dots in the Poincaré section. These orbits merely sample the countable infinity of periodic orbits predicted by Birkhoff and Smale.

Two periodic orbits from this infinite set are plotted as state-space trajectories in Fig. 19.10. Both trajectories describe a pendulum simply swinging back and forth in response to the periodic drive torque, but the period of solution A is one drive cycle, while that of solution B is three drive cycles. If you carefully trace both trajectories, you'll find that the path of B encircles that of A over the course of three drive cycles. Thus, as in the horseshoe of Chapter 18, the periodic trajectories of the homoclinic tangle are generally entwined in a complex braid.

Figure 19.9 samples two countably infinite sets of trajectories associated with a tangle: the homoclinic and periodic trajectories. However,

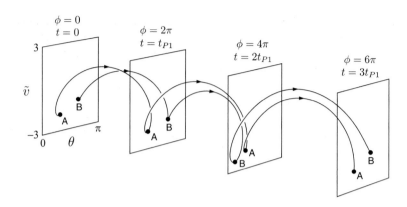

Fig. 19.10 Two of a countable infinity of periodic solutions associated with the homoclinic tangle of Fig. 19.9, plotted over three drive cycles in the full θ–$\tilde{v}$–ϕ state space. Solution A, with initial conditions $(\theta, \tilde{v}) = (0.99387, -1.88111)$, has a period of one drive cycle, and solution B, with initial conditions $(\theta, \tilde{v}) = (0.39598, -2.39534)$, has a period of three cycles.

as Smale revealed by application of symbolic dynamics, a horseshoe also implies an uncountable infinity of nonperiodic trajectories within the same region of state space. Because the nonperiodic trajectories aren't countable, they far outnumber their homoclinic and periodic cousins. There is effectively one nonperiodic trajectory for every irrational number between 0 and 1, and a typical one has an element of randomness equivalent to an infinite string of coin tosses. Thus, a homoclinic tangle is probably more complex than Poincaré or Birkhoff ever dared to imagine. All of the trajectories, homoclinic, periodic, and nonperiodic alike, are confined to the eight shaded areas of Fig. 19.9, but within this region we find a complex tangle of paths that never cross but are entwined with unimaginable complexity. This tangle is the mathematical heart of chaotic motion.

Due to the work of Poincaré, Birkhoff, and Smale, the complexity of a homoclinic tangle is beyond question, although it remains surprising when we consider the simplicity of the underlying equations of motion. In particular, Eqs. (16.6)–(16.8) for the driven pendulum describe a flow that is everywhere continuous, so neighboring trajectories necessarily remain neighbors in the short term. However, the smoothness of the flow does not prevent the existence of a saddle orbit with an inset and outset, nor the intersection of the inset and outset, nor any of the more surprising things that this intersection implies. Poincaré, Birkhoff, and Smale demonstrated that we can begin with something whole and smooth yet end with chaos.

We might think of the homoclinic tangle as a kind of tumor, a knot of intertwined and improbable trajectories confined to a limited region of state space. In this metaphor, the particular tangle explored in this chapter would be diagnosed as benign. Like the horseshoe of the previous chapter, the chaos implied here by the infinity of nonperiodic trajectories has a fragility that would make it difficult to observe in a real pendulum. Indeed, all trajectories confined to the horseshoe of Fig. 19.8 are highly unstable. As with Smale's horseshoe of the last chapter, the slightest noise or offset in initial conditions will quickly send the pendulum from one of the homoclinic, periodic, or nonperiodic solutions of the horseshoe off toward a quasiperiodic attractor in another part of state space. Thus, for almost all initial conditions, the pendulum specified in Fig. 19.1 ends up in rapid rotation, driven by the constant component of torque $\tilde{\tau}_0$, as if the oscillatory torque $\tilde{\tau}_1 \sin(\omega t)$ didn't exist. Thus, it would be virtually impossible to observe anything chaotic in a real pendulum with the parameters assumed in Fig. 19.1.

But don't be disappointed that our homoclinic tangle shows up only on the computer and not in the real world. In the following sections, we'll take a look at tangles responsible for the pendulum's more malignant forms of chaos.

19.7 Heteroclinic tangle

Our next examples of tangles will be of a second type, which Poincaré called "heteroclinic." The two types of tangles are most easily compared using cartoon sketches of their Poincaré sections. First consider the sketch of a homoclinic tangle shown in Fig. 19.11. While not topologically identical to the tangle of Fig. 19.6, this sketch shows the essential elements of any homoclinic tangle: an inset and outset of a saddle orbit S that cross first at a homoclinic point H^0, and infinite sets of further crossings that are revealed as the outset is extended forward in time (H^1, H^2, etc.) and as the inset is extended backward in time (H^{-1}, H^{-2}, etc.). As shown in Fig. 19.11, the extensions of the inset and outset take the form of ever wilder oscillations as they approach the saddle, intersecting each other at secondary homoclinic points and creating what Poincaré described as a trellis, web, or net.

A heteroclinic tangle is similar to a homoclinic one but involves the intersection between the insets and outsets of two different saddle orbits. A Poincaré section of this more complicated situation is sketched in Fig. 19.12, where the two saddle orbits are S_1 and S_2. Here the outset of S_1 intersects the inset of S_2 at the point H_a^0, which defines a trajectory that is in both sets. Because H_a^0 is in the inset of S_2, the trajectory

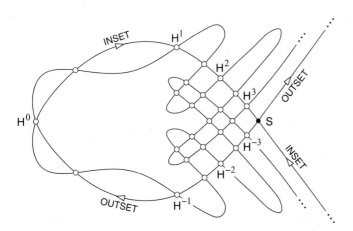

Fig. 19.11 Cartoon of the Poincaré section of a homoclinic tangle associated with saddle orbit S. Points of one homoclinic trajectory H are labeled as H^n for times (iterations) from $n = -3$ to 3.

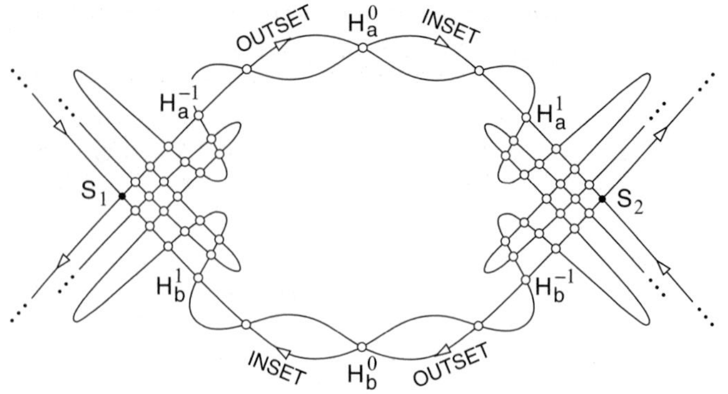

Fig. 19.12 Cartoon of the Poincaré section of a heteroclinic tangle associated with saddle orbits S_1 and S_2. Points of two heteroclinic trajectories H_a and H_b are labeled as H_a^n and H_b^n for times from $n = -1$ to 1.

approaches S_2 as time goes forward, and because H_a^0 is in the outset of S_1, it approaches S_1 as time goes backward. Thus, one forward iteration takes H_a^0 to H_a^1 and a backward iteration takes it to H_a^{-1}. The trajectory H_a is thus doubly asymptotic, but with different asymptotes for forward and backward times. Poincaré's term "heteroclinic," literally "opposite inclining," thus describes H_a perfectly.

In Fig. 19.12, a second heteroclinic trajectory H_b results from an intersection of the outset of S_2 with the inset of S_1 and defines a path from S_2 to S_1. As a result, the inset and outset of S_1 form a trellis near S_2 and those of S_2 form a trellis near S_1. Is this a situation ripe for chaos? Yes. As Poincaré realized, the double trellis structure implies a countable infinity of doubly asymptotic trajectories. And in 1935 Birkhoff proved that a heteroclinic tangle also yields a countable infinity of periodic orbits. Finally, although a heteroclinic tangle doesn't include a horseshoe as such, symbolic dynamics can be applied to prove the existence of an uncountable infinity of nonperiodic trajectories analogous to those of the horseshoe. Thus, heteroclinic and homoclinic tangles exhibit the same sorts of entwined trajectories, and both lead to chaotic motion that can be either dynamically fragile or robust.

19.8   Fractal basin boundary

As our first tangle with experimentally observable consequences, we consider a driven pendulum with parameters that lead to 14 saddle orbits and two attracting orbits, all with a period of one drive cycle. As you might imagine, the insets and outsets of all the saddle orbits create a messy tangle in state space, with homoclinic and heteroclinic points scattered everywhere.

Figure 19.13 gives some idea of this mess, with the insets plotted in (a) and the outsets in (b). The 14 saddles orbits, shown as unlabeled open circles, correspond to unstable periodic motion in which the pendulum advances by -2, -1, 0, $+1$, or $+2$ full revolutions during each drive

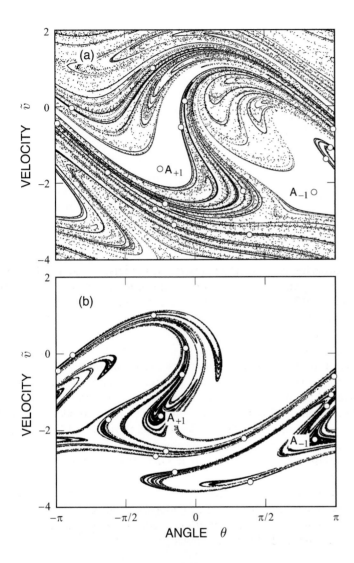

Fig. 19.13 Poincaré sections of the driven pendulum showing (a) insets and (b) outsets of 14 saddle orbits (unlabeled open circles), all with a period of one drive cycle. Two attracting orbits, labeled A_{+1} and A_{-1}, advance by $+1$ and -1 revolutions during each drive cycle. The pendulum parameters are $\rho = 0.1$, $\omega = 0.8$, $\tilde{\tau}_0 = 0$, and $\tilde{\tau}_1 = 1.2$.

cycle. The two attractors, labeled A_{+1} and A_{-1}, represent stable motion in which the pendulum advances by $+1$ or -1 rotation per drive cycle. The small dots in Fig. 19.13(a) are all points on the inset of one or another of the saddles. The insets should appear as curves in this Poincaré section, but we haven't plotted enough points to fill out every curve. Similarly, the small dots in Fig. 19.13(b) are points on the outset curves of the saddle orbits. Comparing frames (a) and (b) leads to the immediate conclusion that the inset and outset curves of the various saddles intersect at a large number of points, implying the existence of many homoclinic and heteroclinic trajectories and tangles.

The flow depicted in Fig. 19.13 is certainly complex, but, as in the simpler homoclinic tangle discussed earlier, the wealth of entwined trajectories yields only dynamically fragile sorts of chaotic motion. The problem is that the slightest noise or offset in initial conditions will push the pendulum off any of the homoclinic, heteroclinic, saddle, or nonperiodic trajectories, and it will quickly move to one of the stable attractors, either A_{+1} or A_{-1}. Indeed, as Fig. 19.13(b) shows, A_{+1} and A_{-1} are embedded in the outsets of all the saddle orbits, where they gather in trajectories that wander away from any saddle. Thus, in our quest for dynamically robust chaotic motion, we have again come up empty handed.

There are, however, experimental consequences to the tangle of Fig. 19.13. To observe the tangle, we only need to map the basins of attraction of A_{+1} and A_{-1}. A real experiment would be difficult, so a computer simulation will have to suffice. Suppose we select initial conditions $(\theta, \tilde{v})$ from a grid of points that spans the range of θ and $\tilde{v}$ in Fig. 19.13. Then, using each $(\theta, \tilde{v})$ as the starting point, we can observe the pendulum's motion and determine which attractor it approaches. If all the grid points leading to A_{+1} are colored black and those leading to A_{-1} are colored white, we obtain Fig. 19.14, a map of the two basins of attraction.

The complexity of these basins contrasts with the simplicity of those we saw earlier in Figs. 15.9 and 15.11. However, as in Chapter 15, the boundary between the basins of A_{+1} and A_{-1} is still formed by the insets of saddle orbits. The difference is that the insets now have a fractal geometry because they are part of a system of tangles. Thus, the complexity of the boundary between the basins of A_{+1} and A_{-1} reflects the complexity of the insets in Fig. 19.13(a), and our numerical experiment reveals an essential element of the tangle's structure. At last, we have a tangle with an experimentally observable consequence, a fractal basin boundary,[2] even if it isn't dynamically robust chaotic motion.

[2] The Mandlebrot set, perhaps the most famous fractal, also depicts two basins of attraction separated by a fractal basin boundary.

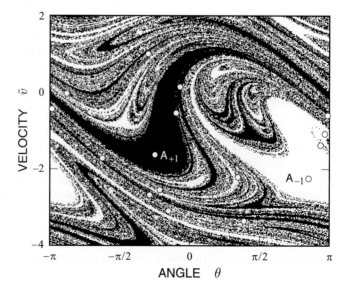

Fig. 19.14 Poincaré section of the basins of attraction for the two attractors of Fig. 19.13. Black areas are attracted to A_{+1} and white areas to A_{-1}.

19.9   Robust chaos

Actually, the tangle of Fig. 19.13 is very close to producing the kind of chaotic attractor with which we are now thoroughly familiar. In fact, reducing the amplitude of the periodic drive torque from $\tilde{\tau}_1 = 1.2$ to 1.1 suffices. This modest change eliminates the attracting orbits A_{+1} and A_{-1} along with two of the saddle orbits, leaving a dozen period-1 saddles and a complex system of tangles. With the periodic attractors eliminated, we might wonder where the outset trajectories of the saddles will go. Is there a quasiperiodic rotation of the pendulum that becomes the destination of the outsets? No, in the present example the constant component of torque is zero ($\tilde{\tau}_0 = 0$), so there is nothing to keep the pendulum rotating quasiperiodically. Surprisingly, in the absence of other attractors, the outset trajectories themselves become the new attracting set.

The resulting state-space topology is revealed in Fig. 19.15, which plots the insets and outsets of the dozen saddle orbits. As in Fig. 19.13, the many intersections between insets and outsets yield a rich structure of homoclinic and heteroclinic trajectories and tangles and an uncountable infinity of nonperiodic trajectories. However, the chaos inherent in these trajectories is now easily observed, because the pendulum is drawn to the chaotic region from virtually any point in state space. Thus, instead of the fragile chaos found at $\tilde{\tau}_1 = 1.2$, at 1.1 we have chaotic motion that is completely robust in the presence of noise. Moreover, the outset trajectories of Fig. 19.15(b) don't just approach the attracting set asymptotically, they are a part of it. Indeed, the outsets of Fig. 19.15(b) alone clearly depict the strange attractor, the epitome of chaotic motion. Thus, we have come to the end of the journey begun by Poincaré when he first realized that insets and outsets can intersect, and we have acquired a mathematical inkling of how chaos happens.

However, the global stability of a chaotic attractor remains a little mysterious. Certainly the inset trajectories of Fig. 19.15(a) suggest stability. By definition, these trajectories all approach one of the saddle orbits embedded in the attractor, so they support the notion of a converging flow. Nevertheless, it's surprising that all of the trajectories in the surrounding area are attracted to an inherently unstable region of state space. In fact, nothing in the work of Poincaré or Smale suggests that the unstable trajectories which they investigated might ever become the core of an attracting set. But, even if we don't completely understand the global stability of strange attractors, they are an experimental fact. Cartwright and Littlewood had bumped into a strange attractor in 1945, but it wasn't until Yorke began to circulate Lorenz's 1963 paper that the wider scientific community was awakened to the possibility of dynamically robust chaotic motion. As discussed previously, at this point a mathematical backwater suddenly became mainstream science.

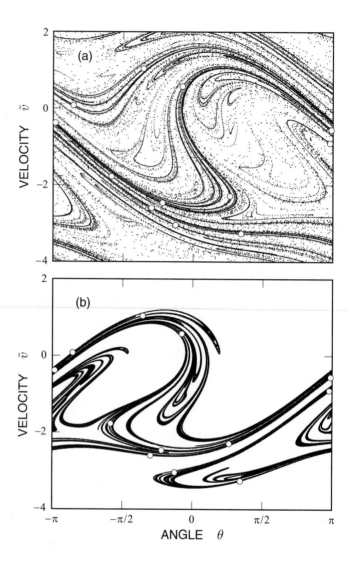

Fig. 19.15 Poincaré sections of the driven pendulum showing (a) insets and (b) outsets of 12 saddle orbits (open circles) with periods of one drive cycle. The pendulum parameters are $\rho = 0.1$, $\omega = 0.8$, $\tilde{\tau}_0 = 0$, and $\tilde{\tau}_1 = 1.1$.

19.10   Paradox lost

Thanks to Poincaré, Birkhoff, and Smale, we can now picture chaos in terms of the topology of state space. In this picture, chaos is a product of saddle orbits, insets and outsets, homoclinic and heteroclinic points, tangles, stretching and folding, and infinite symbol sequences. As these concepts were introduced, we noted how they help explain the paradox of chaos. By way of summary, we'll take a final look at the mechanism of this strange phenomenon.

Perhaps the most rudimentary form of chaos occurs in the 1-D shift map introduced in Chapter 14. Here chaos results from the iterated actions of stretching (multiplication by 2) and folding (taking the fractional part) applied to a number Y_i between 0 and 1. Stretching assures that two trajectories beginning with nearly the same Y_0 quickly diverge, which explains the butterfly effect, while folding keeps Y_i between 0 and 1, which explains the global stability of chaotic motion. When Y_i is expressed in binary, we can view these actions as shifting the radix point one digit to the right (stretching) and discarding the ones digit (folding). In this picture, Y_i can be thought of as a sequence of the symbols 0 and 1 that orchestrates the motion. The radix point moves one digit to the right with each iteration, and folding occurs when the digit following the radix point is 1. Because symbols are destroyed in this process, the shift map can't be inverted to reconstruct previous values of Y_i, but it provides an intuitive picture of chaos in terms of stretching and folding. However, the shift map can't explain the origin of chaotic randomness, which is simply built into the initial Y_0: periodicity results if Y_0 is rational and aperiodicity if it's irrational.

Our picture of chaos in 1-D is completed by the logistic map, $X_{i+1} = 4X_i(1 - X_i)$, even though its trajectories exactly parallel those of the shift map. The factor of 4 in the logistic map insures stretching on average, and the parabolic shape of the map provides the folding needed to keep X_i between 0 and 1. Most important, chaotic randomness in the logistic map results for almost all X_0, whether rational or irrational. Thus, the map itself generates the infinite random symbol sequence that must exist in the parallel world of the shift map to account for chaotic motion.

Many elements of chaos in 1-D maps carry over to 2-D maps and 3-D flows, where the Smale horseshoe is the prototype. As described in Chapter 18, the 2-D horseshoe map acts on an area of state space by stretching it in one direction, compressing it in another direction, and finally folding and repositioning it to overlap its original location. Thus, unsurprisingly, the horseshoe map involves stretching and folding on each iteration. Moreover, trajectories that remain forever within the original area can be described by a symbol sequence comprising 0's and 1's. For the horseshoe map this sequence is bi-infinite ($\ldots 101{:}011 \ldots$) with a colon indicating the present time. Unlike the shift and logistic maps, the horseshoe map is invertible, and time can move either forward or backward, depending on whether the colon is shifted to the right or left. In either case, the state-space point follows a folded trajectory if the colon steps over a 1, but not if it steps over a 0, just as in the shift map. As Smale demonstrated, there are horseshoe trajectories corresponding to every possible bi-infinite symbol sequence, implying the existence of a countable infinity of periodic trajectories and an uncountable infinity of aperiodic trajectories, all unstable. Thus, the horseshoe map provides the raw material for chaotic motion.

As we've learned in this chapter, horseshoes are found in 3-D flows in association with periodic saddle orbits. A saddle orbit incorporates both stability and instability in the form of a 2-D sheet of trajectories that approach the saddle (inset) and a sheet of trajectories that leave the saddle (outset). If the inset and outset of a single saddle orbit cross, then their intersection defines a homoclinic trajectory that approaches the saddle orbit when we follow it either forward or backward in time. As Poincaré realized in 1889, a homoclinic trajectory further implies a tangled state-space topology in which the inset and outset sheets are folded in an infinite number of ever closer pleats that intersect to create an infinite number of homoclinic trajectories. Birkhoff later showed that a homoclinic tangle also implies an infinite number of unstable periodic orbits, and Smale showed that it implies a horseshoe map. A single saddle orbit can thus give rise to the topology of a chaotic flow, a fantastic jumble including countable infinities of homoclinic trajectories and periodic orbits and an uncountable infinity of intertwined aperiodic trajectories.

Chaotic trajectories exists wherever we find a homoclinic or heteroclinic tangle. But often the chaos is dynamically fragile, and the slightest perturbation will divert the motion to another part of state space. In this case, the insets associated with the tangle may form a fractal basin boundary between separate state-space attractors. On the other hand, for reasons not entirely understood, it may happen that the outsets of the tangle are themselves an attracting set, and chaotic motion is dynamically robust. This is the kind of motion discovered by Sellner with his Tilt-A-Whirl, van der Pol and Ueda with their driven oscillators, and Lorenz with his weather model. This is the easily observed chaos that has forced science to come to grips with the astonishing topological complexity of three-dimensional state-space flows.

19.11   Stability of the Solar System

The revised version of Poincaré's prize winning paper, published more than a century ago, left open the possibility that our Solar System is dynamically unstable. So, which is it, stable or unstable? Today we have a much more complete answer to this question, but not a definitive one.

At present, our best window on the question of stability comes from computer simulations of Solar-System dynamics. In principle, such simulations are straightforward. The Sun and eight planets are assumed to be point masses, all of which attract one another through Newton's law of gravitation. Given their positions and velocities at one instant, it's a simple matter to calculate the acceleration and determine the positions and velocities at a slightly later instant in time, just as we did in Chapter 3. Of course, we now have to keep track of three position coordinates and three velocity components for each mass, so there are a total of 27 state variables, but the computer program is no different in principle from that of Fig. 3.2. In practice, of course, maintaining accuracy over long time periods is difficult, but through heroic efforts several groups of scientists have calculated the motion of the Solar System more than one billion years into the future.

The findings of these scientists confirm Poincaré's fear that the Solar System might be unstable. In particular, several independent calculations show that the Solar System has a positive Liapunov exponent of approximately $\lambda = 1/(4 \times 10^6$ years). That is, the separation between two simulations with nearly identical initial conditions grows by a factor of $e = 2.72$ every 4 million years. Expressed in more mundane terms, an uncertainty of 15 meters in our knowledge of the Earth's position today would make it impossible to predict where the Earth will be in its orbit 100 million years from now. But most important, a positive Liapunov exponent allows us to conclude that the Solar System really is chaotic in the technical sense of the word.

Does this chaotic instability have any dramatic long-term consequences? The Solar System has been around for 4.5 billion years, so it's remarkable to find an instability that grows exponentially on a time scale of only 4 million years. But the billion-year simulations haven't revealed any future cataclysmic event, so perhaps the 4-million-year time constant isn't as alarming as it sounds. On the other hand, the simulated Solar System does slowly change through a mechanism that astronomers refer to as resonance. A resonance acts like the periodic push that we give a child's swing, slowly but inevitably increasing the amplitude of the motion. Planetary resonances have similar long-term effects, gradually changing orbital parameters like eccentricity. Over hundreds of millions of years planetary orbits tend to evolve in one direction and then back again, making it impossible to know for sure what will happen in the very long term. However, one researcher has concluded that the planet Mercury might be ejected from the Solar System within the next 5 to 10 billion years. Although humanity is unlikely to survive long enough to see such an event, it would be dramatic confirmation of our present conception of chaos in the Solar System.

Further reading

Discovery of homoclinic points

- Andersson, K. G., "Poincaré's discovery of homoclinic points", *Archive for History of Exact Sciences* **48**, 133–147 (1994).

- Béguin, F., "Poincaré's memoir for the Prize of King Oscar II: Celestial harmony entangled in homoclinic intersections", in *The Scientific Legacy of Poincaré*, edited by Charpentier, É., Ghys, É., and Lesne, A. (American Mathematical Society, Providence, 2010) chapter 8.

- Barrow-Green, J., "Oscar II's prize competition and the error in Poincaré's memoir on the three body problem", *Archive for History of Exact Sciences* **48**, 107-131 (1994).

- Barrow-Green, J., *Poincaré and the Three Body Problem* (American Mathematical Society, Providence, 1996).

○ Diacu, F. and Holmes, P., "A great discovery—and a mistake", in *Celestial Encounters: The Origins of Chaos and Stability* (Princeton University Press, Princeton, 1996) pp. 3–50.

○ Ekeland, I., "Poincaré and beyond", in *The Best of All Possible Worlds: Mathematics and Destiny* (University of Chicago Press, Chicago, 2006) chapter 5.

- Gray, J., "Poincaré, topological dynamics, and the stability of the solar system", in *The Investigation of Difficult Things: Essays on Newton and the History of the Exact Sciences*, edited by P. M. Harman and A. E. Shapiro (Cambridge University Press, Cambridge, 1992) pp. 503–524.

- Goroff, D. L., "Henri Poincaré and the birth of chaos theory", in *New Methods of Celestial*

Mechanics, Part I (American Institute of Physics, New York, 1993) pp. I1–I107.

- Holmes, P., "Poincaré, celestial mechanics, dynamical-systems theory and 'chaos'", *Physics Reports* **193**, 137–163 (1990).

○ Peterson, I., "Prophet of chaos", in *Newton's Clock: Chaos in the Solar System* (Freeman, New York, 1993) chapter 7.

- Poincaré, H., "Sur le problème des trois corps et les équations de la dynamique", *Acta Mathematica* **13**, 1–270 (1890).

- Poincaré, H., *Les Méthodes Nouvelles de la Méchanique Céleste* III (Gauthier-Villars, 1899); *New Methods of Celestial Mechanics*, Part III, English translation, edited by D. L. Goroff (American Institute of Physics, Woodbury, New York, 1993).

○ Stewart, I., "The last universalist", in *Does God Play Dice? The New Mathematics of Chaos,* 2nd edition (Blackwell, Malden, Massachusetts, 2002) chapter 4.

○ Szpiro, G. G., "An Oscar for the best script", in *Poincaré's Prize: The Hundred-Year Quest to Solve One of Math's Greatest Puzzles* (Dutton, New York, 2007) chapter 4.

Fixed-point theorems

- Birkhoff, G. D., "Proof of Poincaré's geometric theorem", *Transactions of the American Mathematical Society* **14**, 14–22 (1913).

- Birkhoff, G. D., "On the periodic motions of dynamical systems", *Acta Mathematica* **50**, 359–379 (1927).

- Birkhoff, G. D., "Nouvelles recherches sur les systèmes dynamiques", *Memoriae Pontifical Academia Scienze Novi Lyncaei*, S. 3, **1**, 85–216 (1935).

- Poincaré, H., "Sur un théorèm de géométrie", *Rendiconti Circolo Matematico di Palermo* **33**, 375–407 (1912).

○ Shinbrot, M., "Fixed-point theorems", *Scientific American* **214**(1), 105–110 (January 1966).

Stability of the Solar System

○ Chown, M., "Chaotic heavens", *New Scientist* **181**(2436), 32–35 (28 February 2004).

○ Frank, A., "Crack in the clockwork", *Astronomy* **26**(5), 54–59 (May 1998).

○ Hartley, K., "Solar system chaos", *Astronomy* **18**(5), 34–39 (May 1990).

○ Laughlin, G., "The solar system's extended shelf life", *Nature* **459** 781–782 (2009).

○ Parker, B. R., "Is the solar system stable?" in *Chaos in the Cosmos: The Stunning Complexity of the Universe* (Plenum, New York, 1996) chapter 11.

○ Peterson, I., "Digital orrery" and "Celestial disharmonies", in *Newton's Clock: Chaos in the Solar System* (Freeman, New York, 1993) chapters 10–11.

○ Soter, S., "Are planetary systems filled to capacity?" *American Scientist* **95**, 414–421 (2007).

○ Thuan, T. X., "Chaos in the cosmic machinery, and uncertainty in determinism", in *Chaos and Harmony: Perspectives on Scientific Revolutions of the Twentieth Century* (Oxford University Press, Oxford, 2001) chapter 3.

Topology of chaos

○ Abraham, R. H. and Shaw, C. D., "Chaotic Behavior" and "Global Behavior", in *Dynamics: The Geometry of Behavior*, 2nd edition (Addison-Wesley, Reading, 1992) Parts 2 and 3.

- Gilmore, R. and Lefranc, M., *The Topology of Chaos: Alice in Stretch and Squeezeland* (Wiley, New York, 2002).

Part VI
Conclusion

Chaos goes to work

20

Now that we've probed the mathematical depths of chaos, it's time to put our knowledge to practical use. There are several possible applications to consider. For example, chaos can be used simply to generate random numbers or signals. Or, when chaos arises naturally, we can use our understanding to reduce or eliminate unwanted consequences. Or, our goal might be to discover the predictable element hidden within a chaotic situation. All of these scenarios have been explored as scientists have sought to apply their new understanding of chaotic systems. Each application generally exploits one of the three basic properties of chaos: short-term predictability, long-term randomness, or sensitivity to small perturbations. Let's look first at how chaotic randomness can be put to work.

20.1   Randomness

In Chapter 7 we learned that typical chaotic motion is highly correlated on short time scales but statistically random over longer periods. Moreover, Smale demonstrated that the random element is as random as you can get, like a series of coin tosses.

Using chaotic motion as a source of randomness is perhaps its simplest application and, as we've discussed in previous chapters, first occurred long before the nature of chaos was fully understood. Thus, many devices for gambling based on chaos are centuries old, including dice, the roulette wheel, and the quincunx. More recently, our fascination with random motion has led to entertainments ranging from kinetic sculptures like the Lava Lamp and the Space Circle to carnival rides like the Tilt-A-Whirl and the Octopus.

On the scientific side, we previously noted that in 1947 Ulam and von Neumann proposed using the logistic map as a source of random numbers in numerical simulations. Because they are simple and efficient, deterministic random number generators based on chaotic maps remain a mainstay of scientific computing. However, such algorithms sometimes fail to meet certain statistical tests for randomness and always begin to repeat after some large number of iterations.

More recently, researchers have obtained random numbers from diode lasers operated in a chaotic regime. Producing more than 1 billion random bits per second, such schemes have passed standard statistical tests for randomness and, because they rely on microscopic motion as their ultimate source of noise, the bits are certain never to repeat or exhibit long-term predictability.

Another area in which chaotic randomness seems ripe for exploitation is in the transmission of encrypted messages. In one scheme, the message is transmitted as a small addition to the output of a chaotic circuit. This conceals the message, so that an eavesdropper hears only noise. If the intended recipient has a similar chaotic circuit, however, he can monitor the corrections needed to keep his circuit synchronized with that of the sender and recover the message as the correction signal. Without a copy of the chaotic circuit, a would-be spy would seem to be at a total loss. However, there is always a battle of wits between code maker and code breaker, and in this case the code breaker has the upper hand. If the spy is clever enough to guess that chaos has been used to hide the message, he may be able to reconstruct the chaotic circuit from information available in the short-term correlations of the signal. Then the spy can build his own decoder and eavesdrop at will. Thus, chaotic encryption doesn't allow messages to be transmitted with complete security, but it may find application in less demanding situations.

20.2   Prediction

An ability to predict the future is helpful in most human activities, and understanding chaos can improve our prediction skills. In particular, simply knowing that apparently random events can be due to chaos tells us not to give up too soon in looking for patterns. Although short-term correlations aren't always obvious when we look at deterministic chaos, we may be able to tease out some regularity and make useful predictions in situations that appear hopelessly confused.

A somewhat artificial example of discovering a pattern within chaos is provided by Lorenz's 1963 paper on chaotic convection. The equations of motion for this system, Eqs. (10.1)–(10.3), are sufficiently complex that it would be difficult to reconstruct them from a sample waveform like that of Fig. 10.3. However, Lorenz suggested another approach. To simplify the situation, he ignored the continuous waveforms associated with the state variables, x, y, and z, and focused his attention only on the successive maximum values of z. If we denote the ith maximum by $z_{\max}(i)$ and use the full convection equations to determine 100 successive maxima, we obtain the plot shown in Fig. 20.1.

As we've come to expect in chaotic systems, Fig. 20.1 reveals both an element of randomness and also some systematic behavior. On the other hand, we wouldn't be surprised if Fig. 20.1 represented earthquake activity or stock market trends rather than chaotic convection. With these alternatives in mind, discovering any pattern hidden in this figure takes on an importance beyond its humble origins.

Lorenz cracked the problem of prediction for his convection cell using a simple strategy. Plotting the $(i+1)$th maximum of z against the ith maximum, he discovered the amazingly regular curve shown in Fig. 20.2. If successive values of $z_{\max}(i)$ were uncorrelated, we would expect to find

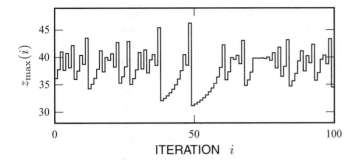

Fig. 20.1 One hundred successive maximum values of the Lorenz variable z computed from Eqs. (10.1)–(10.3) with $\sigma = 10$, $r = 28$, and $b = 8/3$.

a cloud of points scattered over the plane of $z_{\max}(i)$ and $z_{\max}(i+1)$. Instead, we find that $z_{\max}(i+1)$ is almost completely determined by $z_{\max}(i)$, as if the Lorenz system were no more than a simple 1-D iterated map. We know that the map is only an approximation because Eqs. (10.1)–(10.3) can be used to calculate previous values of x, y, and z (using negative values for Δt), while the map isn't invertible. The approximate nature of the map is also revealed by the inset in Fig. 20.2 which shows that the calculated points don't fall on a single curve. Nevertheless, the 1-D map is an excellent approximation to the 3-D flow of Eqs. (10.1)–(10.3) and can provide useful predictions for the next several values of $z_{\max}$.

The best thing about a map like $z_{\max}(i+1)$ versus $z_{\max}(i)$, usually called a return map, is that it can often be constructed without any special knowledge about the system. Although we used the Lorenz equations to create Fig. 20.2, a return map can be plotted for any

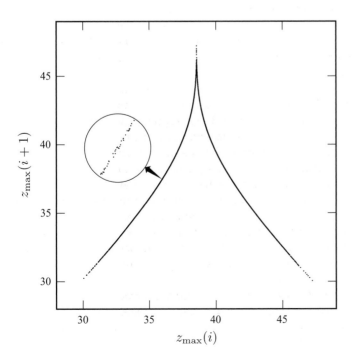

Fig. 20.2 The next maximum value of the Lorenz variable z as a function of the preceding maximum for the parameters of Fig. 20.1. Points for 10,000 successive maxima are plotted. The inset reveals scatter in a section of the curve magnified by a factor of 100.

source of apparently random data without knowing what equations (if any) might lie behind it. For example, the dripping faucet mentioned in chapter 1 was explored by Robert Shaw, a graduate student in physics at Santa Cruz and a member of the Dynamical Systems Collective. Shaw simply recorded the times T_i between successive water drops and plotted T_{i+1} versus T_i. For certain flow rates, he discovered a roughly parabolic return map similar to the logistic map. With this discovery, Shaw immediately knew that his dripping faucet harbored short-term correlations as well as random behavior, and he could easily use the return map to predict the next interval between drops.

The value of predicting the future becomes clear if we pretend that $z_{\max}(i)$ in Fig. 20.1 represents the maximum price in dollars of XYZ Corporation stock on day i. In this case, the figure becomes a 100-day price history of the stock. Suppose that on day 39, with the price at $32.06, we don't know for sure what will happen next. But, assuming that the return map of Fig. 20.2 is a good predictor, we can use it to forecast an upward trend in price beginning at $32.06. So let's buy 100 shares of XYZ and keep an eye on the predictions of the return map. At first we see our investment grow, but on day 49 we'd surely sell our shares at $46.22, because the return map predicts a sharp drop in price the following day. Of course, buying low and selling high is the goal of every investor, and the return map allows us to do just that. We made an imaginary profit of $1,416 or 44% in just 10 days. Unfortunately, finding a stock like XYZ with a price so accurately predicted by a return map is highly unlikely.

Nevertheless, in 1991 Doyne Farmer and Norman Packard left comfortable positions at the Los Alamos National Laboratory and the University of Illinois to form a new company premised on the belief that financial markets include exploitable "islands of predictability." As former members of the Dynamical Systems Collective, Farmer and Packard were well aware of Shaw's demonstration of the predictability of chaotic dripping. As graduate students they also spent considerable effort attempting to win big at Las Vegas using Newtonian mechanics to predict the motion of a roulette ball. By 1991 Farmer and Packard were recognized experts in nonlinear dynamics and chaos, and they were ready for a new challenge. Thus, under the name "Prediction Company," they began to develop sophisticated numerical techniques for detecting patterns in the ebb and flow of money in a variety of commodity and securities markets. And this wasn't an ivory-tower exercise. With capital supplied by Union Bank of Switzerland (UBS), Farmer and Packard put their predictions to the test on the trading floor. Were they successful predictors? While its financial records are not public, 14 years after it was founded, Prediction Company was sold to UBS for an undisclosed sum. Reading between the lines suggests that Farmer and Packard were indeed successful.

Forecasting based on the recognition of chaotic behavior can also have social benefits. Mathematician George Sugihara obtained his doctorate under Robert May and went on to work in a number of areas involving chaos and complexity, including the economics of marine ecosystems. Based on a study of the population dynamics of sardines published in 2006, Sugihara has suggested that the long-term viability of the fishing industry may depend on recognizing the essential instability of fish populations and their propensity for chaotic fluctuation. He has proposed that predictions based on nonlinear models be used in adjusting fishing harvests to eliminate boom and bust cycles and insure a continuing supply of this valuable resource. In this case, an entire economic community may benefit from understanding the short-term predictability of a chaotic system.

20.3 Suppressing chaos

In Chapter 12, we remarked that riders of the Tilt-A-Whirl, Octopus, and other chaotic carnival rides often discover that they can control their car's motion by throwing their weight around at crucial moments. Such active intervention can make a wild ride even wilder, and it relies on the fact that chaos is extremely sensitive to small perturbations. The rider's lunge is exponentially amplified by the ride itself, quickly diverting the car to an entirely new trajectory. Although ignorant of Liapunov exponents, ride fans discovered that chaotic systems can be controlled with small forces long before dynamical systems experts arrived at the same conclusion. On the other hand, the experts have proposed a wider range of applications, and we'll devote the remainder of this book to exploring a few of them.

The object of controlling a chaotic system is often simply the elimination of chaos itself. An engineer who runs into chaos might do this by adding some viscous damping or otherwise modifying her design. But when modification of the basic mechanism isn't practical, she can suppress chaos by careful application of the same small perturbations used by Tilt-A-Whirl enthusiasts. This approach to controlling chaos was first suggested in 1990 by three scientists at the University of Maryland, Edward Ott, Celso Grebogi, and James Yorke, and it involves what engineers call feedback.

One of the simplest feedback mechanisms is the thermostat that controls the temperature inside your house. In the winter, the thermostat turns on the furnace when the temperature falls say a degree below the thermostat setting. The thermostat shuts off the furnace again when the temperature rises a degree above the set point. By monitoring the temperature and turning the furnace on and off, a thermostat thus keeps your house within a narrow range of temperature suited to people and cats.

Ott, Grebogi, and Yorke (OGY) realized that feedback can also be used to confine the trajectory of a chaotic system to one of the periodic saddle orbits embedded in its attracting set, thereby suppressing the chaos. Although a saddle orbit is unstable, in the absence of noise the system will follow the saddle orbit forever. So, with only a small amount of noise, small corrections should suffice to keep the system on a selected orbit. By constantly monitoring the trajectory for deviations from the saddle orbit and applying a judicial kick now and then, we can thus hope to convert chaotic motion into stable periodic motion.

The feedback stabilization suggested by OGY is analogous to the technique you use to balance on one foot. In principle, it shouldn't require any work at all to maintain balance, but you always begin falling to one side or the other, and your foot must constantly twitch from side to side, applying the corrections needed to keep you upright. It's a tricky business, but something a foot can learn to do.

Let's see how feedback can be applied to bring order to chaos in the driven pendulum. In particular, we'll stabilize the saddle orbit at $(\theta, \tilde{v}) = (-0.241, 0.572)$ in the chaotic attractor of Fig. 19.15(b). The region of state space including this saddle orbit is redrawn at a magnified scale in Fig. 20.3. Here the saddle is labeled S_{+1}, as it represents motion in which the pendulum advances by exactly one revolution during each drive cycle, and we've included portions of its inset and outset.

Suppose that the pendulum is currently at the nearby point $P_0 = S_{+1} + (0.0008, 0.0004)$, and we want to move the pendulum closer to the saddle by applying feedback. Without feedback, the pendulum will move to P_1 after one drive cycle, making a quick getaway from the saddle along its outset. To thwart this getaway, OGY suggested altering one of the system parameters during the next iteration of the map (in our example, the next drive cycle) so as to move the trajectory to the saddle's inset. This is a smart strategy because, once on the inset, the system will naturally move toward the saddle orbit.

In the driven pendulum, the logical choice of feedback parameter is the constant component of the applied torque, $\tilde{\tau}_0$. We want to apply just the right amount of torque over the next drive cycle to take the

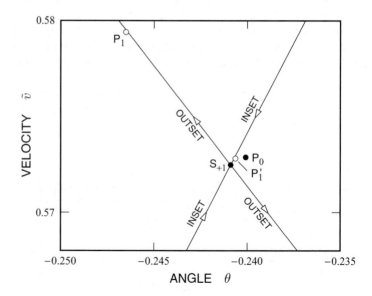

Fig. 20.3 Poincaré section of state space for the driven pendulum showing a saddle orbit S_{+1} and its inset and outset. This saddle is part of the chaotic attractor in Fig. 19.15(b) and corresponds to periodic motion in which the pendulum angle advances by one full rotation during each drive cycle. The nearby point P_0 would normally return as P_1 after one drive cycle. However, in the presence of feedback designed to stabilize S_{+1}, P_0 returns as P'_1. The pendulum parameters are $\rho = 0.1$, $\omega = 0.8$, $\tilde{\tau}_0 = 0$, and $\tilde{\tau}_1 = 1.1$.

pendulum from P_0 to a point on the inset of S_{+1}. Fortunately, in the neighborhood of a saddle orbit the flow is particularly simple because the system's response is almost linear. Thus, it isn't difficult to estimate that a feedback torque of $\tilde{\tau}_0 = 1.73 \times 10^{-4}$ acting over one drive cycle will do the trick. And indeed, solving the equations of motion reveals that with this additional torque the pendulum moves from P_0 to P'_1, a point nearly on the inset, rather than to P_1. Because P'_1 is not precisely on the inset, it is necessary to repeat this process on several additional cycles, always measuring the pendulum's position in state space relative to S_{+1} and using this information to estimate the feedback torque to be applied on the next cycle. However, only a few cycles are required to move the pendulum to the periodic orbit and suppress further chaotic behavior.

In practice, it doesn't make sense to apply feedback to the pendulum until it approaches the saddle orbit that the feedback is designed to stabilize. When this close approach is left to chance, we obtain a result like that in Fig. 20.4. Here we plot the net rotation of the pendulum on each of 100 successive drive cycles. The first 75 cycles show the pendulum's natural chaotic motion, but after the 75th cycle the pendulum ends up close to S_{+1} and feedback is applied thereafter. With feedback in place, the saddle orbit is stabilized, and the pendulum now rotates exactly one revolution during every drive cycle. Thanks to feedback control, chaotic motion has been replaced by periodic motion.

The real test of such stabilizing feedback is how well it works in the presence of noise. Balancing on one foot in a calm environment is good, but can you recover your balance when someone gives you a push? To test our feedback technique, we consider what happens when the pendulum is subject to environmental noise in the form of a randomly applied torque. Figure 20.5 shows how our stabilized pendulum copes with noise. In frame (a) we see the random torque, which never amounts to more than about 3% of the torque required to raise the pendulum to its horizontal position. In frame (b) we see the feedback torque, which is computed at the end of each drive cycle from the pendulum's current position and velocity. In principle this is the torque required to bring the pendulum back to the saddle's inset over the next drive cycle. Of course, the pendulum will be buffeted by additional noise over the next cycle, so this estimate never quite works out, just like the twitching of your foot

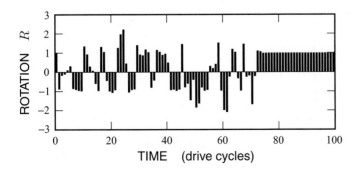

Fig. 20.4 The net rotation R in revolutions during 100 successive drive cycles of a driven pendulum. Stabilizing feedback is applied after the 75th drive cycle, when the pendulum comes near the saddle orbit S_{+1} of Fig. 20.3.

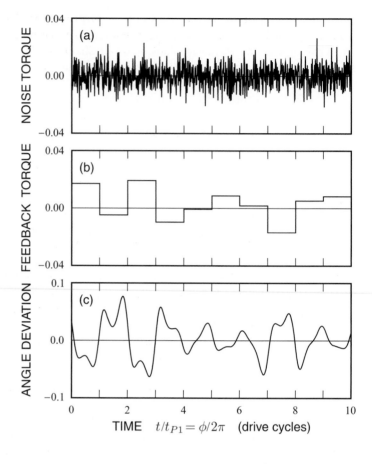

Fig. 20.5 Feedback stabilization of the saddle orbit S_{+1} of Fig. 20.3 in the presence of a random noise torque. The random torque (a), feedback torque (b), and deviation of the pendulum angle from the saddle orbit (c) are plotted as a function of time. The noise torque has a white spectrum and a Gaussian distribution. The deviation angle is in radians and the torques are normalized to the torque required to support the pendulum in a horizontal position.

never quite allows you to achieve a perfect balance. Finally, in frame (c) we see the deviation of the pendulum's angle from the saddle orbit. While this deviation sometimes approaches 0.1 radian (a few degrees), the feedback doesn't allow the pendulum to move too far from the saddle orbit even in a noisy environment. Thus, feedback really can suppress chaos by locking a system to a periodic saddle.

Although feedback stabilization may seem entirely theoretical, it has been successfully applied in a number of cases. One example concerns a laser system that uses a nonlinear crystal to double the frequency of the light generated, producing green light from an infrared source. In this case, adjusting the components to maximize the output invariably led to chaotic fluctuations in the light intensity. With feedback stabilization, however, experimenters were able to replace the chaos with a periodic variation in intensity, and thereby make the laser useful at higher output levels.

Another application of feedback control concerns the chaotic beating of the heart during ventricular fibrillation, a kind of heart attack. Pioneering experiments were performed on part of a rabbit heart treated with a drug to induce a chaotic arrhythmia. The experiments demonstrated that application of OGY feedback can suppress the chaos and restore a regular beat. Based on these results, smart pacemakers may one day monitor the heart and control it with small pulses applied at appropriate times rather than simply providing a regular cadence of strong pulses.

20.4 Hitchhiker's guide to state space

We now know how to stay put on any of the unstable saddle orbits embedded in a chaotic attractor: even in the presence of noise, an occasional judicially applied nudge does the trick. As it happens, it's also easy to move from one saddle orbit to another. In fact, such transitions require only an infinitesimal nudge if we make use of the heteroclinic trajectories introduced in the last chapter. As you'll recall, a heteroclinic trajectory approaches one saddle as time goes backward and another as time goes forward, providing a natural bridge between two saddle orbits. Aside from the nudge required to initiate motion on the heteroclinic trajectory, no further action is required to complete the transition. A heteroclinic trajectory is the ultimate free ride.

For an example of a heteroclinic trajectory, we return again to the chaotic attractor of the driven pendulum introduced in Fig. 19.15(b). In particular, we'll look for a heteroclinic trajectory linking the saddle orbits S_0 and S_{+1} shown in Fig. 20.6. These saddles represent different motions in that the pendulum simply oscillates about $\theta = 0$ on S_0 and advances by one full revolution during each drive cycle on S_{+1}. To find a trajectory linking these motions, we've plotted portions of the outset of S_0 and the inset of S_{+1}. The points (open circles) where the outset and inset intersect define trajectories that approach S_0 as time goes

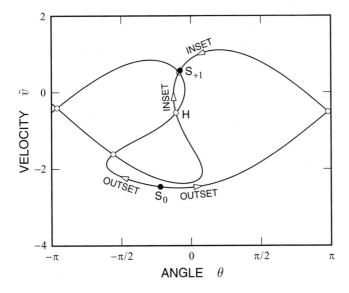

Fig. 20.6 Poincaré section showing two saddle orbits, S_0 and S_{+1}, of the driven pendulum excerpted from the chaotic attractor of Fig. 19.15(b). Open circles show heteroclinic points defined by intersections between the outset of S_0 and the inset of S_{+1}.

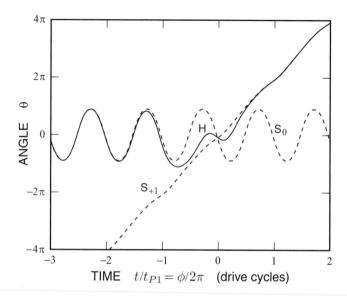

Fig. 20.7 Pendulum angle as a function of time for the heteroclinic trajectory H (solid curve) linking the periodic saddle orbits S_0 and S_{+1} (dashed curves) of Fig. 20.6.

backward and approach S_{+1} as time goes forward, exactly the kind of paths we're seeking. Given such an intersection, we can easily reconstruct the complete link, and, based on the intersection at H in Fig. 20.6, we obtain the trajectory H plotted in Fig. 20.7.

Figure 20.7 begins to reveal a magic aspect of chaotic attractors. Here the pendulum angle is plotted as a function of time for the saddle orbits S_0 and S_{+1} (dashed lines) and for the heteroclinic trajectory H (solid line) connecting them. The remarkable thing is that the transition from oscillatory motion about $\theta = 0$ to motion in which θ steadily advances, requires only an infinitesimal nudge and is completed for practical purposes within a span of three drive cycles. Thus, thanks to their predictability and extreme sensitivity, chaotic systems can be induced to radically change their motion without applying significant force.

But Fig. 20.7 only begins to suggest the possibilities for controlling a chaotic pendulum. The link from S_0 to S_{+1} is one of an infinity of heteroclinic trajectories that link the saddle orbits of the attractor. In Fig. 20.8, we show links from S_0 to S_{+2}, S_{-1}, and S_{-2} as well as S_{+1}. Thus, with a slightly different nudge, the pendulum can be induced to change from simple oscillatory motion to rotary motion in which it advances by $+2$, $+1$, -1, or -2 revolutions per drive cycle. Moreover, there are an infinity of saddle orbits embedded in the attractor, with periodicities of any integral number of drive cycles, and Fig. 19.15 suggests that the outset of any one of these saddles probably intersects the insets of all the other saddles. In this case, there are one or more heteroclinic trajectories linking every saddle with every other saddle. Thus, not only can a nudge move the pendulum from S_0 to S_{+1}, but

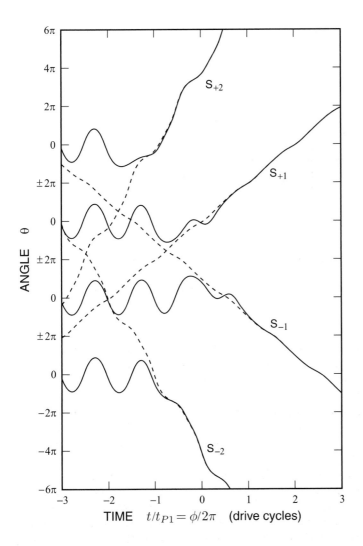

Fig. 20.8 Pendulum angle as a function of time for four heteroclinic trajectories (solid lines) linking the saddle orbit S_0 (not shown) to the saddle orbits S_{+2}, S_{+1}, S_{-1}, and S_{-2} (dashed lines) of the chaotic attractor of Fig. 19.15(b). For clarity, the heteroclinic trajectories are offset vertically from one another by 4π.

a second nudge can move it from S_{+1} to S_{-2}, and a third nudge can move it from S_{-2} to S_{+2} or back to S_0. In effect, with proper knowledge we can hitchhike from any saddle to any other saddle in short order. Thus, the possibilities for controlling motion on a chaotic attractor are endless.

Very likely, this kind of control explains the addiction of some fans of chaotic carnival rides. As Richard Chance of Chance Manufacturing once told me, devotees of the Zipper, often sporting Kamikaze-like headbands, will ride the machine over and over, keeping track of how many times they can flip their car in a single ride. And expert skiers may have a similar addiction to skiing moguls. According to Edward Lorenz, a mogul field can also lead to chaotic behavior and extreme sensitivity to small forces. What could be more fun than speeding down a slope while controlling your motion with the slightest shift in weight?

20.5   Space travel

Unfortunately, a free ride from one unstable orbit of the driven pendulum to another has no apparent practical value. Setting aside thrill seeking at amusement parks and ski areas, we might ask whether there is any real use to jumping between unstable orbits via heteroclinic trajectories. In the realm of outer space, the answer to this question is definitely "yes." In recent years, space scientists have discovered that such trajectories help make possible what's been called the "interplanetary transport network," a kind of celestial highway system that allows spacecraft to make long journeys with modest amounts of fuel. In this network, unstable orbits act like roundabouts, with their insets serving as incoming highways and their outsets as outgoing highways. On such a roundabout, only a nudge is needed to select a new direction of travel and a possible free ride to another roundabout. By hopping from roundabout to roundabout, a spacecraft can sometimes travel to a distant part of the Solar System with almost no fuel—thanks to the chaotic nature of celestial dynamics. What a bargain!

To get the flavor of this mode of space travel, we first turn back the clock to 1750. Around this year Leonhard Euler, the mathematician mentioned earlier in connection with the exponential function, discovered a curious property of three-body systems. In particular, when the third body is of negligible mass and one of the larger bodies orbits the other in a circle, there are three special points in space where the small mass can remain stationary with respect to the positions of the larger masses. These points form the hubs of the roundabouts in the celestial highway system.

Taking the large masses as the Earth and Moon, we've plotted Euler's stationary points in Fig. 20.9. Here, the Earth and Moon are at the points E and M, the x axis connects them and rotates with the Moon, and Euler's stationary points are L_1, L_2, and L_3. A spacecraft at L_1 circles the Earth while maintaining its position between the Earth and Moon. A spacecraft at L_2 or L_3 also circles the Earth in synchrony with the Moon, maintaining its position either on the opposite side of the Moon from the Earth or on the opposite side of the Earth from the Moon. As if three stationary points weren't enough, in 1772 the Italian-French mathematician Joseph Lagrange (1736–1813) discovered two more such points, L_4 and L_5 in Fig. 20.9, and all five are now called Lagrange points.

Lagrange points are the locations at which a spacecraft experiences no force in the rotating coordinate system. That is, the gravitational pull of the Earth and Moon and the centrifugal force of rotation all balance one another for a small mass at a Lagrange point. In the Earth–Moon system, L_4 and L_5 are stable equilibria, and they probably collect space dust like belly buttons collect lint. In contrast, L_1, L_2, and L_3 are all unstable. Thus, a spacecraft positioned at one of these points will inevitably drift away unless the craft monitors its position and occasionally fires its thrusters to correct for external perturbations.

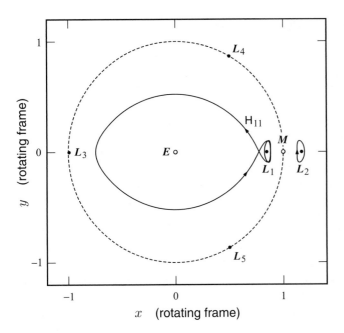

Fig. 20.9 The stationary or Lagrange points (L_1–L_5) of the Earth–Moon (E-M) system in a coordinate system that rotates with the Moon. The Moon's orbit is assumed to be circular and distances are measured in terms of its radius. Also shown are bean-shaped orbits around L_1 and L_2 for a spacecraft with a mass much smaller than that of the Earth or Moon. The homoclinic trajectory H_{11} shows how the spacecraft can circle the Earth without fuel along a path beginning and ending on the L_1 orbit. All of the orbits and trajectories shown here lie in the plane of the Moon's orbit.

We now focus on L_1 and L_2, which are the hubs of useful roundabouts in the celestial highway system. If a permanent base were ever established on the Moon, L_1 would be a convenient transfer point for people and supplies moving between the Earth and the Moon, and L_2 could be used as a stepping stone to points beyond. Surprisingly, a spacecraft need not be positioned exactly at L_1 or L_2, but can occupy any of various orbits around these stationary points. Examples of such orbits are shown in Fig. 20.9 by the bean-shaped paths that encircle L_1 and L_2. Like L_1 and L_2 themselves, these orbits are unstable, so a spacecraft must make occasional corrections to maintain its orbit. However, these orbits are keys to the celestial highway system: they are the roundabouts that allow a spacecraft to head off in new directions.

If we restrict ourselves to motion within the orbital plane of the Moon, the state of our spacecraft at any instant is specified by its x and y coordinates in the rotating frame and its velocity components v_x and v_y in the x and y directions. The craft thus inhabits a four-dimensional state space, providing more than enough dimensions to allow intersections between the insets and outsets of unstable orbits like those at L_1 and L_2. As usual, such intersections lead to homoclinic or heteroclinic trajectories and chaotic motion. Here our primary interest is in heteroclinic trajectories because they are the roads in the celestial highway system, connecting the unstable orbits that serve as roundabouts. Thus, to discover highways in space, scientists are keen to understand the insets and outsets of the unstable Lagrange orbits.

As it happens, the insets and outsets of the orbits around L_1 and L_2 take the form of tubes in the 4-D state space, with the inset trajectories spiraling around an inset tube as they approach the orbit and the outset trajectories spiraling around an outset tube as they leave the orbit. For the Earth–Moon system, these inset and outset tubes intersect to produce both homoclinic and heteroclinic trajectories. Although a 4-D graph isn't possible, we can easily plot the projection of such trajectories in the x–y plane. For example, Fig. 20.9 shows a homoclinic trajectory labeled H_{11} that begins and ends on the orbit circling L_1. Of course, we've previously met homoclinic trajectories, but it's still astounding that a spacecraft orbiting L_1 can, with the slightest nudge at the right instant, be set on a path taking it fully around the Earth (relative to the Moon) and back to its initial orbit. In many ways the craft's motion is like that of a frictionless pendulum poised at its balance point and given a tap to move it full circle, but it's astounding nonetheless.

For celestial transport, however, heteroclinic trajectories are key because they allow a spacecraft to move between unstable orbits. Figure 20.10 shows how such trajectories connect orbits circling L_1 and L_2. When the two orbits have the same energy, as they do here, the outset of one can intersect the inset of the other, creating a connecting trajectory. In Fig. 20.10 we find H_{12}, a heteroclinic trajectory going from an orbit

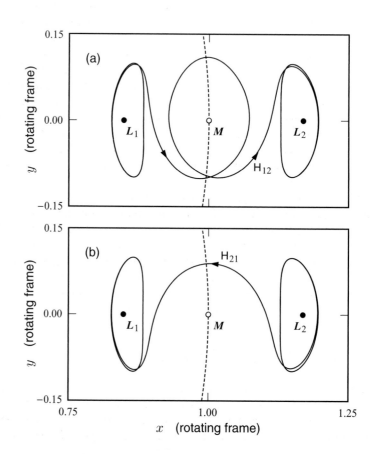

Fig. 20.10 Heteroclinic trajectories connect orbits around the L_1 and L_2 points of the Earth–Moon system. H_{12} allows motion from the L_1 orbit to the L_2 orbit (a) and H_{21} allows motion in the reverse direction (b).

about L_1 to an orbit about L_2, and H_{21}, which provides a return path. Thus, H_{12} and H_{21} allow free rides back and forth between the unstable orbits circling L_1 and L_2 and demonstrate the basic operation of the celestial highway system. In this case, space travel is almost fuel free.

But the interplanetary transport network extends far beyond the Moon. The Lagrange points that serve as hubs for roundabouts exist for all pairs of large orbiting bodies in the Solar System. Thus, as in the Earth–Moon system, there are five Lagrange points for each of the planets paired with the Sun, and we can draw a diagram like Fig. 20.9 with the Earth and Moon replaced by the Sun and any planet. The Lagrange points of the Sun–Earth system have proven to be useful locations for spacecraft. For example, the solar observatory SOHO was positioned at L_1, between the Earth and the Sun, where it has a good view of the Sun, and the WMAP probe was placed at L_2, opposite the Earth from the Sun, in order to observe the cosmic microwave background radiation.

The host of Sun–planet and planet-moon Lagrange points in the Solar System provides many possibilities for fuel-efficient transport. However, most hops are not as simple as hopping between the Lagrange points of a single orbiting pair. In particular, a trip from a Lagrange point of the Sun–Earth system to one of the Earth–Moon system is complicated by the fact that the rotating frames of these systems are entirely different, and the inset and outset tubes of the Lagrange-point orbits in the two systems continually rotate with respect to each other. Thus, while there are low-fuel and even no-fuel paths between the L_1 and L_2 points of the Earth–Moon system and those of the Sun–Earth system, intersections between the relevant inset and outset tubes only occur about once a month. Nevertheless, heteroclinic trajectories between these points do exist, and someday we may find it useful to move a spacecraft stationed at the Sun–Earth L_2 point back to the Earth–Moon L_1 point, repair it, and return it to the Sun–Earth L_2 point, using almost no fuel. What an amazing free ride that would be!

Celestial heteroclinic trajectories have one major disadvantage: they generally require much longer flight times than more direct routes. However, because travel time is not always important for unmanned probes and fuel-efficient trips allow larger payloads, we can expect Lagrange points to be used ever more frequently as stepping stones to remote parts of the Solar System. Space travel is thus one realm in which the heteroclinic trajectories underpinning chaotic motion provide a real-world practical advantage.

20.6   Weather modification

The American essayist Charles Warner (1829–1900) once joked that "Everybody complains about the weather, but nobody does anything about it." Of course, his joke relies on the weather being entirely beyond human influence. Today, however, meteorologists believe that something might eventually be done about the weather. The chaotic nature of weather, first understood by Edward Lorenz, is the obvious key to controlling it. Because chaotic systems are sensitive to small perturbations, meteorologists may someday discover a practical means of bringing rain to an area of drought or defusing a tropical storm before it develops into a hurricane.

Is weather modification a real possibility? In 1977 one of Lorenz's students, Ross Hoffman, proposed writing a dissertation that would answer precisely this question. At the time, Lorenz saw the topic as too ambitious, however, and Hoffman ultimately chose another thesis project. But Hoffman was not completely dissuaded, and two decades later he began to reexamine the question of weather modification. In the meantime, advances in numerical methods and computer power had revolutionized the science of weather forecasting, making Hoffman's task much easier.

To more fully explain the problem of weather modification, we return to the butterfly effect. According to Lorenz, the flap of a butterfly's wing might eventually have a significant effect on the weather. However, it would probably require months or years before a butterfly's tiny disturbance grew large enough to make a perceptible difference. On the other hand, we can only forecast the weather a few days in advance, so the perturbation required to change a foreseeable event is much larger than a butterfly can provide. That is, by the time we can predict that a tropical storm will turn into a hurricane, it's too late for butterflies to be of use. What Hoffman wanted to know was the size and kind of perturbation required to modify the weather of the near future that we can predict with confidence.

In a 2004 article, Hoffman considered two hurricanes from 1992, Iniki and Andrew, that devastated the Hawaiian island of Kauai and the east coast of Florida. These hurricanes were complex dynamical events, but they were subsequently modeled in detail using the latest in forecasting software. By experimenting with the models, Hoffman sought to determine what kind of perturbations might have significantly reduced the impact of each storm. Using a trial-and-error approach, he left the equations of motion untouched but altered the initial conditions in various ways and computed the new outcomes. In the case of Iniki, he discovered that a small increase (less than $2°$ Celsius) in the temperature profile of the hurricane a few hours before landfall was enough to make it miss Kauai by some 60 miles. Similarly, altering the temperature profile of Andrew by up to $3°$ Celsius reduced its intensity a few hours later from Category 3 to Category 1. Such modifications would have dramatically reduced the devastation of both hurricanes.

However, altering the temperature profile of a hurricane over many square miles is no simple task. Among other possibilities, Hoffman has suggested using a large mirror in orbit about the Earth either to direct extra sunlight toward a hurricane or to block sunlight from reaching it. But the efficacy of this and other approaches to perturbing the weather remains to be determined. On the other hand, our ability to model the weather continues to improve, giving us ever better long-range forecasts. And the further in advance we know what the weather will be, the smaller the perturbation needed to modify it. Thus, over the coming decades, creating the nudges needed to control the weather is likely to become an easier task. Warner's joke isn't out of date yet, but someday it may well be.

20.7   Adaptation

In our previous examples of chaos at work, we have gone from simple systems like the driven pendulum described by just three state variables, to spacecraft whose trajectories are realistically predicted only when dozens of variables are included, to the Earth's atmosphere, which is modeled by equations involving millions of state variables. Using such models, we can accurately predict the motion of a spacecraft or make weather forecasts that are informative if not entirely reliable. However, as our final example, we turn to systems that are so complex that realistic modeling is presently beyond the realm of possibility. In particular, we consider various processes associated with living organisms that involve adaptation of one kind or another. These include the genetic adaptation of species to changing environments, the adaptive response of the immune system to new microbial threats, and the creativity that the brain applies to problem solving. In each of these areas, nature itself has apparently discovered a way to benefit from chaotic dynamics.

Often called complex adaptive systems, these products of the living world are among the most fascinating of any in the human environment. Who wouldn't like to better understand the mechanism of evolution or intelligent thought? But we can't expect to model living organisms by beginning with the dynamics of molecules and working our way up to cells, tissues, organs, and entire animals. Instead, the best we can do is explore highly simplified models that might capture the essence of an adaptive process we hope to understand. Thus, whereas we employed realistic equations to describe the dynamics of spacecraft and weather systems, to investigate adaptive processes we turn to toy models, and the conclusions we reach are necessarily speculative. Nevertheless, toy models may provide valuable insights into the mechanisms by which life adapts.

The thread that I wish to follow begins with John von Neumann and Stanislaw Ulam, the Los Alamos mathematicians who we first met in Chapter 11. Intrigued by the electronic computers that he had helped to create, in the late 1940s von Neumann began to wonder exactly what such machines could and couldn't do. In particular, he asked if a machine could ever build a copy of itself. To answer this question, von Neumann needed a mathematical universe that was governed by a few simple rules yet complex enough to allow the possibility of a self-reproducing machine. Ulam suggested that von Neumann consider a class of universes, now called cellular automata, that have proven to display a rich variety of dynamical behaviors.

Ulam's idea was to divide both space and time into discrete units, with space covered by a regular grid of cells and time advancing in steps that we'll call ticks or generations. Associated with each cell is a simple robotic mechanism that can be in one of several states. We'll suppose that the state of a robot at any time is displayed by the color of its cell. All of the robots in the cellular universe are identical, and all robots update their states simultaneously with each tick of the clock. The pattern of cell colors thus changes tick by tick, but in a completely regimented way. Each robot determines its next state or cell color according to a rule that takes into account only its present color and those of its neighboring cells. All robots use the same rule, and the next color pattern depends only on the present pattern, so the dynamics of Ulam's universe is completely deterministic.

Although the rigid mechanics governing cellular automata might suggest an ultimately boring universe, von Neumann discovered just the opposite. Indeed, his discoveries are truly astonishing. For a cellular automaton defined on a two-dimensional grid with 29 possible states for each cell, von Neumann was able to show the existence of an initial color pattern that could generate an identical pattern in an adjacent region of space. Thus, in 1952 von Neumann demonstrated for the first time that machines are capable of reproducing themselves. Accordingly, reproduction, one of the defining characteristics of living organisms, is not unique to the biosphere. In addition, von Neumann was able to prove that a universal computer, capable of performing any computation, could be constructed within his cellular universe. Thus, while cellular automata are no more than toy models of our real universe, they are unexpectedly potent sources of complex behavior. Strangely, von Neumann had not yet written about his work on cellular automata at the time of his death in 1957, and it was published only posthumously in 1966.

For a simple example of a cellular automaton, we turn to an invention of the British mathematician John Conway (born 1937), which he dubbed "Life" for the lifelike quality of its dynamics. Introduced in 1970, Life became a craze among scientists enthralled by the patterns it generated on their computer screens. The Life automaton, like that of von Neumann, uses a 2-D grid of cells, but each cell can only be in one of two states, either dead or alive. And the rule determining whether a cell will be dead or alive in the next generation is especially simple. If a cell is presently alive then it will remain so in the next generation only if either two or three of its eight neighbors is presently alive. If a cell is presently dead, it will spring to life in the next generation only if exactly three of its neighbors are presently alive. There are no more rules to Conway's Life.

Although it's difficult to capture the complex dynamics of the Life automaton in static drawings, Fig. 20.11 gives a taste of what can

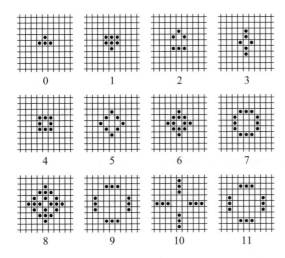

Fig. 20.11 The dynamics of the Life cellular automaton. Four living cells at generation 0, indicated by black dots, develop into an oscillating pattern known as "traffic lights" nine generations later.

happen. Here we show how an initial pattern of four living cells at generation 0 (indicated by dots) changes over the next 11 generations. Of course, the history captured here results merely from applying the rule given above to each cell of each successive generation. For example, in going from generation 0 to 1, we see that the four original living cells all survive because each is surrounded by two or three living neighbors. And three new cells spring to life in generation 1 because each had exactly three living neighbors in generation 0. Can you see how the same rules explain generation 2?

The Life history of Fig. 20.11 ends with generation 11, which happens to repeat generation 9. From the deterministic property of Life, we can conclude that patterns 9 and 10 will continue to alternate forever, an oscillation known to fans of Life as "traffic lights." But can Life produce more interesting behavior? Surprisingly, Conway sketched proofs indicating that the Life automaton supports the construction of both universal computers and self-replicating machines. Indeed, Conway hypothesized, "It's probable, given a large enough Life space, initially in a random state, that after a long time, intelligent self-replicating animals will emerge and populate some parts of the space." How could the simple rules of Life lend it such fantastic properties? If Conway is correct, there is much more to Life than we would naively expect from its simplicity. Is chaos part of the explanation?

In 1983, while working at the Institute for Advanced Study in Princeton, the British physicist and mathematician Stephen Wolfram (born 1957) established a more general perspective on cellular automata. Conway had discovered Life by investigating a few simple automaton rules, but Wolfram wanted to understand the range of behavior across a broad spectrum of rules. In principle, there are an infinite number of rules to be considered. However, even if we limit ourselves to 2-D automata with two states per cell (dead or alive) and choose the next state based on the states of the cell itself and its eight neighbors, then the cell's robot has $2^9 = 512$ possible inputs, each with two choices for the cell's next state, or a total of $2^{512} \approx 10^{154}$ possible rules.

Wolfram could sample only a tiny fraction of these rules, but he soon concluded that the dynamical behaviors of cellular automata always fall into one of four classes. In class I behavior, the automaton quickly goes to a completely uniform grid, with all cells dead or all cells alive, no matter what the initial pattern. In class II behavior, the final grid is also reached quickly, but it may include static arrangements of cells or patterns that repeat periodically. Class III automata represent the opposite extreme, in which most initial grids lead to chaotic behavior reminiscent of the constant agitation of the molecules in a liquid or gas. Finally, class IV automata occupy the middle ground, where the final grid usually includes only static and periodic patterns, but it appears after a long-lived chaotic transient. As Wolfram noted, the Life automaton is the epitome of class IV, allowing complex behavior that falls between the rigidity of classes I and II and the complete randomness of class III. He also observed that class IV automata are by far the most interesting, in part because they allow the possibility of constructing a universal computer.

Although class IV behavior falls logically between classes II and III, the rules governing the dynamics of cellular automata are so numerous and diverse that in 1983 it wasn't clear whether rules from the same class are otherwise related. Not long after Wolfram published his classification scheme, however, Christopher Langton (born 1949) discovered a parameter that helps to sort rules into behavior classes. At the time, Langton was a graduate student in computer science at the University of Michigan and was fascinated by what he called "artificial life," basically von Neumann's idea that life might be simulated by computers. Often denoted Λ, Langton's parameter is easily calculated for any cellular automaton, and it often provides a useful guide to the behavior class of an automaton rule, with the Λ values of class IV behavior nestled between those of classes II and III, just as one might expect. Examples of automata from these three classes are animated in Experiment 26 of the Dynamics Lab.

Langton introduced his Λ parameter in a 1986 paper entitled "Studying artificial life with cellular automata." Later, the fact that this parameter places class IV behavior between classes II and III led Langton to propose an interesting hypothesis, an idea now commonly referred to as "life at the edge of chaos." Stated in today's language, Langton proposed that complex adaptive systems, including living organisms, may exist at the edge of chaos, where dynamical behavior is neither too ordered nor too chaotic and universal computation is possible. Certainly, life relies on ordered behavior for self-replication, but it also requires the ability to adapt to environmental change. Thus, evolution may guide life to the edge of chaos, precisely because organisms that are too chaotic have a lower chance of reproducing and organisms that follow too strict a regimen are unable to adapt to new situations. The edge of chaos is the natural place for life to exist when it comes to dynamics.

Although the edge-of-chaos idea has a strong intuitive appeal, scientific acceptance requires a degree of proof that has yet to be established for this hypothesis. Charles Darwin spent years collecting evidence for his theory of evolution by natural selection before publishing *On the Origin of Species*, and even then his idea was considered speculative. Today life at the edge of chaos is similarly speculative, but it's intriguing to think that life on Earth and human intelligence may owe their existence in part to the adaptive and innovative capacities inherent in long-lived chaotic transients.

20.8 *Terra incognita*

Today we look back with nostalgia on the rapid-fire discoveries of the 1960s, 70s and 80s that made chaotic dynamics a vibrant and exciting field of research. The days of surprise and delight are now gone, but they have left in their wake a legacy of knowledge that will forever remain a part of the scientific edifice. Scientists have a new tool in their kit that will help explain many curious phenomena yet to be observed. Today, when a nonlinear system acts strangely, it's perfectly natural to suspect that chaos is at play.

All the same, we shouldn't think that nothing remains to be discovered in nonlinear dynamics. Although the Poincaré–Bendixson theorem gives us a complete catalog of the topologies possible in a two-dimensional state space, no such catalog exists for flows in spaces of higher dimension. Poincaré and his followers contributed wonderfully to our knowledge by uncovering the amazing intricacies of chaotic tangles, but we still stand on the edge of vast unknown regions when it comes to higher dimensions. Who can say what topological monsters remain to be discovered when dozens or millions of state variables come into play? Perhaps a new theorem will one day allow us to wring the last bit of predictability out of Earth's weather or discover a new trajectory through some labyrinth of nature.

As a final speculation, I wonder if the paradigm of chaos might have something to say about the human condition. Near the end of the movie that takes his name, Forrest Gump says, "I don't know if we each have a destiny or if we're all just floating around accidental like on a breeze, but I think maybe it's both—maybe both happening at the same time." This poetic thought might once have seemed beyond science, but the existence of predictable random motion suggests that what seems to be destined and what seems to be accidental may in fact be one and the same thing.

Further reading

Adaptation

- Lewin, R., *Complexity: Life at the Edge of Chaos* (Macmillan, New York, 1992).
- Robson, D., "Disorderly genius", *New Scientist* **202**(2714), 34–37 (27 June 2009).
- Solé, R. and Goodwin, B., *Signs of Life: How Complexity Pervades Biology* (Basic Books, New York, 2000).
- Waldrop, M. M., *Complexity: The Emerging Science at the Edge of Order and Chaos* (Simon and Schuster, New York, 1992).

Cellular automata

- Gardner, M., "The game of life", in *Wheels, Life and Other Mathematical Amusements* (Freeman, New York, 1983) chapters 20–22.
- Poundstone, W., *The Recursive Universe: Cosmic Complexity and the Limits of Scientific Knowledge* (Contemporary Books, Chicago, 1985).
- Wolfram, S., *A New Kind of Science* (Wolfram Media, Champaign, Illinois, 2002).

Controlling chaos

- Ditto, W. and Munakata, T., "Principles and applications of chaotic systems", *Communications of the ACM* **38**(11), 96–102 (November 1995).
- Ditto, W. L. and Pecora, L. M., "Mastering chaos", *Scientific American* **269**(2), 78–84 (August 1993).
- Hunt, E. R. and Johnson, G., "Keeping chaos at bay", *IEEE Spectrum* **30**(11), 32–36 (November 1993).
- Ott, E., Grebogi, C., and Yorke, J. A., "Controlling chaos", *Physical Review Letters* **64**, 1196–1199 (1990).
- Ott, E. and Spano, M., "Controlling chaos", *Physics Today* **48**(5), 34–40 (May 1995).
- Schöll, E. and Schuster, H. G., editors, *Handbook of Chaos Control*, 2nd edition, (Wiley–VCH, Weinheim, Germany, 2008).

Economic prediction

- Bass, T. A., *The Predictors* (Holt, New York, 1999).
- Berreby, D., "Chaos hits Wall Street", *Discover* **14**(3), 76–84, (March 1993).
- Lewin, R., "Making maths make money", *New Scientist* **134**(1816), 31–34 (11 April 1992).
- Raeburn, P., "Chaos and the catch of the day", *Scientific American* **300**(2), 76–78 (February 2009).
- Savit, R., "Chaos on the trading floor", *New Scientist* **127**(1729), 48–51 (11 August 1990).

Encryption

- Brooks, M., "Spies, lies and butterflies", *New Scientist* **188**(2526), 32–35 (19 November 2005).
- Lesurf, J., "A spy's guide to chaos", *New Scientist* **133**(1806), 29–33 (1 February 1992).

Space travel

- Belbruno, E., *Fly Me to the Moon: An Insider's Guide to the New Science of Space Travel* (Princeton University Press, Princeton, 2007).
- Klarreich, E., "Navigating celestial currents", *Science News* **167**, 250–252 (2005).
- Koon, W. S., Lo, M. W., Marsden, J. E., and Ross, S. D., "Heteroclinic connections between periodic orbits and resonance transitions in celestial mechanics", *Chaos* **10**, 427–469 (2000).
- Ross, S. D., "The interplanetary transport network", *American Scientist* **94**, 230–237 (2006).
- Smith, D. L., "Next exit 0.5 million kilometers", *Engineering and Science* **65**(4), 6–15 (2002).
- Stewart, I.,"Ride the celestial subway", *New Scientist* **189**(2544), 32–36 (25 March 2006).
- G. Taubes, "Surfing the solar system", *Discover* **20**(6), 88–93 (June 1999).

Weather modification

- Hoffman, R. N., "Controlling hurricanes", *Scientific American* **291**(4), 68–75 (October 2004).
- Mullins, J., "Raising a storm", *New Scientist* **175**(2353), 29–33 (27 July 2002).

Bibliography

The following sources on chaos are intended for general audiences; they do not require advanced mathematics.

Books

Abraham, R. H. and Shaw, C. D., *Dynamics: The Geometry of Behavior*, 2nd edition (Addison-Wesley, Reading, 1992).

Bass, T. A., *The Predictors: How a Band of Maverick Physicists Used Chaos Theory to Trade Their Way to a Fortune on Wall Street* (Henry Holt, New York, 1999).

Belbruno, E., *Fly Me to the Moon: An Insider's Guide to the New Science of Space Travel* (Princeton University Press, Princeton, 2007).

Bird, R. J., *Chaos and Life: Complexity and Order in Evolution and Thought* (Columbia University Press, New York, 2003).

Briggs, J. and Peat, F. D., *Turbulent Mirror: An Illustrated Guide to Chaos Theory and the Science of Wholeness* (Harper and Row, New York, 1989).

Çambel, A. B., *Applied Chaos Theory: A Paradigm for Complexity* (Academic, San Diego, 1993).

Diacu, F. and Holmes, P., *Celestial Encounters: The Origins of Chaos and Stability* (Princeton University Press, Princeton, 1996).

Ekeland, I., *Mathematics and the Unexpected* (University of Chicago Press, Chicago, 1988).

Gleick, J., *Chaos: Making a New Science* (Viking, New York, 1987).

Gribbin, J., *Deep Simplicity: Bringing Order to Chaos and Complexity* (Random House, New York, 2004).

Hall, N., editor, *Exploring Chaos: A Guide to the New Science of Disorder* (Norton, New York, 1994).

Kellert, S. H., *In the Wake of Chaos: Unpredictable Order in Dynamical Systems* (University of Chicago Press, Chicago, 1993).

Lewin, R., *Complexity: Life at the Edge of Chaos* (Macmillan, New York, 1992).

Lorenz, E. N., *The Essence of Chaos* (University of Washington Press, Seattle, 1993).

Parker, B. R., *Chaos in the Cosmos: The Stunning Complexity of the Universe* (Plenum, New York, 1996).

Peak, D. and Frame, M., *Chaos under Control: The Art and Science of Complexity* (Freeman, New York, 1994).

Peterson, I., *Newton's Clock: Chaos in the Solar System* (Freeman, New York, 1993).

Ruelle, D., *Chance and Chaos* (Princeton University Press, Princeton, 1991).

Sardar, Z. and Abrams, I., *Introducing Chaos* (Icon Books, Cambridge, 1998).

Schroeder, M., *Fractals, Chaos, Power Laws: Minutes from an Infinite Universe* (Freeman, New York, 1991).

Smith, L., *Chaos: A Very Short Introduction* (Cambridge University Press, Cambridge, 2007).

Solé, R. and Goodwin, B., *Signs of Life: How Complexity Pervades Biology* (Basic Books, New York, 2000).

Stewart, I., *Does God Play Dice? The New Mathematics of Chaos*, 2nd edition (Blackwell, Malden, Massachusetts, 2002).

Waldrop, M. M., *Complexity: The Emerging Science at the Edge of Order and Chaos* (Simon and Schuster, New York, 1992).

Williams, G. P., *Chaos Theory Tamed* (Joseph Henry Press, Washington, 1997).

Articles

Appell, D., "Celestial swingers", *New Scientist* **71**(2302), 36–39 (4 August 2001).

Balibar, S., "Cyclists and butterflies", in *The Atom and the Apple: Twelve Tales from Contemporary Physics* (Princeton University Press, Princeton, 2008) chapter 9.

Bass, T. A., "Predicting Chaos: Norman Packard", in *Reinventing the Future: Conversations with the World's Leading Scientists* (Addison Wesley, Reading, Massachusetts, 1994) pp. 198–216.

Berreby, D., "Chaos hits Wall Street", *Discover* **14**(3), 76–84 (March 1993).

Berry, M., "Quantum physics on the edge of chaos", *New Scientist* **116**(1587), 44–47 (19 November 1987).

Brooks, M., "Spies, lies, and butterflies", *New Scientist* **188**(2526), 32–35 (19 November 2005).

Brown, J., "Where two worlds meet", *New Scientist* **150**(2030), 26–30 (18 May 1996).

Buchanan, M., "Breakout!", *New Scientist* **187**(2514), 34–37 (27 August 2005).

Burger, E. B. and Starbird, M., "Chaos reigns: Why we can't predict the future", in *Coincidences, Chaos, and All That Math Jazz: Making Light of Weighty Ideas* (Norton, New York, 2005) chapter 2.

Carlson, S., "Falling into chaos", *Scientific American* **281**(5), 120–121 (November 1999).

Chown, M., "Chaotic heavens", *New Scientist* **181**(2436), 32–35 (28 February 2004).

Coveney, P., "Chaos, entropy and the arrow of time", *New Scientist* **127**(1736), 49–52 (29 September 1990).

Cromer, A., Zahopoulos, C., and Silevitch, M. B., "Chaos in the corridor", *The Physics Teacher* **30**, 382–383 (1992).

Crutchfield, J. P., Farmer, J. D., Packard, N., and Shaw, R. S., "Chaos", *Scientific American* **255**(6), 46–57 (December 1986).

Davies, P., "Chaos", in *The Cosmic Blueprint: New Discoveries in Natures's Creative Ability to Order the Universe* (Simon and Schuster, New York, 1988) chapter 4.

Davies, P., "Chaos frees the universe", *New Scientist* **128**(1737), 48–51 (6 October 1990).

Dewdney, A. K., "Probing the strange attractions of chaos", *Scientific American* **257**(1), 108–111 (July 1987).

Dewdney, A. K., "Leaping into Lyapunov space", *Scientific American* **265**(3), 178–180 (September 1991).

Dewdney, A. K., "The edge of chaos: Unpredictable systems", in *Beyond Reason: 8 Great Problems that Reveal the Limits of Science* (Wiley, Hoboken, 2004) chapter 4.

Ditto, W. L. and Pecora, L. M., "Mastering chaos", *Scientific American* **269**(2), 78–84 (August 1993).

Ditto, W. and Munakata, T., "Principles and applications of chaotic systems", *Communications of the ACM* **38**(11), 96–102 (November 1995).

Ehrlich, R., "Three mechanical demonstrations of chaos", *The Physics Teacher* **28**, 26–29 (1990).

Ekeland, I., "From computations to geometry", "Poincaré and beyond", and "Pandora's box", in *The Best of All Possible Worlds: Mathematics and Destiny* (University of Chicago Press, Chicago, 2006) chapters 4–6.

Esbenshade, D. H., "Computer-specific initial conditions and chaos", *The Physics Teacher* **32**, 40–41 (1994).

Flake, G. W., "Chaos", in *The Computational Beauty of Nature: Computer Explorations of Fractals, Chaos, Complex Systems, and Adaptation* (MIT Press, Cambridge, Massachusetts, 1998) pp. 137–227.

Ford, J., "What is chaos, that we should be mindful of it?", in *The New Physics*, Davies, P. C. W., editor (Cambridge University Press, Cambridge, 1989) pp. 348–371.

Frank, A., "Crack in the clockwork", *Astronomy* **26**(5), 54–59 (May 1998).

Gedzelman, S. D., "Chaos rules: Edward Lorenz capped a century of progress in forecasting by explaining unpredictability", *Weatherwise* **47**(4), 21–26 (August/September 1994).

Gleick, J., "Solving the mathematical riddle of chaos", *New York Times Magazine*, 10 June 1984.

Goldberger, A. L., Rigney, D. R., and West, B. J., "Chaos and fractals in human physiology", *Scientific American* **262**(2), 43–49 (February 1990).

Goldstein, G., "Francis Moon: Coming to terms with chaos", *Mechanical Engineering* **112**(1), 40–47 (January 1990).

Graham-Rowe, D., "Engines of chaos", *New Scientist* **200**(2680), 40–43 (1 November 2008).

Gutzwiller, M. C., "Quantum chaos", *Scientific American* **266**(1), 78–85 (January 1992).

Hartley, K., "Solar system chaos", *Astronomy* **18**(5), 34–39 (May 1990).

Hofstadter, D. R., "Strange attractors: Mathematical patterns delicately poised between order and chaos", *Scientific American* **245**(5), 22–43 (November 1981).

Holden, A., "Chaos is no longer a dirty word", *New Scientist* **106**(1453), 12–15 (25 April 1985).

Hooper, J., "Connoisseurs of chaos", *Omni* **5**(9), 85 (June 1983).

Hunt, B. R. and Yorke, J. A., "Maxwell on chaos", *Nonlinear Science Today* **3**, 1–4 (1993).

Jensen, R. V., "Classical chaos", *American Scientist* **75**, 168–181, (1987).

Kanigel, R., "The coming of chaos", *Johns Hopkins Magazine* **39**(3), 36–42 (June 1987).

Kauffman, S., "Edge of chaos", in *At Home in the Universe: The Search for the Laws of Self-Organization and Complexity* (Oxford University Press, Oxford, 1995) pp 86–92.

Kauffman, S., "Candidate law 1: The dynamical edge of chaos", in *Investigations* (Oxford University Press, Oxford, 2000) pp 160–188.

Killian, A. M., "Playing dice with the solar system", *Sky and Telescope* **78**, 136–140 (1989).

Kostelich, E., "Symphony in chaos", *New Scientist* **146**(1972), 36–39 (8 April 1995).

Lesurf, J., "Chaos on the circuit board", *New Scientist* **126**(1723), 63–66 (30 June 1990).

Lesurf, J., "A spy's guide to chaos", *New Scientist* **133**(1806), 29–33 (1 February 1992).

Mancuso, R. V. and Somerset, E. M., "Change of the state of a diode and chaos", *The Physics Teacher* **35**, 31–33 (1997).

Marchese, J., "Forecast: Hazy", *Discover* **22**(6), 44–51 (June 2001).

Mathews, R., "Don't blame the butterfly", *New Scientist* **171**(2302), 24–27 (4 August 2001).

May, R., "The chaotic rhythms of life", *New Scientist* **124**(1691), 37–41 (18 November 1989).

McAuliffe, K., "Get smart: Controlling chaos", *Omni* **12**(5), 42 (February 1990).

McRobie, A. and Thompson, M., "Chaos, catastrophes and engineering", *New Scientist* **126**(1720), 41–46 (9 June 1990).

Mitchell, M., "Dynamics, chaos, and prediction", in *Complexity: A Guided Tour* (Oxford University Press, Oxford, 2009) chapter 2.

Monastersky, R., "Forecasting into chaos", *Science News* **137**, 280–282 (1990).

Mullin, T., "Turbulent times for fluids", *New Scientist* **124**(1690), 52–55 (11 November 1989).

Mullins, J., "Raising a storm", *New Scientist* **175**(2353), 28–33 (27 July 2002).

Murray, C., "Is the solar system stable?", *New Scientist* **124**(1692), 60–63 (25 November 1989).

Neff, J. and Carroll, T. L., "Circuits that get chaos in sync", *Scientific American* **269**(2), 120–122 (August 1993).

Oliver, D., "A chaotic pendulum", *The Physics Teacher* **37**, 174 (1999).

Page, D., "Formula for fun: The physics of chaos", *Funworld* **12**(3), 42–46 (March 1996).

Pagels, H., "Order, complexity, and chaos" and "Life can be so non-linear", in *The Dreams of Reason: The Computer and the Rise of the Sciences of Complexity* (Simon and Schuster, New York, 1988) chapters 3 and 4.

Palmer, T., "A weather eye on unpredictability", *New Scientist* **124**(1690), 56–59 (11 November 1989).

Percival, I., "Chaos: A science for the real world", *New Scientist* **124**(1687), 42–47 (21 October 1989).

Peterson, I., "The dragons of chaos", in *The Mathematical Tourist: Snapshots of Modern Mathematics* (Freeman, New York, 1988) chapter 6.

Peterson, I., "Chaos in the clockwork: Uncovering traces of chaos in planetary orbits", *Science News* **141**, 120–121 (1992).

Peterson, I., "Chaos in spacetime", *Science News* **144**, 376–377 (1993).

Peterson, I., "Cavities of chaos: Sorting out quantum chaos in the lab", *Science News* **147**, 264–265 (1995).

Peterson, I., "Complete chaos", in *The Jungles of Randomness: A Mathematical Safari* (Wiley, New York, 1998) chapter 7.

Porter, M. A. and Liboff, R. L., "Chaos on the quantum scale", *American Scientist* **89**, 532–537 (2001).

Rockmore, D., "God created the natural numbers ... but, in a billiard hall?", and "Making order out of (quantum) chaos", in *Stalking the Riemann Hypothesis: The Quest to Find the Hidden Law of Prime Numbers* (Pantheon, New York, 2005) chapters 12 and 13.

Rosenfeld, J., "The butterfly that roared", *Scientific American Presents* **11**(1), 22–27 (Spring 2000).

Ruelle, D., "Strange attractors", *The Mathematical Intelligencer* **2**, 126–137 (1980).

Savit, R., "Chaos on the trading floor", *New Scientist* **127**(1729), 48–51 (11 August 1990).

Schwartzschild, B. M., "Chaotic orbits and spins in the solar system", *Physics Today* **38**(9), 17–20 (September 1985).

Scott, S., "Clocks and chaos in chemistry", *New Scientist* **124**(1693), 53–59 (2 December 1989).

Smith, D., "How to generate chaos at home", *Scientific American* **266**(1), 144–146 (January 1992).

Stewart, I., "Portraits of chaos", *New Scientist* **124**(1689), 42–47 (4 November 1989).

Stewart, I., "Does chaos rule the cosmos?", *Discover* **13**(11), 56–63 (November 1992).

Stewart, I., "Christmas in the house of chaos", *Scientific American* **267**(6), 144–146 (December 1992).

Stewart, I. and Golubitsky, M., "Icons of chaos", in *Fearful Symmetry: Is God a Geometer?* (Blackwell, Oxford, 1992) chapter 9.

Svoboda, E., "Hurtling toward chaos", *Discover* **26**(7), 20–21 (July 2005).

Taubes, G., "The mathematics of chaos", *Discover* **5**(9), 30–39 (September 1984).

Thuan, T. X., "Chaos in the cosmic machinery, and uncertainty in determinism", in *Chaos and Harmony: Perspectives on Scientific Revolutions of the Twentieth Century* (Oxford University Press, Oxford, 2001) chapter 3.

Tritton, D., "Chaos in the swing of a pendulum", *New Scientist* **111**(1518), 37–40 (24 July 1986).

Vivaldi, F., "An experiment with mathematics", *New Scientist* **124**(1688), 46–49 (28 October 1989).

Videos

Chaos: A New Science, New Dimension Media, 1990.

Peitgen, H.-O., Jürgens, H., Saupe, D., and Zahlten, C., *Fractals: An Animated Discussion with Edward Lorenz and Benoît Mandelbrot*, W. H. Freeman, New York, 1990.

Suzuki, D., host, *Chaos, Science and the Unexpected*, The Nature of Things, Canadian Broadcasting Corporation, 1990.

Taylor, J., *The Strange New Science of Chaos*, Nova, WGBH Television, Boston, 1989.

Tsonis, A., host, *Chaos and Fractal Forms: Irregularity in Nature*, The Science Bag, University of Wisconsin, 1990.

Index

Note: Indexed items with page numbers followed by "n." appear in the footnote of the page indicated.